ATOM PROBE TOMOGRAPHY

ATOM PROBE TOMOGRAPHY

Put Theory Into Practice

Edited by

WILLIAMS LEFEBVRE-ULRIKSON
FRANÇOIS VURPILLOT
XAVIER SAUVAGE

ELSEVIER

Amsterdam • Boston • Heidelberg • London
New York • Oxford • Paris • San Diego
San Francisco • Singapore • Sydney • Tokyo

Academic Press is an imprint of Elsevier

Academic Press is an imprint of Elsevier
125 London Wall, London EC2Y 5AS, UK
525 B Street, Suite 1800, San Diego, CA 92101-4495, USA
50 Hampshire Street, 5th Floor, Cambridge, MA 02139, USA
The Boulevard, Langford Lane, Kidlington, Oxford OX5 1GB, UK

Notices

Knowledge and best practice in this field are constantly changing. As new research and experience broaden our understanding, changes in research methods, professional practices, or medical treatment may become necessary.

Practitioners and researchers must always rely on their own experience and knowledge in evaluating and using any information, methods, compounds, or experiments described herein. In using such information or methods they should be mindful of their own safety and the safety of others, including parties for whom they have a professional responsibility.

To the fullest extent of the law, neither the Publisher nor the authors, contributors, or editors, assume any liability for any injury and/or damage to persons or property as a matter of products liability, negligence or otherwise, or from any use or operation of any methods, products, instructions, or ideas contained in the material herein.

British Library Cataloguing-in-Publication Data
A catalogue record for this book is available from the British Library

Library of Congress Cataloging-in-Publication Data
A catalog record for this book is available from the Library of Congress

ISBN: 978-0-12-804647-0

Publisher: Matthew Deans
Acquisition Editor: Christina Gifford
Editorial Project Manager: Heather Cain
Production Project Manager: Lisa Jones
Designer: Greg Harris

Typeset by TNQ Books and Journals
www.tnq.co.in

CONTENTS

CONTRIBUTORS

D. Blavette
Groupe de Physique des Matériaux, University and INSA of Rouen, Normandie University, France

I. Blum
Groupe de Physique des Matériaux, University and INSA of Rouen, Normandie University, France

F. Cuvilly
Groupe de Physique des Matériaux, University and INSA of Rouen, Normandie University, France

G. Da Costa
Groupe de Physique des Matériaux, University and INSA of Rouen, Normandie University, France

F. Danoix
Groupe de Physique des Matériaux, University and INSA of Rouen, Normandie University, France

J. Houard
Groupe de Physique des Matériaux, University and INSA of Rouen, Normandie University, France

W. Lefebvre-Ulrikson
Groupe de Physique des Matériaux, University and INSA of Rouen, Normandie University, France

C. Pareige
Groupe de Physique des Matériaux, University and INSA of Rouen, Normandie University, France

L. Rigutti
Groupe de Physique des Matériaux, University and INSA of Rouen, Normandie University, France

X. Sauvage
Groupe de Physique des Matériaux, University and INSA of Rouen, Normandie University, France

A. Vella
Groupe de Physique des Matériaux, University and INSA of Rouen, Normandie University, France

F. Vurpillot
Groupe de Physique des Matériaux, University and INSA of Rouen, Normandie University, France

PREFACE

Atom probe tomography belongs to the very few characterization techniques enabling the investigation of matter at the atomic scale. In addition, the 3D reconstruction routinely accessible with this technique confers its unique capabilities, which justifies why this instrument stepped in more and more fields of research. Initially dedicated to metallurgy, applications of this technique are nowadays numerous in micro- and nanoelectronics, and applications are also now opened to other fields like biology, soft matter, and geoscience. Theory and applications of atom probe tomography are now taught in universities. This book is first dedicated to graduate students and PhDs and also to researchers having a start with this technique or willing to improve their skill and knowledge. In this book, a special emphasis is given on basic physical and technological concepts for a better understanding of the ultimate capabilities and limitations of atom probe tomography.

The idea of this book naturally arose from the annual international school of Atom Probe Tomography organized by the "Groupe de Physique des Matériaux" at the University of Rouen (France) since 2008. Each chapter in this book is the transcription of a lecture given in that school with a similar educational approach. During the past years, the constant feedback of attendees (beginners or experienced users) was of great help to design this pedagogical content now listed in a series of chapters covering the most relevant aspects of the technique. The first chapter is a general presentation of the context, the early developments, and the basic concepts of atom probe. The theory about the field ion emission mechanisms is introduced to the reader in "Chapter 2, Field Ion Emission Mechanisms." The third chapter is dedicated to field ion microscopy that can be seen as the historical ancestor of Atom Probe Tomography. Sample preparation, which is a key issue prior to any analysis, is presented in "Chapter 4, Atom Probe Sample Preparation." The fifth chapter deals with time-of-flight mass spectrometry and the final aim which is quantitative composition measurements. The design of detectors and the resulting limitations in detection capabilities are described in "Chapter 6, Atom Probe Tomography: Detector Issues and Technology." This is followed by, in the seventh chapter, the presentation of 3D reconstruction procedures from basic to advanced approaches. The specific mechanisms and practical consequences of laser-assisted field evaporation are described in "Chapter 8, Laser-Assisted Field Evaporation."

The ninth chapter is dedicated to data mining, from basic processing like identification of clusters to more advanced statistical approaches. The last two chapters are more application oriented with a special emphasis on correlative microscopy in combination with the more classical transmission electron microscopy (Chapter 9: Data Mining) or photoluminescence studies (Chapter 10: Correlative Microscopy by (Scanning) Transmission Electron Microscopy and Atom Probe Tomography).

This book being the natural output of ongoing series of Atom Probe Schools organized every year by the "Groupe de Physique des Matériaux" at the University of Rouen, the authors would like thank at first all attendees who actively contributed to the great success of this school. We hope that in the present book they could find a hint of the great atmosphere that we shared with them. We also would like to thank all invited speakers who accepted to deliver some lectures. Their contributions provided for sure some useful and unique inspiration to this book: B. Gault, D.J. Larson, N. Kruse, L. Renaud, A. Shabli, A. Sharik, D. Raabe, B. Geiser, T. Prosa, D. Haley, T. Visard de Bocarmé, C. Oberdorfer, and F. De Geuser. We also would like to thank our colleagues who contributed to the Atom Probe School but do not appear in the chapter author list: Marie-Ange Derue, Agnès Dalle, Germain Martigny, Christine Vurpillot, Caroline Jorry, Philippe Pareige, Auriane Etienne, Bertrand Radiguet, Sébastien Duguay, Rodrigue Lardé, and Alain Bostel.

Williams Lefebvre-Ulrikson
François Vurpillot
Xavier Sauvage

LIST OF ABBREVIATIONS

1D	One-dimensional
2D	Two-dimensional
3D	Three-dimensional
ADLD	Advanced DLD
AP	Atom Probe
APT	Atom Probe Tomography
BF-TEM	Bright Field TEM
BSE	Backscattered electrons
CCD	Charge-Coupled Device (related to cameras and detection systems)
CFD	Constant Fraction Discriminator (related to detection systems)
Cs-corrector	Hardware correction of spherical aberration in STEM
CTEM	Conventional TEM
DC	Direct Current (related to voltage)
DF-TEM	Dark Field TEM
DLD	Delay Line Detector
EBSD	Electron Backscattered Diffraction
EDS	Energy Dispersive X-ray Spectroscopy (related to electron microscopy)
EELS	Electron Energy Loss Spectroscopy (related to TEM)
FDTD	Finite Difference Time Domain (related to numerical simulation)
FEG	Field Emission Gun (related to electron microscopy)
FIB	Focused Ion Beam
FIM	Field Ion Microscopy
FWHM	Full Width at Half Maximum (related to mass resolution)
FWTM	Full Width at Tenth Maximum (related to mass resolution)
GB	Grain Boundary
HAADF	High-Angle Annular Dark Field (related to STEM)
HR-STEM	High-Resolution STEM
HRTEM	High-Resolution TEM
HV	High Voltage
IBID	Ion Beam Induced Deposition
IR	Infrared (related to laser illumination)
KNN	Kth nearest neighbor
La-APT	Laser APT
LAWATAP	Laser-Assisted Wide-Angle Tomographic Atom Probe
LEAP	Local Electrode Atom Probe
MCP	Microchannel Plates (related to detectors)
ML	Monolayer
MQW	Multi Quantum Well
MRP	Mass Resolving Power (related to detection systems)
MSD	Maximum Separation Distance
PL	Photoluminescence
POSAP	Position-Sensitive Atom Probe

QCSE	Quantum Confined Stark Effect
QD	Quantum Dot
QW	Quantum Well
RDF	Radial Distribution Function
S/N	Signal to Noise ratio
S/TEM	Refers to TEM and STEM
SE	Secondary Electrons
SEM	Scanning Electron Microscopy
SIMS	Secondary Ion Mass Spectroscopy
SRP	Spatial Resolving Power (related to detection systems)
STEM	Scanning Transmission Electron Microscopy
STM	Scanning Tunneling Microscopy
TDC	Time to Digital Converter (related to detection systems)
TEM	Transmission Electron Microscopy
TKD	Transmitted Kikuchi Diffraction
TOF	Time of Flight
TRP	Time Resolving Power (related to detection systems)
UV	Ultraviolet (related to laser illumination)
VCA	Virtual Crystal Approximation (related to semiconductors)
μPL	Microphotoluminescence

CHAPTER ONE

Early Developments and Basic Concepts

D. Blavette, X. Sauvage
Groupe de Physique des Matériaux, University and INSA of Rouen, Normandie University, France

INTRODUCTION

Since the 1990s, atom probe tomography (APT) technique has reached a recognized position among the most important analytical microscopy techniques. APT is able to map out the spatial distribution of atomic species in three dimensions (3D). With a unique combination of chemical sensitivity and 3D capabilities with a spatial resolution near the atomic scale it provides unique information about nanoscale structures. In this introduction chapter, the overall fundamentals and performances of APT its main fields of application in material science, the historical background, and modern challenges are briefly described.

ATOM PROBE TOMOGRAPHY IN MATERIALS SCIENCE TODAY

The Unique Information Provided by Atom Probe Tomography

Imaging materials in 3D at the atomic scale has not only been a dream but also a challenge. This has been of vital importance for fundamental researches dealing with phase transformations or segregation of solute atoms to lattice defects. The characterization of microstructures at the ultimate scale is also of great importance for materials of industrial importance as it enables to better control the genesis and development of microstructures and to optimize processing routes and properties. Both functional and mechanical properties of materials depend not only on microscale features like grain sizes, precipitates, but also on nanoscale features such as clusters of solutes, chemical gradients, nanoprecipitates, and segregations on structural defects.

Scanning electron microscopes (SEM) have high imaging resolutions down to about 10 nm thanks to Field Emission Gun but their analytical capability based on X-ray emission is, however, still limited. This is the result

Atom Probe Tomography
ISBN: 9780128046470
http://dx.doi.org/10.1016/B978-0-12-804647-0.00001-2

of the scattering of electrons in matter that gives rise to a significant spread of the signal up to about 1 µm. This limitation can be easily overcome in the transmission electron microscope (TEM) or scanning-TEM where the sample thickness is typically less than 100 nm. In TEM, the limitation of the spatial resolution comes mainly from the aberrations of lenses used to focus the electron beam. During the past 10 years, the implementation of spherical aberration correctors in commercial instruments pushed the routine analytical capability of TEMs down to the atomic scale [1]. However, the chemical sensitivity of analytical TEM, based on electron energy loss spectroscopy or energy dispersive X-ray spectroscopy is typically limited to about one atomic percent in routine experiments. Besides, data and images are provided through the projection in two dimensions across the electron-transparent sample and part of the 3D information might be lost. Recent developments in TEM tomography [2] make it also possible to access part of this 3D information but hardly with a near-atomic scale resolution. Secondary ion mass spectroscopy overcomes electron microscopy techniques for the chemical sensitivity by several orders of magnitude (Fig. 1.1) but at the expense of

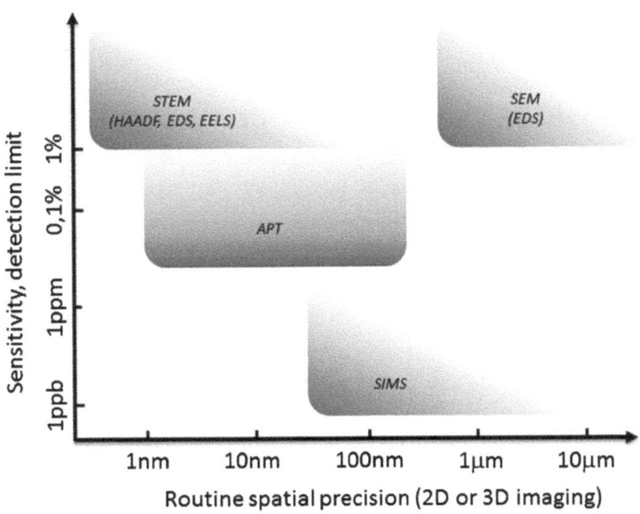

Figure 1.1 Range of routine spatial precision in two-dimensional (2D) or three-dimensional (3D) imaging mode and chemical sensitivity of major analytical microscopy techniques showing that the atom probe tomography (APT) combines near-atomic resolution with sensitivity close to few atomic parts per million. *EDS*, Energy dispersive X-ray spectroscopy; *EELS*, electron energy loss spectroscopy; *SEM*, scanning electron microscopy; *SIMS*, secondary ion mass spectroscopy; *STEM*, scanning transmission electron microscope.

the spatial resolution as the imaging capability is typically limited by a resolution of about 50 nm [3].

In the landscape of analytical microscopy (Fig. 1.1), APT exhibits unique capabilities: it combines 3D information, near-atomic resolution (0.1 nm in depth <1 nm at the sample surface) with a chemical sensitivity down to a few atomic parts per million in the best cases. The typical analyzed volumes are 50 × 50 × 200 nm^3 (Fig. 1.2). Compared to TEM, APT is certainly less versatile. First, it is a destructive technique. Measurements cannot be repeated on the same sample. It is also difficult to relate the chemical information to structure. Besides, the field of view or the region of interest is more limited. However, the unique capabilities of APT to investigate materials in 3D at a near-atomic scale, including semiconductors and oxides, have been leading to a growing community both in the academic world and industry.

Fields of Application

Field ion microscopy (FIM), invented by E.W. Müller in the early 1950s, can be viewed as the precursor of modern APT. Both are based on high electric fields. It has provided major contributions in surface science and physical metallurgy thanks to its atomic-scale resolution (Fig. 1.3) As early as the 1960s it has been used to image crystalline defects like grain boundaries (GBs) [4] or vacancies [5]. Phase contrast also made it possible to investigate at least qualitatively precipitation, voids, irradiation defects, segregation to lattice defects or surface diffusion, or reactions in metallic alloys [6].

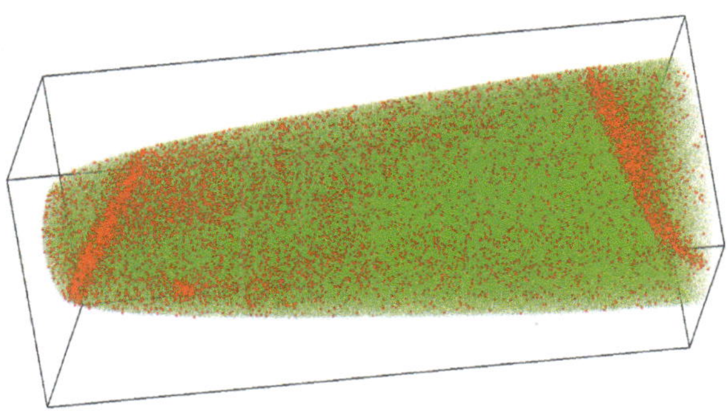

Figure 1.2 Three-dimensional reconstruction of a volume analyzed by atom probe tomography in a steel showing both the carbon segregation to grain boundaries (two are displayed) and a nanoscale carbide [Fe, green; C, red; 45 × 45 × 120 nm^3].

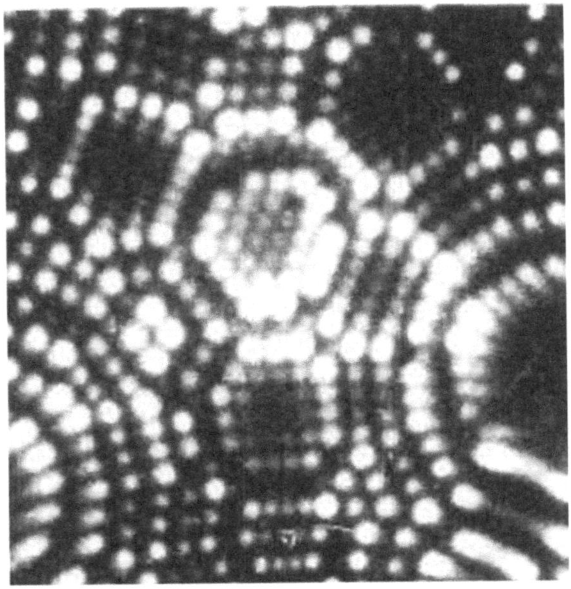

Figure 1.3 Field ion microscopy image of the surface of a PtAu alloy. Each dot is the image of a single atom. The intersection of crystalline atomic planes with the hemispherical surface of the sample gives rise to the specific ring contrast. *With courtesy of C. Martin, GPM, France.*

The early applications of atom probe in the 1970s were mainly limited to good conductors of electricity, that is, metallic alloys. Metallic glasses [7], steels [8], nickel-based superalloys [9], and aluminum alloys [10] were among the first materials investigated. For the first time it was possible to reveal composition fluctuations at the nanoscale and to get some insights into the early stages of precipitation.

One of the major breakthroughs came in the late 1980s and early 1990s with the implementation of a position-sensitive detector on the instrument [11,12]. This innovation gave new dimensions to the instrument. The one-dimensional (1D) atom probe of E.W. Müller, which gave only 1D composition profiles, became a 3D atom probe (or an atom probe tomograph). The nanoscale composition fluctuations were then imaged in 3D, revealing the shape, the distribution, and the morphology of nanoscale clusters or decorated defects (solute segregation). At the beginning of the 21st century, the APT technique started to spread widely in material science laboratories, becoming a complementary tool to TEM.

A second important breakthrough has been done with the replacement of electric pulses by ultrafast laser pulses. This opened up APT to poorly conductive materials [13] like semiconductors [14], functional materials [15], oxides [16], ceramics and even some polymers [17], and biological materials [18,19]. This innovation has transformed APT into a unique technique in nanosciences and microelectronics (nanowires, magnetic multilayers for spintronic including tunnel junctions containing highly resistive oxide layers). This evolution has been possible thanks to the parallel development of focused ion beam (FIB) milling techniques [20].

Combining Atom Probe Tomography and Other Microscopy Techniques

SEM combined with FIB (SEM-FIB) is a technique commonly used in combination with APT for the specimen preparation stage [20]. One of the main advantages is the imaging capabilities of the instrument for site-specific specimen preparation. Electron back scattered diffraction may be used, for example, to select specific GBs and to correlate segregations (to be revealed by APT) with the misorientation. In the past 5 years, correlative microscopy approaches combining information from TEM and APT have been proposed [21,22]. There are typically two very different objectives: (1) using TEM imaging capability to understand or correct APT 3D volume reconstruction artifacts (if any); (2) using TEM to get an accurate picture of crystalline structures, defects and orientation relationships, and APT to access the 3D information on the spatial distribution of solute elements around lattice defects. In the latter case, two options can be considered. APT samples can be observed in TEM prior to analysis so as to identify and characterize defects, but this requires specific specimen holders, and a needle-shaped sample with a nonuniform thickness is not optimal for analytical TEM. Thus, in case of extended features like GBs, the easiest and optimal procedure might be to prepare with FIB one APT sample and one TEM sample along the region of interest in close locations. In case of high-density objects like nanoscale precipitates, randomly prepared TEM and APT samples can be analyzed separately and compared on the basis of statistics.

BASIC CONCEPTS

Seeing Atoms With Field Ion Microscopy

The history of APT begins in the early 1950s with the invention of a new imaging technique by E.W. Müller, the FIM [23]. The physical principle of this instrument is based on the field ionization of a rare gas in the vicinity of the sample surface. FIM provides a picture of the distribution of the electric field over the sample surface. A high electric field is required for field ionization, typically in the range of 30–60 V/nm. Such a field results from a positive high voltage V applied to the sample that is prepared with a very specific shape: a sharp tip with an end radius R typically in a range of 20–50 nm. The resulting electric field F writes:

$$F = V/k_f R \tag{1.1}$$

where k_f is the field factor, typically in a range of 2–10. Gas atoms ionize preferentially over atomic protrusions on the tip surface where the local field is higher. They are then accelerated by the electric field up to a screen located at a distance L of few centimeters ahead where they generate a magnified image of the atomic-scale distribution of the electric field at the tip surface (Fig. 1.4). FIM image is close to the stereographic projection of the sample surface and the magnification simply writes as:

$$\text{Mag} = L/\alpha R \tag{1.2}$$

where α is the image compression factor. The value of α is between 1 and 2. A typical magnification of about 1 million is routinely reached. The first instrument developed by E.W. Müller was extremely simple. It was made of a glass vacuum chamber containing the sample cooled by liquid nitrogen and

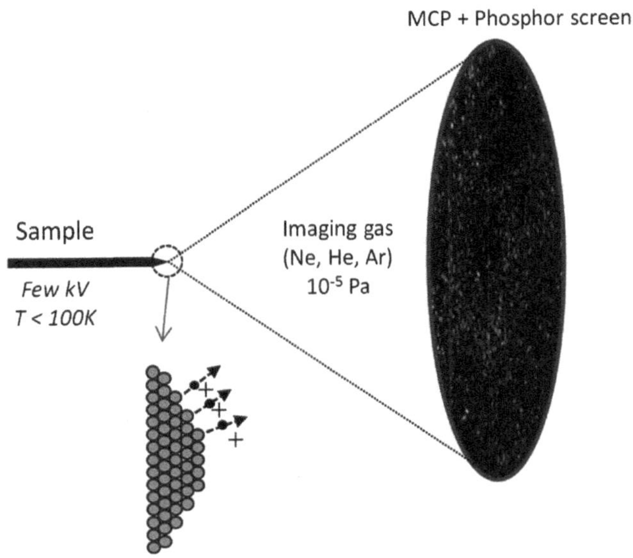

Figure 1.4 Schematic representation of the field ion microscope. *MCP*, Microchannel plate.

a phosphor screen. This was the first microscope able to provide an atomic resolution image of a surface. However, the main limitation of this early microscope was the extremely weak intensity of the FIM images. A great breakthrough was made with the implementation of charge amplifiers (microchannel plates) to the imaging system. This dramatically increased the brightness of the FIM images.

Using this instrument, E.W. Müller discovered a second physical phenomenon, the field evaporation of surface atoms. When the electric field is increased and high enough to break atomic bonds at the surface, atoms leave the surface as positive ions and are accelerated by the electric field, they are "field evaporated." This way, it is possible to evaporate the sample atom by atom, and thus to explore the volume of the sample, atomic layer by atomic layer. This concept will be further exploited by E.W. Müller in the design of the first atom probe. In these early times of development, it was discovered that phase contrast may appear in FIM images. It was quickly understood that this contrast was driven by the different evaporation fields of phases. High (respectively low) evaporation field precipitates appear in bright (respectively dark) contrast in FIM images. This made it possible to image nanoscale precipitates and characterize microstructures.

Field evaporation is of capital importance as this makes it possible to explore the sample in depth and as a consequence in the three dimensions of space. It is now possible to easily build 3D reconstructions of the evaporated volume in FIM using digital recording of successive images collected during the evaporation of the sample. This technique, called 3D-FIM [24], provides images with an atomic-scale resolution and a relatively large field of view as compared to APT (Fig. 1.5).

Analyzing Materials Thanks to Time-of-Flight Mass Spectroscopy

One of the main limitations of FIM has been that this microscope is unable to identify the chemical nature of imaged atoms. The invention of the atom probe by Müller, Panitz, and McClean in 1969 [25] has been a decisive breakthrough as it made it possible to solve this issue. The principle is based on the field evaporation of surface atoms and their identification of atoms by time-of-flight mass spectrometry. Light (respectively heavy) species have a larger (respectively smaller) velocity and consequently a smaller time of flight (t). Surface atoms are evaporated using electric pulses (V_p) that are

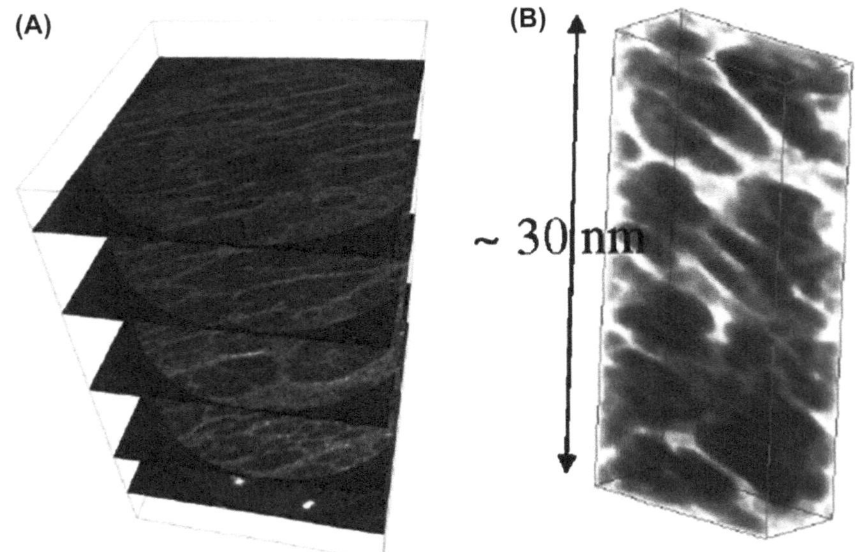

Figure 1.5 Three-dimensional field ion microscopy data collected on an Al-Sm alloy. (A) Stack of three-dimensional images showing the two-phase mixture (Sm-rich phase is brightly imaged), (B) three-dimensional reconstruction generated from the three-dimensional field ion microscopy images. *Reproduced with permission from Vurpillot F, Gilbert M, Deconihout B. Surf Interface Anal 2007;39:273–7.*

superimposed to the steady state voltage V_0. These pulses give the origin of times and are repeated periodically (typically 100 kHz).

The measurement of time of flight (t) gives the mass to charge ratio (M/n) of evaporated ions. Writing that the initial energy $W = neV$ (n is the charge state and e the electron charge) is equal to the final kinetic energy leads to:

$$neV = \frac{1}{2}Mv^2 \tag{1.3}$$

where $V = V_0 + V_p$, M is the mass of atom, $v = L/t$ the velocity of ions, L is the flight distance.

A mass spectrum can be built and a careful interpretation of various peaks permits the identification of chemical species (Fig. 1.6). It is clear that a higher mass resolution reduces overlaps and leads to a more reliable identification of peaks in spectrum. A high mass resolution $m/\Delta m$ (>1000) requires ultrashort pulses (1 ns) and a time precision better that the nanosecond. It is worth keeping in mind that electric pulses generate energy deficits: atoms may field evaporate on the falling edge of pulses leading to

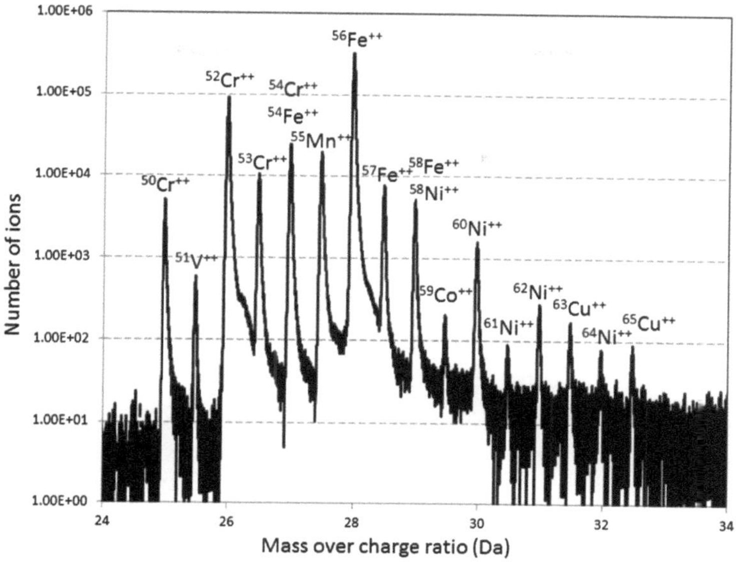

Figure 1.6 Portion of mass spectrum collected from an industrial stainless steel. *Courtesy of F. Danoix.*

a potential energy that is smaller. This contributes to degrade the mass resolution. Major innovations have been brought about to overcome this problem. Energy deficit compensation systems that were subsequently implemented to instruments made it possible to both improve the mass resolution and reduce the background noise in mass spectra, increasing the sensitivity threshold [26]. The use of local electrode to field evaporate specimens in modern local electrode atom probes (LEAP™) has also been a decisive breakthrough [33].

It is worth mentioning that in the early "1D atom probes," atoms were collected from a very small area of the sample surface (typically a few nanometers in diameter compared to 50 nm at present). Thus, the chemical information was collected atomic layer after atomic layer, in a cylinder of 1 nm or so in diameter [27]. This is the way composition profiles were generated. The second limitation of such instrument was ion pile-ups on detector (when two ions are evaporated on the same pulse and reach the detector in the same time window). To reduce the resulting artifacts in composition measurements, multiple anodes detectors were designed in the 1980s [28]. At that time, the Atom Probe community was small and only a few groups were able to set up such instruments where ultravacuum, low temperatures (20–80K), ultrafast electronics (<1 ns), and computer-aided experiments were necessary.

Experiments also provided essential information for a better understanding of field evaporation mechanisms. Let us mention the postionization theory developed by Haydock and Kingham [29]. According to this theory, atoms first field evaporate as singly charged ions. In the early steps of their travel close to the tip, ions subsequently may lose extra electrons depending on the electric field. The higher the field, the higher the charge state of ions. Charge state distributions of ions are today used to assess the electric field during experiments.

At that time, the major advantage of atom probe compared to other analytical tools was its quantitativity in composition measurements, high sensitivity, and high spatial resolution. However, by essence the lateral resolution was not the ultimate resolution reachable as it was given by the size of the analyzed area ($\mathrm{d}a = \mathrm{d}D/\mathrm{Mag}$ with $\mathrm{d}D$ the detector size). At the same time, the analyzed region was extremely small (1−5 nm) so as to preserve the lateral spatial resolution. To overcome this issue, Panitz designed the imaging atom probe, the precursor instrument of today's 3D atom probes. In this instrument, a screen, placed at 10 cm from the tip, is activated at a chosen time-of-flight interval. The image thus provides the two-dimensional (2D) distribution of the selected chemical species related to this time-of-flight interval. The main shortcoming of this 2D imaging technique is that only one selected species can be imaged and composition data cannot be derived from images.

The First Three-Dimensional Atom Probes

An outstanding breakthrough occurred in 1988 with the design of the first position-sensitive atom probe by the Oxford group (A. Cerezo and G.D.W. Smith). This stimulated the French laboratory (University of Rouen) and US laboratory (Oak Ridge National Laboratory) to quickly enter the adventure. The first two prototypes (Oxford 1988 [11], Rouen 1993 [12]) were based on different detector designs. These instruments rely on the implementation of a time-resolved position-sensitive detector (Fig. 1.7). This detector should be able to simultaneously provide the time of flight of ions and the position of their impact. This was the key problem to solve in the early developments of the 3D atom probes. Besides, since the evaporation of two or more atoms on the same pulse cannot be avoided, quantitative measurements also require a multihit capability with a time separation as low as a few nanoseconds. These considerations led to the implementation or the design of various detectors. The Oxford team used a commercial wedge and strip detector and the ONRL team an optical device, whereas the Rouen team

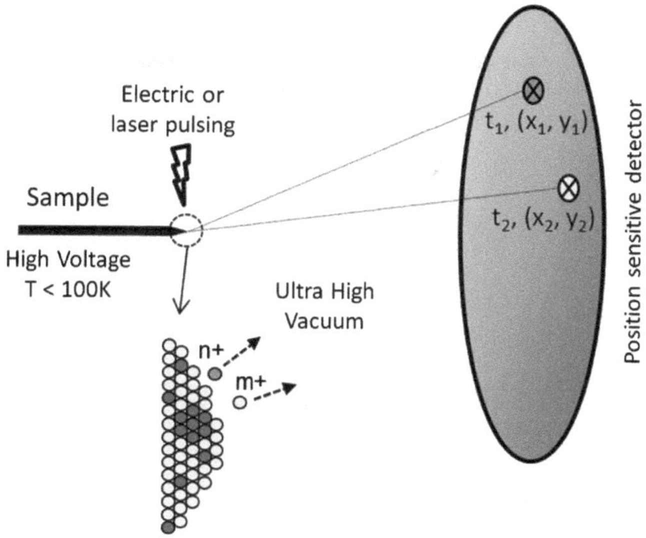

Figure 1.7 Schematic representation of the atom probe tomography experimental setup.

designed a 10 × 10 multianode detector, which was in the 1990s the most efficient device to deal with multihits [30].

Once these detectors were successfully implemented on atom probes and the first data collected, then came the next challenges related to 3D reconstructions and data processing. Projection laws and reconstruction procedures were inspired from those used in FIM [31]. The lateral positions of atoms at the tip surface (x,y) were simply derived from the position of ion impacts (X,Y) using simple relationships (Fig. 1.7):

$$x = X/\text{Mag} \tag{1.4}$$

$$y = Y/\text{Mag} \tag{1.5}$$

The magnification of APT is again given by Eq. (1.2). The third coordinate can be derived from the cumulative number of detected ions that is proportional to the number of evaporated layers and to the analyzed depth consequently. These simple geometrical reconstruction procedures are still used. More accurate and sophisticated procedures are described in "Chapter 7, Three-Dimensional Reconstruction in Atom Probe Tomography: Basics and Advanced Approaches." When atoms are evaporated in a regular sequence from the sample surface, these procedures provide reconstructions with a routine spatial resolution of about 0.2 nm along the analysis direction and about 0.5 nm in lateral directions. This later can be considerably degraded close to interfaces that separate regions with different evaporation fields due to local magnification effects and trajectory aberrations. The designers of the early APT instruments (Fig. 1.7) had also to face a major issue: the compromise between mass resolution and field of view. Indeed, a high mass resolution requires a long flight length, whereas a large field of view in this projection microscope requires a detector as close as possible from the sample. In the late 1990s, the typical sample area "seen" by the detector was $10 \times 10\ nm^2$ (as compared to $50 \times 50\ nm^2$ today) and the mass resolution (full width at half maximum) was $m/\Delta m \sim 300$. The typical pulse rate was lower than 1 kHz, ie, almost three orders of magnitude lower than at present.

To enlarge the field of view, and thus to increase the size of the analyzed volume in a single experiment, it has been necessary to decrease the magnification and thus the flight length (Eq. (1.2)). However, reducing the flight length and consequently times of flight has a direct negative impact on the mass resolution (Eq. (1.3)). Therefore, reflectrons (electrostatic mirrors [32]) and local electrodes (LEAP IMAGO™ [33]) have been implemented at the beginning of the 21st century, leading to a dramatic improvement of the capability of instruments. Faster acquisition systems combined with high-frequency pulsers also helped to reduce acquisition times. New detectors have also been developed, and modern atom probes are now all equipped with so-called delay-line detectors [34] that are able to combine excellent multihit capability and good time separation limit.

The Challenge of Data Mining

The large field of view of modern atom probes combined with fast acquisition systems allows collecting in routine mode several millions of ions per hour with the final data set typically containing up to 50 millions of atoms. Thus, the first challenge of data mining is to technically deal with such a huge amount of data. There are two main softwares that have been developed and that are routinely used in the community, the CAMECA commercial IVAS software and the "GPM3D-soft" designed by the APT Rouen team. The second challenge is to extract quantitative information, beyond the simple visualization of the data. The processing of 3D data has mainly started on the basis of procedures developed for the early "one-dimensional" atom probes [35]. Then, lots of routines and procedures have been proposed and developed, especially for the statistical study of clustering or the automatic identification of nanoscale precipitates. There is, however, a third challenge for data mining. The phase contrast that made the old FIM so attractive and powerful to visualize nanoscale composition fluctuations or nanoscale particles arises from spatial variations of the evaporation field. Since these variations are not known a priori, it is impossible to implement them in the 3D reconstruction procedure and it typically leads to the so-called local magnification effects. Such local magnification effects may lead to severe image distortions and also may significantly affect the lateral resolution, especially at interfaces between two phases or where chemical gradients are sharp (segregations). Then, for nanoscale precipitates, the local composition might be biased by ions coming from the surrounding matrix. Models to simulate ion trajectories have been proposed and they turned out to be of great help for the understanding and sometimes for the correction of aberrations and image distortions [36–40].

Laser Atom Probe Tomography, Opening the Technique to New Fields

In the early development of atom probes, the field evaporation of atoms from sample surfaces was controlled by electric field pulses generated by high-voltage pulses. It intrinsically limited the technique to rather good electrical conductors, ie, mainly metallic alloys. Laser pulses superimposed on standing voltage instead of electric pulses have been first implemented on 1D atom probe in the 1980s by Kellogg and Tsong [41]. In 2005, femtosecond lasers have been implemented on 3D atom probes and the first quantitative laser atom probes were designed [13]. Under certain conditions, a much better mass resolution as compared to electric mode was obtained. But the main breakthrough with laser APT has been to open the technique to a wider range of materials including bad conductors like semiconductors and oxides. Since then, the APT is an advanced analytical microscope not only for metallurgists but also for scientists in the field of nanoscience, functional materials, and microelectronics.

The physical mechanisms of laser-enhanced field evaporation have been the subject of numerous debates (electric field rectification and enhancement against laser-induced temperature increase). It is now well accepted that the laser pulse induces a sharp temperature increase that promotes field evaporation of surface atoms. However, the limited thermal conductivity of the tip leads to a gradual decrease of sample temperature. Field evaporation keeps on proceeding well after laser pulses leading to enlarged mass peaks. The degradation of the mass resolution depends on many factors, including the shank angle and conductivity of sample, the wavelength of laser, beam intensity and focalization.

CONCLUDING REMARKS, THE SPECIFICITIES OF ATOM PROBE TOMOGRAPHY

In the analytical microscopy landscape, APT stands in a unique place and not only because of its unique combination of chemical sensitivity and spatial resolution. Contrary to TEM, for example, where samples can be observed and analyzed in various conditions and as long as necessary (if care is taken to avoid contamination and irradiation damages), APT is intrinsically a destructive technique. Then, as the sample is destroyed during the analysis, it means that the conditions used (temperature, laser wave length and energy, pulse fraction, pulse repetition rate, and detection rate) should be settled carefully before starting the experiment as they are of critical importance. Another specificity of APT is that the sample is also the microscope lens. Indeed, trajectories are almost exclusively determined by the tip shape (radius and shank angle), the crystallographic directions of investigated phases, and their local chemical composition. Thus, special care should be taken during the specimen preparation and data interpretation. But, the other important aspect is that the shape of the sample also changes during the analysis. Due to the shank angle, as soon as few atomic layers are evaporated, the tip radius increases. This is why it is necessary to continuously increase the standing voltage, to keep constant the electric field to ensure an almost constant evaporation rate. Such an evolution of the tip radius leads to a constant change of the magnification that cannot be directly controlled by the operator. This is basically why most of APT-reconstructed volumes exhibit a cone shape when the full data set is displayed. Last specificity, successful analyses typically last a few hours, and are fully automatically controlled. But the analysis time is usually less than that required for accurate data processing, which starts with the mass spectrum calibration and interpretation, then the 3D reconstruction and at last the visualization and data mining.

REFERENCES

[1] Bleloch A, Lupini A. Mater Today 2004;7(12):42–8.

[2] Möbus G, Inkson BJ. Mater Today 2007;10(12):18–25.

[3] Christien F, Downing C, Moore KL, Grovenor CRM. Surf Interface Anal 2012;44: 377–87.

[4] Brandon DG, Ralph B, Ranganathan S, Wald MS. Acta Metall 1964;12(7):813–21.

[5] Speicher CA, Pimbley WT, Attardo MJ, Galligan JM, Brenner SS. Phys Lett 1966; 23(3):194–6.

[6] Brenner SS, Camus PP, Miller MK, Soffa WA. Acta Metall 1984;32(8):1217–27.

[7] Piller J, Haasen P. Acta Metall 1982;30(1):1–8.

[8] Babu SS, Hono K, Sakurai T. Metall Mater Trans A 1994;25(3):499–508.
[9] Wanderkaa N, Glatzel U. Mater Sci Eng A 1995;203(1–2):69–74.
[10] Ringer SP, Hono K, Sakurai T. Metall Mater Trans A 1995;26(9):2207–17.
[11] Cerezo A, Godfrey TJ, Smith GDW. Rev Sci Instrum 1988;59:862.
[12] Blavette D, Deconihout B, Bostel A, Sarrau JM, Bouet M, Menand A. Rev Sci Instrum 1993;64:2911.
[13] Gault B, Vurpillot F, Vella A, Gilbert M, Menand A, Blavette D, et al. Rev Sci Instrum 2006;77:043705.
[14] Cojocaru-Mirédin O, Cadel E, Vurpillot F, Mangelinck D, Blavette D. Scr Mater 2009;60:285–8.
[15] Rigutti L, Blum I, Shinde D, Hernández-Maldonado D, Lefebvre W, Houard J, et al. Nano Lett 2014;14:107–14.
[16] Roussel M, Talbot E, Gourbilleau F, Pareige P. Nanoscale Res Lett 2011;6:164.
[17] Zhang Y, Hillier AC. Anal Chem 2010;82:6139–47.
[18] Gordon LM, Tran L, Joester D. ACS Nano 2012;6(12):10667–75.
[19] Narayan K, Prosa T, Fu J, Kelly TF, Subramaniam S. J Struct Biol May 2012;178(2): 98–107.
[20] Larson DJ, Foord DT, Petford-Long AK, Liew H, Blamire MG, Cerezo A, et al. Ultramicroscopy September 1999;79(1–4):287–93.
[21] Arslan I, Marquis EA, Homer M, Hekmaty MA, Bartelt NC. Ultramicroscopy 2008; 108:1579–85.
[22] Herbig M, Choi P, Raabe D. Ultramicroscopy 2015;153:32–9.
[23] Mueller EW. The field ion microscope. Zeit F Phys 1951;131:136–42.
[24] Vurpillot F, Gilbert M, Deconihout B. Surf Interface Anal 2007;39:273–7.
[25] Müller EW, Panitz J, Mc Lane SB. The atom probe field ion microscope. Rev Sci Instrum 1968;39:83–8.
[26] Poschenrieder WP. Int J Mass Spectrom Ion Phys 1972;9:357–73.
[27] Blavette D, Bostel A. Acta Metall 1984;32:811–6.
[28] Blavette D, Bostel A, Sarrau JM. J de Phys 1986;47(C2):473–7.
[29] Haydock R, Kingham DR. Surf Sci 1981;103:239.
[30] Blavette D, Bostel A, Sarrau JM, Deconihout B, Menand A. Nature 1993;363:432–5.
[31] Bas P, Bostel A, Deconihout B, Blavette D. Appl Surf Sci 1995;87:298.
[32] Kelly TF, Camus PP, Larson DJ, Holzman LM, Bajikar SS. Ultramicroscopy 1996; 62(1–2):29–42.
[33] Bemont E, Bostel A, Bouet M, Da Costa G, Chambreland S, Deconihout B, et al. Ultramicroscopy 2003;95(1–4):231–8.
[34] Da Costa G, Vurpillot F, Bostel A, Bouet M, Deconihout B. Rev Sci Instrum 2005;76: 013304.
[35] Blavette D, Sarrau JM, Bostel A, Gallot J. Rev Phys Appl 1982;17:435–40.
[36] Vurpillot F, Bostel A, Blavette D. Appl Phys Lett 2000;76:3127.
[37] Oberdorfer C, Schmitz G. Microsc Microanal 2011;17:15.
[38] De Geuser F, Lefebvre W, Danoix F, Vurpillot F, Forbord B, Blavette D. Surf Inter Anal 2007;39:268.
[39] Vurpillot F, Cerezo A, Blavette D, Larson DJ. Microsc Microanal 2004;10:384.
[40] Philippe T, Gruber M, Vurpillot F, Blavette D. Microsc Microanal 2010;16:643.
[41] Kellog G, Tsong TT. J Appl Phys (USA) 1980;51(2):1184.

CHAPTER TWO

Field Ion Emission Mechanisms

F. Vurpillot
Groupe de Physique des Matériaux, University and INSA of Rouen, Normandie University, France

INTRODUCTION

The specimen in atom probe tomography (APT) or field ion microscopy (FIM) is a field emitter, or more generally a field ion emission source. This means that these two techniques exploit the colossal static electric field that is produced at the end of a sample to emit positively charged particles. Understanding how these techniques work requires describing not only how a high and static electric field may affect the nature of surface atoms of materials of interest but also how it is possible to generate it.

Field-induced emission is an extremely large and spread mechanism, exploited in various techniques and applications such as electron field emission, transmission electron microscopy (TEM), or liquid metal ion source for scanning electron microscopy coupled with focused ion beam (SEM-FIB). Much of the source physics was established long ago when FIM was being developed by Müller and coworkers. Therefore, most of the basics can be found in Refs. [1−7]. We have focused this chapter on principles useful for their applications in APT.

The general idea of all these techniques is to generate an electric field so strong that it will approach or overcome the electrostatic field existing in the core of atoms present at the surface of the specimen of interest. A sufficient negative electric field (with respect to the specimen surface) may induce the emission of negative particles (electrons), without help of any external excitation (eg, temperature, photo absorption). This is the principle of the field electron microscopy (FEM). A sufficient positive electric field may generate the controlled emission of positive particles from the near surface, from external atoms to the material (field ionization, FI in the FIM) or may ionize atoms of the material surface itself [field evaporation (FEV) in atom probe]. The apparent simplicity of these processes, which are tuned at the macroscopic scale only by the amplitude of the voltage bias applied

Atom Probe Tomography
ISBN: 9780128046470
http://dx.doi.org/10.1016/B978-0-12-804647-0.00002-4

to the sample, is disarming. Nevertheless, the field magnitude that we actually produce at the surface sample is far beyond the values we use to manipulate in classical physical problems. For instance, in the best field effect transistors, where the channel is controlled by an electric field and isolated by a good insulator (a silicon dioxide layer), the dielectric strength can approach a few 10^9 V/m. Above this value, the materials reach a breakdown point, that may induce the physical destruction of the device. In FIM or APT, an electric field 10 times larger is applied, with a total control of the emission current.

This chapter will give an introduction to the physics of how and why this nonstandard electric field is generated. A review of some basic ideas of the electrostatics physics in field emitters will be presented. The consequence of this field will enable to introduce practical considerations on the principal physical mechanisms induced in field emission, ie, field emission, field ionization, field evaporation, and postionization processes.

FIELD GENERATION AT THE SURFACE OF A FIELD EMITTER

Electric Field in the Electrostatic Approximation

Electrostatics is a branch of physics that deals with the phenomena and properties of stationary or slow-moving electric charges with no acceleration. Electrostatic phenomena arise from the forces that electric charges exert on each other. Such forces are described by Coulomb's law. Even though electrostatically induced forces seem to be rather weak, the electrostatic force between, eg, an electron and a proton, that together make up a hydrogen atom, is about 36 orders of magnitude stronger than the gravitational force acting between them. The forces sticking atoms together is in the same range of magnitude. On a field emitter, a sufficient electric field is required to induce forces to enable counterbalance of the attractive forces that maintain the cohesion of atoms from the surface. Surprisingly, the generation of this field is understandable in a classical description of the phenomenon, ie, using standard Maxwell's equations in the electrostatic approximation.

The validity of this approximation is only questionable when considering nonstationary flux of charges at the sample surface of significant amplitude. Considering that at maximum only a few thousand elementary charges are experimentally drained from the sample per second, a static description is in most of the case a very good approximation of the actual field generation mechanism.

Charges and Electric Field Distribution in the Vicinity of a Field Emitter

Electrical charges and field in the vicinity of a field emitter control the erosion mechanism from the sample surface to the detector. A close inspection of these distributions in the geometry of the atom probe (AP) is mandatory to fully understand the basic mechanisms involved. This section describes in the simplest approach the theoretical calculation of these distributions. The reader is invited to follow given references for advanced mathematical treatments, which could have been avoided for reasons of simplicity and space.

Geometrical Description of the Atom Probe Tip in the Atom Probe

The sample of interest is prepared as a sharply pointed needle, with an end apex approximately hemispherical. The end radius is in the range 10–200 nm. The length of the needle is always longer than several microns to produce sufficient enhancement of the local field at the apex of the pointed sample (see Fig. 2.1). A positive electrical bias (voltage V_{DC} in the range 1–30 kV) is applied to the sample. The needle is often placed in front of a grounded counter electrode (named Local Electrode in commercial instruments from CAMECA) [8]. Ions pass through a hole in this

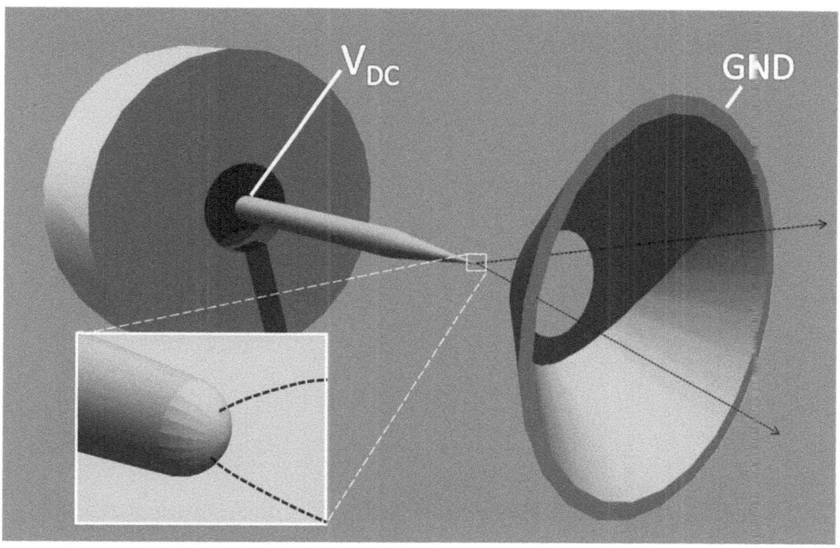

Figure 2.1 Schematic image illustrating the physical geometry of a tip in front of a counter electrode. The sharp needle is biased to a voltage V_{DC} that generates a high electric field at the tip surface, counter electrode is biased to the ground voltage (GND). This tip has submicron dimension (end radius <100 nm).

counter electrode and fly to the ion detector placed a few centimeters away. We will discuss the application of voltage pulses onto this electrode later in the chapter. Note that those ion trajectories can be modified by additional ion optic devices but these devices have normally no influence on the local field at the sample surface and will not be considered hereafter.

The sample must be able to drain the electrical current generated by FEV or field ionization to avoid a significant drop in electric voltage from the tip apex to its base. Since this current induced by FEV is in the subfemtoampere range ($\sim$1–10,000 elemental charges drained per second), and the tip length (the material of interest) is in the micron scale, it was shown that even for highly electrically resistive material, the field distribution around the tip is similar to a metal case (a wide range of materials, from metals to ceramics can be analyzed in APT [9,10]). We will consider in this description the sample as a perfect conductor for the sake of simplicity.

The Sphere's Toy Model: Field, Charge, and Surface Stress on a Field Emitter

In Maxwell's electrostatics, conductive surface charge is treated as a thin uniform layer. The electric field of a conducting sphere biased to a voltage V can be obtained by straightforward application of Gauss' law. Considering a surface in the form of a sphere at radius $r > R$ (Fig. 2.2A), the electric field has the same magnitude at every point of the surface and is directed outward. At the surface of the sphere the amplitude of the electric field is given by

$$F = \frac{V}{R} \tag{2.1}$$

Following Coulomb's law, this field results from the surface density of charge on the sphere

$$\sigma = \varepsilon_0 F \tag{2.2}$$

For $r > R$, the field decreases with r as

$$\vec{F}(\vec{r}) = \frac{V}{R} \times \left(\frac{R^2 \vec{r}}{\|\vec{r}\|^3} \right) \tag{2.3}$$

Considering a sphere with a radius about 100 nm submitted to a voltage of 1 kV, the generated field reaches 10^{10} V/m, which corresponds to a surface charge density of about 0.5 electrons/nm^2. Such field is colossal, three orders of magnitude stronger than the dielectric strength of dry air, and this

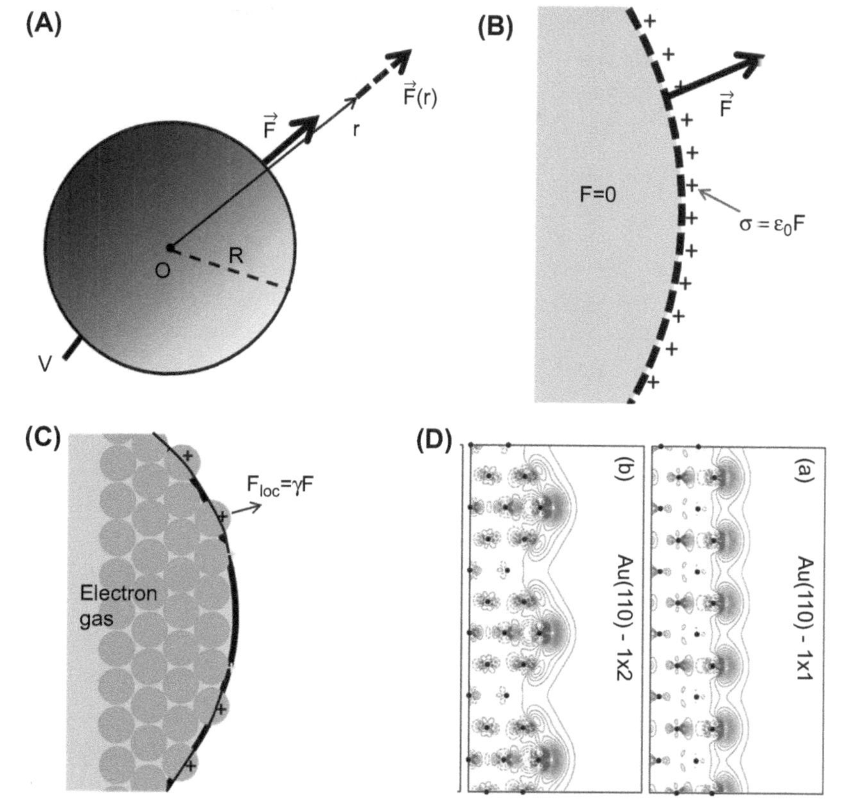

Figure 2.2 Schematic image illustrating the generation of field in the simple sphere model. (A) Gauss' law applied to the spherical geometry, the conductive sphere of radius R is submitted to a voltage V, giving rise to a surface electric field F. (B) At the surface of the sphere, a sheet with density of charge σ gives rise to the external field. (C) On an atomically defined surface, charge tends to be located at the position of protruding surface atoms. The local field is higher than these local charges and an enhancement factor γ of the electric field can be defined locally. (D) Field enhancement and localization of the surface charge is predicted by quantum theory calculation (see Ref. [11]).

field approaches the internal strength present inside atoms (it should be noted that electric field may normally not be considered as a perturbation in atomic physics when the amplitude of field approaches a few percent of an atomic unit of field, ie, an a.u. $= 5.14 \times 10^{11}$ V/m).

The presence of this colossal field at near surface of the material gives rise to drastic modifications on free atoms, molecules, ions, and cations that could surround the considered sphere. It can also strongly affect the surface of the material itself. In a first approach, the surface stress induced by the charge layer directed outward the surface is considered. The force acting on the surface, known as *Maxwell stress*, is induced by the natural interrepulsion of charges present at the surface. Therefore this force always tends to pull on the surface whatever the sign of the charge layer. The intensity of this force per surface unit is

$$f_s = \frac{1}{2}\varepsilon_0 F^2 \quad (2.4)$$

In our simple example, it gives rise to a stress higher than 500 MPa, a value close to the mechanical yield strength of most materials. Outside the tip surface, the intense field is located close to the surface. The field decreases one order of magnitude in a few hundreds nanometers. Sufficient electric field produces direct emission of atoms as charged particles, a phenomenon named as field ionization that will be treated in the following part of this chapter.

The Field Factor and the Mesoscopic Influence of the Tip Shape and Local Environment

The real shape of a field emitter deviates from the spherical approximation. It is generally described as a truncated cone with a cylindrical symmetry terminated by a rounded cap, the average radius of curvature of this cap giving rise to the required field enhancement when a voltage is submitted to the sample (see Fig. 2.1). The presence of the shaft strongly influences the magnitude of the field that may be attained. Nevertheless, the field can be enhanced by placing the sample in front of a grounded electrode (local electrode).

The intensity of the produced field may be evaluated by modeling the tip with a more rigorous geometry. Such kinds of computation have been done by several authors using analytical solutions, and the impact of geometrical features on the local surface field was found. An elegant analytical description of the field distribution around a simple field emitter is found when considering the sample and the counter electrode as two confocal paraboloids of revolution. In this geometry, the conductive sample is a constant voltage surface (V) in front of a grounded surface ($V = 0$) situated at a distance L. In this system of coordinates an analytical solution to the Laplace equation exists; the voltage simply decreases from the tip to the counter electrode (Fig. 2.3A, Ref. [12]). Gomer and some other authors presented typical results of such calculations (see for instance Refs. [13–16]).

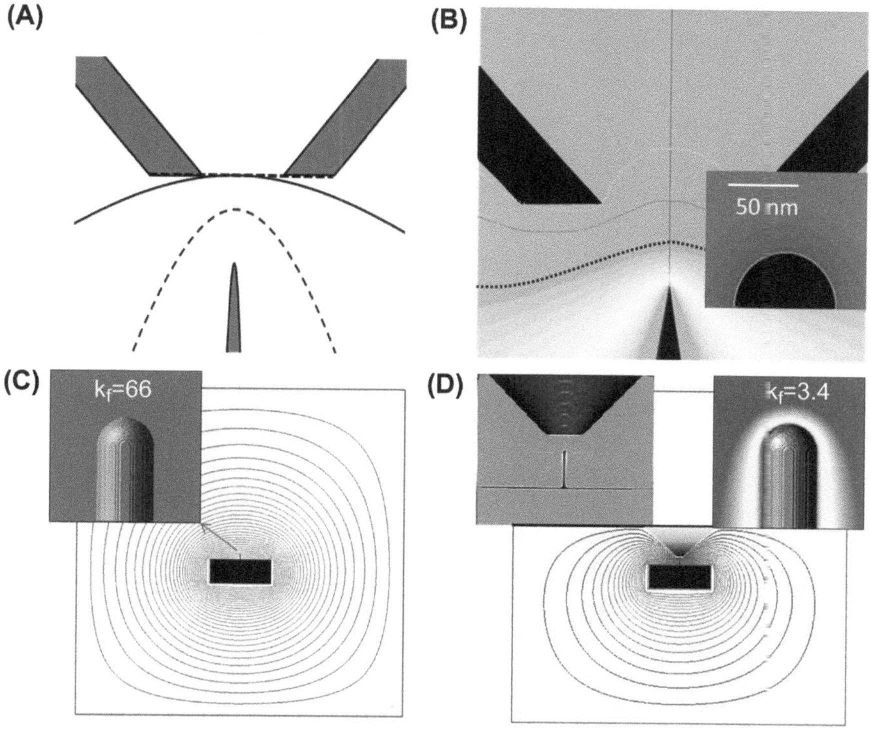

Figure 2.3 Schematic image illustrating the distribution of the voltage around the tip in the (A) paraboloidal geometry (B) with numerical field electromagnetic modeling taking into account the actual geometry of the tip and the counter electrode (C) numerical modeling of the electrostatic distribution around a microtip of 50 μm height on a planar substrate (1 mm size) with a 10-nm end radius. Due to the presence of this base biased at the tip voltage, low field enhancement is measured at the tip apex ($k_f = 66$). (D) Same geometry with a microscale counter electrode at 20 μm from the tip, k_f decreases to 3.4.

For a paraboloidal geometry, electric field and voltage decrease with r, the distance to the focus of the parabola is:

$$V(r) = V\frac{\ln\left(\frac{r}{L}\right)}{\ln\left(\frac{R}{2L}\right)} \text{ and } F(r) = \frac{V}{r\ln\left(\frac{2L}{R}\right)} \tag{2.5}$$

As a result, on the tip axis and at the tip surface (distance $R/2$ to the focal point), field is derived to

$$F = \frac{V}{k_f R} \tag{2.6}$$

with

$$k_f = \frac{1}{2}\ln\left(\frac{2L}{R}\right) \tag{2.7}$$

k_f is named as "*field factor*" and is a semiconstant that accounts for both the tip shape and its electrostatic environment. In the paraboloid approximation, k_f varies as the logarithm of the ratio between the tip to counter electrode L and the end radius R. The field concentrated at the end apex is reduced by a factor between three and eight, depending on the ratio between the tip radius and the tip to counter electrode distance. It is worth noting that the field factor increases slightly with tip blunting, as the radius of curvature increases in AP analysis for instance. Similar values are found when treating the tip in different approximations of the geometry such as the hyperboloidal geometry or the sphere-on-orthogonal cone geometry Eq. (2.5).

To take into account the real tip shape (tip radius and shank angle) as well as the exact geometry of the AP chamber, several investigators have modeled the field distribution around the specimen with finite element methods [17,18]. This kind of calculation is not straightforward since the tip must be precisely defined close to the tip surface (refined mesh) without neglecting the large scale of the AP chamber. A multiscale model is useful in this aim. Using this approach, Gipson showed that surface field may be expressed by an approximated expression:

$$k_f = k_1(R/L) \times \left(1 + \sin^2\frac{\theta}{4}\right)^{\frac{1}{2}} \times e^{k_2(R/L)} \tag{2.8}$$

Some values of k_1 and k_2 were tabulated by Gipson in his paper [15]. Note that the field was shown to vary with the angular position θ onto the tip, which is somewhat surprising since the tip shape is formed by a unique evaporation field (assuming a homogeneous material). The electric field is expected to be constant on the whole apex surface (if we neglect evaporation field variations induced by the sample crystallography). The approach of Gipson neglects the modification of the equilibrium end shape induced by FEV (assuming a hemisphere) and also neglects the fine scale enhancement of the field induced by the presence of atoms at the tip surface.

To overcome these shortcomings, improved modeling approaches have been developed. Using numerical approaches the surface electric field can be calculated with tip shape that develops a surface satisfying the evaporation process and AP chamber geometries. An example of one of this model is shown in Fig. 2.3B.

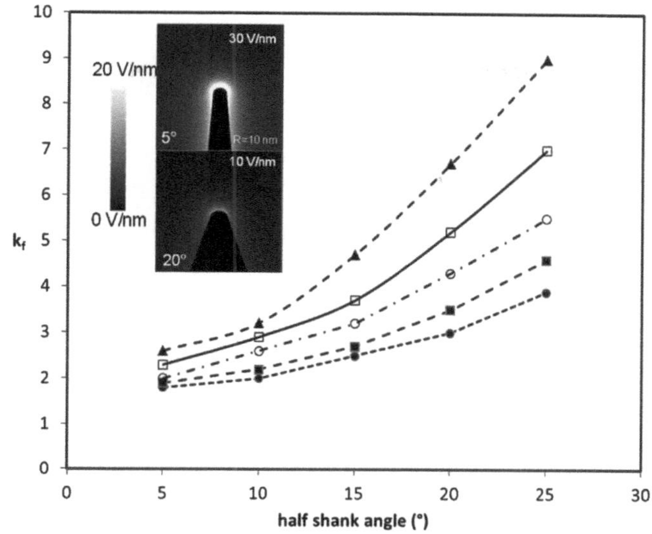

Figure 2.4 Increases in field factor induced by large shank angle tip calculated by finite element method modeling. Spread in field factor decreases when reducing the tip to counter electrode distance for a constant tip radius. Curves calculated for $R = 50$ nm, tip to counter electrode (CE) distances varying between 30 μm (*solid circle, bottom curve*), 60 μm (*solid square*), 120 μm (*open circle*), 300 μm (*open square*), and 1.2 mm (*solid triangle, upper curve*). Inset: field distributions represented with a gray scale color map around two tips with half shank angle of 5° and 20°.

In good agreement with sphere-on-orthogonal cone model, finite element method model shows a strong dependency of the field factor with the shank angle of the tip (when the tip is modeled in this simple geometry, Fig. 2.4). A large shank angle reduces strongly the enhancement of field at the angle. Similar effects exist when using small tips on planar substrates; tip lengths must be adapted in function of the tip to counter-electrode distance. The use of micrometer-size counter electrodes placed at a few micrometers from the apex enables tip of about 100 μm to be analyzed, the field factor being comparable to standard tips. In practice, aperture diameter and aspect ratio of the microelectrode must be adapted to the distance and the size of the tip. Commercial instruments (LEAP) use a standard local electrode with 40 μm diameter aperture with tip to local electrode distance of the same order. Smaller apertures produce micron-scale curvature radius at the rim of the local electrode. This roughness may induce direct field electron emission from the electrode when several kilovolts are applied to the fronted sample. This emission, irradiating the sample substrate, could be sufficient to induce desorption and ionization from the substrate and could increase significantly the noise in the AP chamber. It can also be noted that the field rapidly drops along the shaft (in a quasi-exponential behavior), so that the field is naturally confined at the tip apex.

Microscopic Inspection of the Surface Field

The earlier calculations of the electric field and voltage distribution are based on classical assumptions of a smooth surface of a conductive material considered as a continuous medium. As a result, the charge surface is considered as an infinitely thin layer, and classical electrostatic field can be applied. In reality, looking at the surface composed of atoms, these hypotheses are not strictly correct. The calculation of the local field at the atomic scale is a difficult problem because quantum calculations are mandatory to express the distribution of charges inside the volume of the tip. To understand how the electric field may affect the surface atoms, and the way the tip is eroded during AP analysis, it is important to determine how the electric field spreads inside the sample and what the exact relationship between the curvature at the atomic scale and this field is (which is simply the result of the distribution of surface charges). Presence of an internal field may induce bulk ionic diffusion, for instance [19], or induce variation of potential energy for ions leaving the surface [9,20,21]. In addition, the distribution of charges and the induced field at the surface of the material are important to understand the regularity of the mechanisms of atom erosion by FEV.

The more advanced treatments used the density functional theory (DFT) to calculate how electron density redistributes when a sheet of materials composed of individual atoms is biased to a constant voltage. These models reproducing the atomistic surface of the material ought to give more reliable results. DFT models are limited to a few hundreds of atoms at maximum and are unable to give a complete picture of a biased tip at the quantum level. Nevertheless, these models can mimic a small patch of surface or a cluster of atoms embedded in a high field region and provide a useful understanding of the underlying physics of high field phenomena [22–28].

In addition, these models calculate self-consistently both the electronic charge distribution and the resulting electrostatic field in both metals and insulators. Without entering into the details (the reader is referred to papers dealing with charged surfaces calculated by DFT of W, Mo, Al, Al–Sc, Au, MgO, and ZnO [22–28]), the charge distribution above the surface producing a high electric field can be observed in Fig. 2.2D. The method reduces the surface to a limited number of atoms from a crystalline structure defined with periodic conditions to model an infinite surface, or by a cluster of atoms. Field is created by placing planar layers of negative charge on either side of a central slab, and adding some positive charge to make the whole system neutral overall. In this model, electrical charge inside the simulated cluster of atoms is redistributed following self-consistency quantum calculation.

Surprisingly, the quantum models predict that the equilibrium state of a surface producing a high electric field is rather similar to more classical model treating the charge surface as a geometric surface in the Gauss approach. Charge is found to concentrate on surface atoms at the tip surface. Surface atoms may be considered as partial ions; more protruding atoms building the strongest charge, and therefore being the locations of the highest external field. Inside the volume, the electric field rapidly drops to a very small value in an atom size distance, only residual oscillation contributions arising in the few layers in. This result justifies the use of more rough calculations or modeling approaches, that approximates the surface as atoms partially ionized, or by an atomical representation of the surface of the tip where atoms are modeled at a subatomic scale by a polyhedral representation [29,30]. As a result, the surface charge produces a field that strongly depends on the local curvature of the surface. Each atom producing its own protrusion at the tip surface, the local field is strongly different from the average field above the surface. Classical calculation gives a maximum value of enhancement of field $\gamma = F_{\mathrm{local}}/F$ of three in the case of a hemispherical cap on a planar surface. Actual field enhancements calculated by finite element method are generally limited to 1.5–2 on surface atoms as shown in Fig. 2.5 [30].

Figure 2.5 Distribution of the surface enhancement factor above a tip composed of partial ions submitted to a direct current voltage. Maximum enhancement factor of about 1.5 is found above atoms on edge positions of crystalline terraces. Note also that atoms situated just below these atoms are submitted to field much lower than the average field *F* determined by the average radius of the tip surface.

Under this local field, atoms are submitted to a strong electrostatic stress (local Maxwell stress) that may give rise to surface reconstruction, local surface migration or rearrangement [27] in the first steps of FEV as it will be shown at the end of the chapter. In semiconductors or in insulators, additional penetration of the field in the few layers underneath is classically predicted in the equilibrium state and without any electrical current inside the material. The field just below the surface is indeed the external field dividing by the intrinsic dielectric permittivity of the material. In silicon, it indicates that an internal field of a few volts per nanometer may be present inside the material. Nevertheless, this field rapidly drops in the first nanometers of the surface. More details will be found in the laser—matter interaction section (Chapter 8: Laser Assisted Field Evaporation).

FIELD EMISSION AT THE TIP SURFACE FROM ELECTRON EMISSION TO FIELD EVAPORATION

From Eq. (2.3), it was demonstrated that the main consequence of biasing the tip to a high voltage is the generation of a charged layer at the surface of the rounded apex. The intense field produced by the presence of this charged layer is located close to the surface. In this section, we will consider that in the first nanometers close to the surface, a quasiconstant field region exists and may generate charged particles emission. These charged particles could be electrons that are of interest when considering FEM (with a negative voltage applied to the tip), the ancestor of AP techniques. Note that the theoretical concepts developed in FEM were adapted to the interpretation of FIM and AP. When biasing the tip to a positive voltage, emission of ions by direct ionization of gas atoms in the high field—free space or by emission of ionized atoms from the tip surface could be triggered. These processes are discussed in this section. But note that more advanced mathematical treatments may be found in some more theoretical references [5].

A Few Words on Electron Field Emission

An empirical way to understand field emission processes is to look at the result of an experiment involving a field emitter (a tip) held at a negative voltage in a vacuum chamber. When the tip reaches a sufficient voltage, electron emission can be measured easily and can be of several teraamperes per square meter for a small radius tip (which means a very high density of current at the local surface of the sample). The emission behavior is mostly

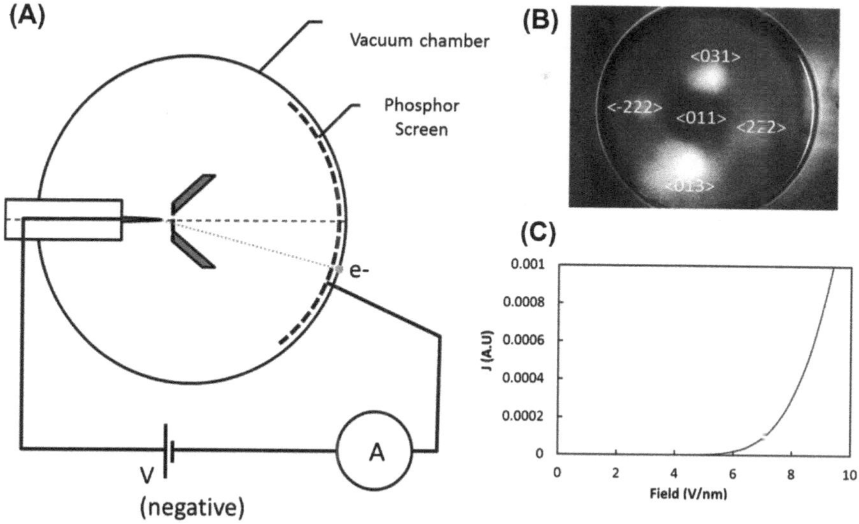

Figure 2.6 (A) Schematic drawing of a field emission microscopy setup. The image of the impacts of electrons on a screen are displayed in (B), the sample is a tungsten wire with end apex oriented along the <011> crystallographic direction. Note the variation of intensity depends on the crystallographic directions. (C) Electron c current exponentially depends on the surface field *F*. *Images courtesy of K. Kadri, GPM.*

field dependent in the conditions we will consider the emitter (a smooth surface at low temperature, in a vacuum environment). Empirical measurements in this case (Fig. 2.6C) show an exponential behavior of the measured current with the applied voltage similar to an electronic diode.

Field emission was occurring in many early experiments where high voltage was applied to sharply pointed cathodes or narrow gaps in evacuated tubes, and also in microelectronic devices, where an isolating barrier separates two materials held at different voltages. The main cause of emission is tunneling of electrons from the material. This process can be understood easily by considering the potential energy of an electron confined inside the material. For a metal, electrons occupy the conduction band, this band having its highest occupied state (Fermi level) at an energetic distance ϕ from the vacuum level electrons. This quantity defines the work function of electron, or the energy that must be given to one electron to quit the material of interest. Applying a field outside the material bends the energy diagram, and creates an energy barrier that can be overcome easily by the electron. The tunneling probability can be calculated, by solving the Schrödinger equation of the electron considered as a wave packet.

The basic solution uses the WKB method or approximation (the WKB method or JWKB, for Jeffreys-Wentzel—Kramers—Brillouin method, is a method for finding approximate solutions to linear differential equations with spatially varying coefficients [31]). In a first approach, if the potential energy diagram of the electron is assimilated as a triangle, the transmission of the barrier for electrons situated at the Fermi level can be fully analytically determined (Fig. 2.7). The transmission of the barrier is given by

$$T_{\mathrm{JBKW}} \propto e^{\left[-2\frac{\sqrt{2m}}{heF}\int_0^{\phi}\sqrt{X}dX\right]} = e^{\left[-4\frac{\sqrt{2m}}{3heF}\phi^{3/2}\right]} \tag{2.9}$$

The exponential dependency of the electronic current to the reciprocal of the applied field F, the fact that no change of current can be detected in function of the temperature (room temperature to cryogenic temperature), and the narrow width of energy distribution of electrons (<0.5 eV) validate the quantum interpretation of the electronic emission of FEM. The dependency of the current as a function of the crystallographic direction

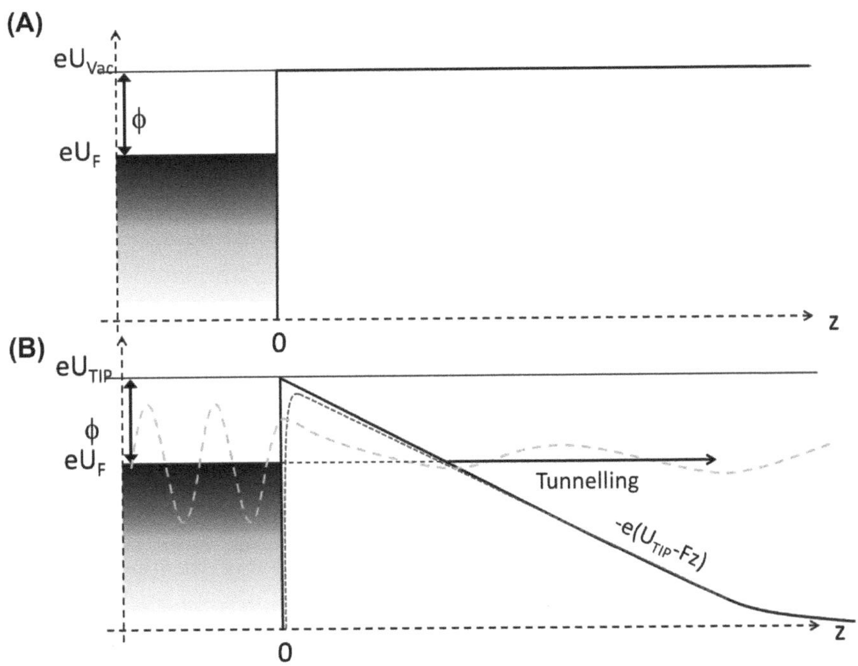

Figure 2.7 Schematic drawing of the energy diagram of electrons close to the material surface without (A) and with (B) applied field. Under sufficient field, the energy barrier, triangular in shape, can be tunneled by electrons of the Fermi level.

(Fig. 2.6B) was interpreted by the variation of the work function and from local field differences. Note that, the quantitative calculation of the emission current as a function of physical parameters is a little bit more complex as the model must take into account the real density of states of electrons in the material, and a more refined model for the energy diagram of electrons close to the surface (including the impact of the attractive force induced by the image charge of the emitted electron in the vacuum). This theory named "Fowler—Nordheim" gives a more correct equation of the electronic current with F and ϕ and may be found in several textbooks [32]. An electric field of a few volts per nanometer induces an electronic current that can be easily measured.

Atoms in High Field: Field Ionization Mechanisms

Reversing the voltage applied to the sample makes emission of electrons impossible, since the energy curves are bent upward. Nevertheless electron tunneling may be possible from atoms situated inside the high field region to the tip surface. In this case, electron tunneling produces direct ionization of the free atoms (or gas atom). This is the main concept used in FIM, where a low-pressure gas is injected close to the emitter submitted to a high positive direct current (DC) voltage (Fig. 2.8). This effect was experimentally evidenced by E.W. Müller in 1951 and theory was developed in the following years [34].

Energy diagram for an electron in a free atom, in the presence of a high positive field is displayed in Fig. 2.9A, and gives a fair picture of the ionization process. The electron of the last occupied state inside the free atom sees the coulomb attractive force of the ion core $U_e(z) = -\frac{e^2}{4\pi\varepsilon_0|z-z_i|}$, with z_i being the atom position. The electron occupies a quantified electronic energy level inside the atom defined by ionization energy I_1 (first ionization energy). The field bends the energy curves and an energy barrier exists in the free space between the tip and the atom. Considering roughly in a first approach this barrier as triangular (Fig. 2.9A) gives a quantum transmission using the same method (WKB) as previously and can be expressed by

$$T(I, F) \propto e^{\left[-4\frac{\sqrt{2m}}{3\hbar eF}I^{3/2}\right]} \tag{2.10}$$

This simple expression is successful in predicting the main features of field ionization, that is the steep rise of the ion current with the field (Fig. 2.8C), and the dependency of the ionization probability with the 3/2 power of the gaseous ionization energy.

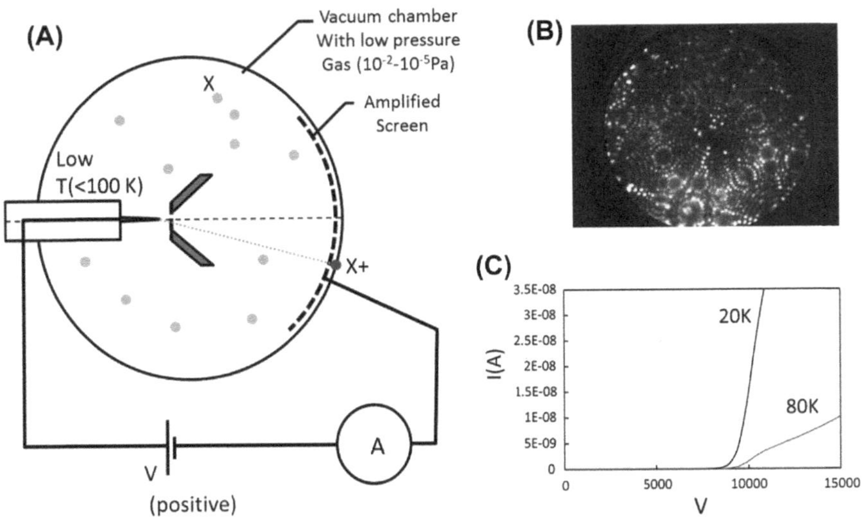

Figure 2.8 (A) Schematic drawing of a field ion microscope. The image of the impacts of ions on an amplified screen are displayed in (B), the sample is a tungsten wire with end apex oriented along the <011> crystallographic direction. (C) Ion current exponentially depends on the surface field *F*, with a different rate regime at high *F* and with a dependency with the sample temperature. This dependency is related to the gas supply function, which is strongly modified by the thermal mobility of gas atoms [33]. *Images courtesy of K. Kadri, GPM.*

Thus considering a constant rate of ionization for a given gas, an actual threshold field of ionization can be determined for all kind of elements. A threshold field was considered taking into account the empirical value of $F_{\mathrm{Th}}(\mathrm{He}) = 44$ V/nm for He determined by Müller in an FIM (Table 2.1); the other values were calculated assuming the ratio $I^{3/2}/F$ being a constant.

This simple calculation, giving good estimates for the field ionization thresholds in the general case for common imaging gases (He, Ne, Ar, Kr, Xe, and H_2), is also interesting to point out a simple observation. The noble gases [helium (He), neon (Ne), argon (Ar), and also krypton (Kr) or xenon (Xe)] have the highest ionization energies, and therefore field of ionization. Metallic atoms have much lower threshold fields, which means that above 10 V/nm any gaseous metallic atom is not stable in the high field region created by a standard AP or FIM sample. It means generally that if an atom is emitted from this sample it will be ionized with a ~100% probability.

Nevertheless ionization could be more difficult considering elements such as C, O, N, or any combination of these atoms on materials submitted

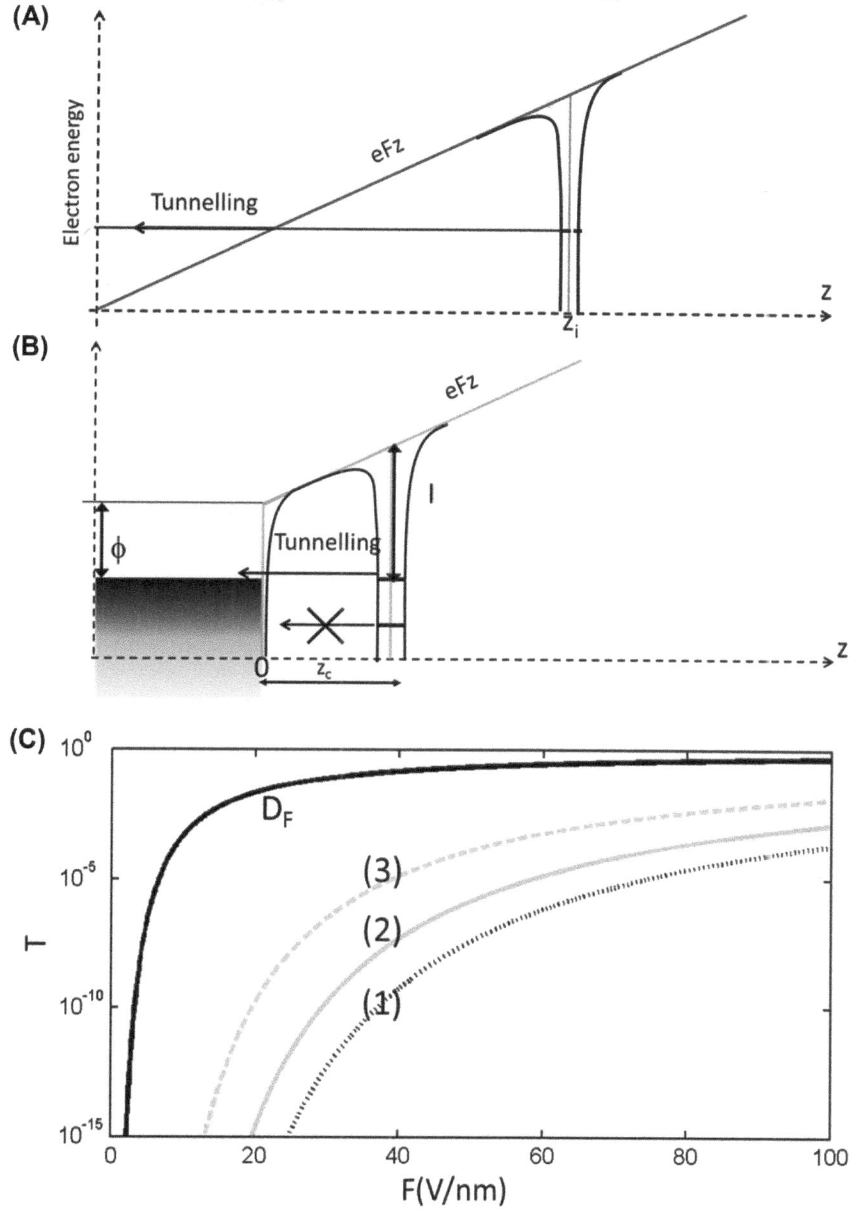

Figure 2.9 Schematic drawing of the energy diagram of electron far (A) or close (B) to the material surface with applied field. The energy barrier mostly triangular in the free space, is deformed close to the tip surface, making the ionization probability higher. Note the presence of a critical distance z_c. Atoms closer than z_c to the surface are not allowed to be ionized. (C) Transparency of a barrier as a function of the applied field in the case of (D_F) electron emission; (1) He ionization transparency with Eq. (2.10); (2) He ionization transparency with Gomer formula; (3) Ne ionization transparency with Gomer formula.

Table 2.1 Threshold Field of Ionization for Some Selected Elements Compared to the Image Field of He on Tungsten (44 V/nm)

Element	I (eV)	Calculated F_{Th} (V/nm)	Experimental Value (V/nm)
He	24.5	44	44
Ne	21.5	36	35
Ar	15.7	23	22
Kr	14.0	19	14.5–15
Xe	12.6	16	11–13
H_2	15.6	22	22
N_2	14.5	20	~17.5 [35]
O_2	13.6	18	
H_2O	12.6	16	
C	11.3	14	
Al	6.0	5	
Si	8.2	8	
Fe	7.9	8	
Ag	7.6	8	
W	8.0	8	

to intermediate fields ($F \sim$ 5–15 V/nm). This difficulty to ionize these elements has raised some questions in the analysis of oxides, carbides, or nitrides, where applied fields, of a few volts per nanometer in some cases, are probably insufficient to fully ionize the emitted C, O, or N atoms [36–38]. This simple equation gives us a rule of thumb for the prediction of the ionization; nevertheless, it fails to predict quantitatively the behavior of ion emission observed experimentally. Indeed, at a relatively low field, the gas experiences greater ionization close to the tip surface (Fig. 2.9B). When the atom is close to the tip, the barrier height is further reduced by the effect of the electron's positive image and the negative image of the ion behind the conductor surface, and the width of the barrier is reduced by the presence of the Fermi level of the material of interest. The ionization probability is maximum when the atom is situated at the distance z_c named critical distance of ionization. This distance is the smallest distance for which ionization of the free atom in front of the material can occur.

If the atom is at a distance smaller than $z_c = \frac{I-\phi}{eF}$, then tunneling is prohibited by the Pauli exclusion principle, because there are no vacant electronic states of appropriate energy within the material to receive the electron.

Several refined models have been developed to give an approximate accurate analytical expression for the optimal tunneling rate (at $z = z_c$). The most well known is the equation provided by Gomer [1]:

$$T(z_c) \propto e^{\left[-6.83\cdot10^{9}\frac{(I-\phi)}{F}\times\left(I-7.60\cdot10^{-5}\frac{F^{1/2}}{I}\right)^{1/2}\right]} \quad (2.11)$$

with I in eV, and F in V/m.

Refined equations may also be found by considering a real three-dimensional (3D) calculation of the tunneling barrier (the coulomb's contribution is spherical and centered on the atom gas) and adding second-order effects such as the Stark effect [39,40]. Moreover, these equations must be combined to the gas supply function to determine a realistic ion current. This implementation is far from straightforward, since the gas pressure close to the apex is enhanced by strong polarization effects (see, for instance, the work of Van Eekelen for more details [33]). Modeling the gas supply function, which depends on the temperature accommodation, thus gives a better agreement to experimental observation, in particular the dependency of the ion current on the tip temperature. The role of gas-adsorbed layers as well as the influence of the field roughness on the emission mechanism lateral variation of the emission onto the field emitter was also pointed out to understand the emission process, but this is discussed in more detail in the FIM section in "Chapter 3, Field Ion Microscopy."

From Field Ionization to Postionization

By analogy with the field ionization mechanism shown previously, it should be possible for ions to undergo further ionization close to the tip, if the field is sufficient. This process is known as post—field ionization or simply postionization. The mechanism is illustrated in Fig. 2.10, considering the energy diagram of an electron in the presence of an atom n times ionized close to a metal surface with applied field F.

The full calculation of the probability of tunneling of the electron situated at the energy state I_{n+1} inside the considered ion was first done by Ernst in 1979 in the case of rhodium and developed by Kingham in 1981 for a list of elements [41,42]. The calculation takes into account the residence time of the ion in the ionization region (at the critical distance of ionization). Inside, the ion is accelerated by the DC field, and rapidly quits the surface where the field is high. As a result, compared to field ionization of a neutral atom, the threshold field of postionization must be higher to counterbalance the fact that the residence time is shorter. To get a correct solution, the transparency of the tunneling barrier is calculated in 3D, using a solution analogous to the

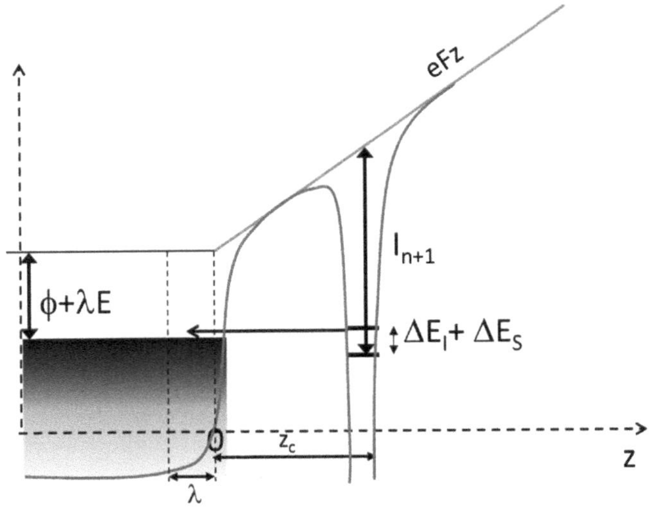

Figure 2.10 Schematic drawing of the energy diagram of electron close to the material surface and confined inside an ion (*n* times ionized) potential well, with applied field *E*. Postionization is predicted quantitatively considering ionization energy from state *n* to $n+1$, Stark effect ΔE_s and ΔE_I image force effect, and field penetration (λ).

JWKB solution in one dimension (note, however, that the 1D solution is fair since tunneling takes place mostly in the region where the barrier is the thinnest, aligned with the field direction). Similar to field ionization, the rate of ionization R is higher at the critical distance of postionization $R(z_c)$ (where the Fermi level is aligned with the current electronic state inside the ion). As shown by Kingham, the ionization rate decreases exponentially with the distance to this critical distance, the ionization zone having a width ($1/\beta$) of a fraction of Angstrom. Using this rate, the probability of secondary ionization is given in a first approach by

$$P_t = 1 - \exp\left(-\frac{R(z_c)}{\beta u_c}\right) \tag{2.12}$$

where u_c is the average speed of the ion on its path at z_c.

The probability represented in Fig. 2.11 shows the existence of a threshold field for which postionization occurs. Kingham, in his paper, adjusted the parameters slightly to fit the experimental behavior of the relative abundance of single- and double-charged rhodium ions plotted against field strength.

As shown in Fig. 2.12, the probability of postionization as a function of the field strength, calculated for common elements, follows a law similar to

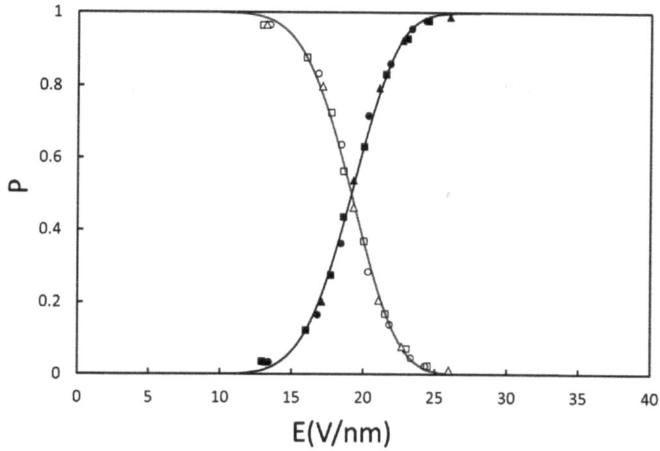

Figure 2.11 Probability of finding an A+ ion (gray) and an A++ ion (black) as a function of the field E at the tip surface, predicted by the postionization theory [$P = \exp(-a \cdot \exp(-b/F))$ for the *gray curve*, with $a = 3300$ and $b = 162$] compared to experimental variations of $N(Si_+)/(N(Si_+) + N(Si_{++}))$ and $1 - N(Si_+)/(N(Si_+) - N(Si_{++}))$ with field E during APT analysis of Si at various field conditions, and different base temperatures (square = 20K; circle = 50K; triangle = 80K). *Courtesy of M. Gilbert, GPM.*

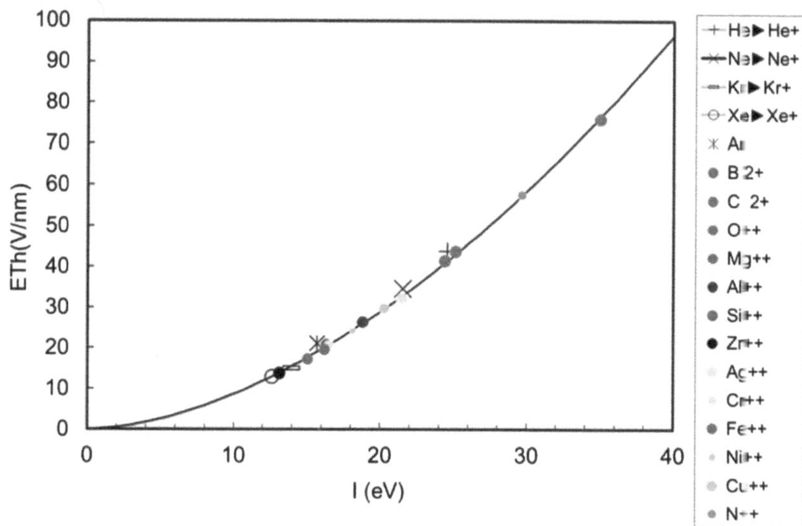

Figure 2.12 Threshold field of postionization of several singly ionized atoms (B, C, O, Mg, Al, Si, Zr, Ag, Cr, Fe, Ni, Cu, N) as a function of the second ionization energy I_2 (in electron volts). These threshold fields are compared to experimental best image fields for noble gas. The same trend as a function of the ionization energy is found (here points are fitted by $Eth = 0.16(I_2)^{1.735}$, close to the $I^{3/2}$ trend).

the field ionization theory. The threshold field of postionization principally depends on the ionization energy of the considered element, with a trend close to the power 3/2 of I with an excellent agreement with most of the experimental data even if a lot of approximations exist in the calculation. For example, no crystallographic dependence is included, such as variation of work function or variations of field strength over the surface at a mesoscopic scale or at a microscopic scale. Large divergences with experimental data were found only in limited cases, where, for instance, dissociation of molecules is observed, or direct emission of multiple-charge ions is possible.

Field Evaporation: Empirical and Mechanical Description of the Mechanism

The colossal external field existing close to the tip affects also the atoms of the material surface itself. This mechanism was used as early as 1941 by Müller to understand and improve the performances of electron field emitters [43]. To increase the emission properties of the tungsten tip, a thin film of a material with a low electron work function could be deposited onto the surface (cesium, barium, or thorium, for instance). Müller studied this emission as a function of the degree of coverage, and found that under a positive electric field, the degree of coverage could be reduced by tearing off the surface atoms. This tearing off effect was found to strongly depend not only on the magnitude of the field, but also on the surface temperature of the tip.

A threshold field for emission of surface atoms could be observed experimentally. This desorption field in the range of several tens of volts per nanometer at low temperature reduces quasi-linearly with the applied temperature. At these ranges of field, atoms are ionized and projected in the chamber (Fig. 2.13). Ion current corresponds to the emission of several atoms per hour to several thousand atoms per second giving subfemtoampere current, impossible to observe with a single phosphor screen. A detector with single ion sensitivity is generally used in this aim.

Empirical Description of the Mechanism

Müller interpreted the field desorption as an evaporation of positive ions of charge ne over a potential energy barrier Q. The mechanism was named FEV, to generalize the tearing off mechanism not only to surface-adsorbed atoms, but also material surface atoms. The energetic barrier Q must be overcome by atoms much heavier than electrons, and therefore quantum tunneling is generally discarded. The rate (per second)

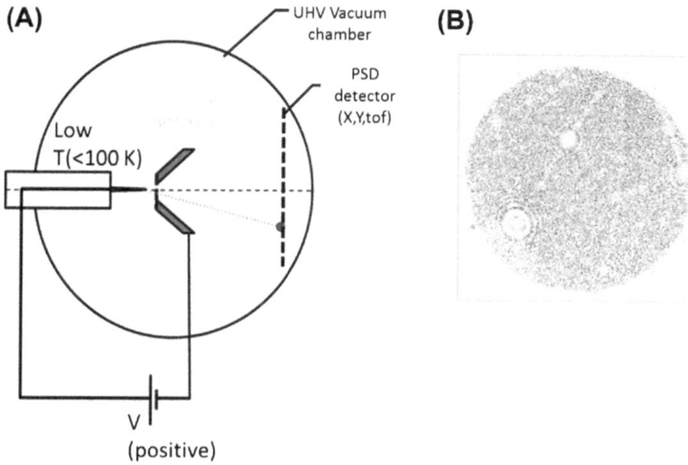

Figure 2.13 Schematic drawing of a field desorption microscope or an atom probe if mass spectrometry of ions emitted is performed (A). The image of the impacts of ions on a position-sensitive detector (PSD) is displayed in (B); the sample is an aluminum wire. The chamber is under ultra high vacuum (UHV).

of FEV for a single atom (K_a) is therefore described empirically by an Arrhenius law with the activation energy Q, Q being a function of E, and a prefactor A:

$$K_a\left(s^{-1}\right) = A\exp\left(-\frac{Q(F)}{k_B T}\right) \tag{2.13}$$

The exponential variation of the evaporation rate with the reciprocal of the applied temperature is reproduced experimentally for a large number of materials, showing the universal nature of the mechanism. An example obtained by G. Kellogg on tungsten is presented in Fig. 2.14. Applied field is shown to reduce the activation barrier from several electron volts to a fraction of an electron volt. With such an energy barrier, application of several tens of K is sufficient to activate the emission of a few atoms per second at the surface of a field emitter in a very controlled manner. Note that the presence of residual gases may affect the observation, and decrease significantly the activation energy; as a result, analyses in AP are carried out under a very low base pressure ($P \sim 10^{-8}$–10^{-10} Pa).

Historical Müller Model

In his first attempt to describe the phenomenon, Müller proposed that the energy barrier was the difference between the binding energy and the

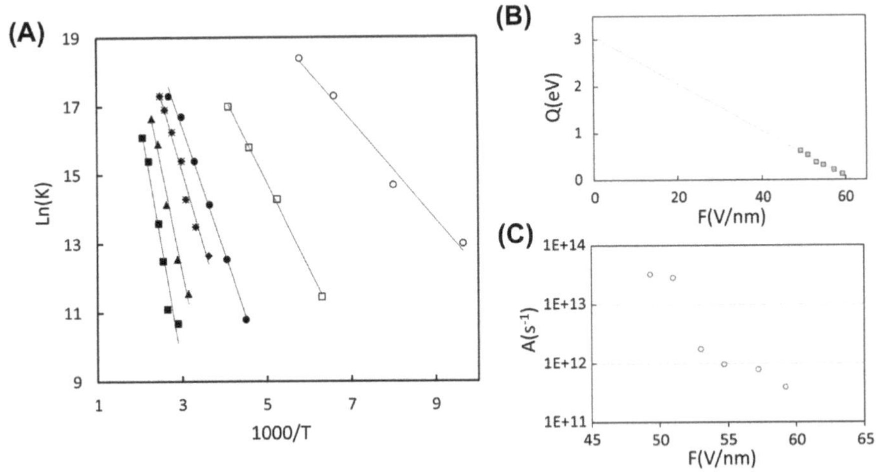

Figure 2.14 (A) Experimental measurements of evaporation rate (K in s^{-1}) as a function of the reciprocal of the temperature T *(data taken from Ref. [44])* for different fields (from 49 to 59.2 V/nm, left to right). Experimental curves are fitted by Arrhenius law. A linear variation of K with F is approximately observed (B) with Eq. (2.17) ($C = 3.05$ eV and $F_{EV} = 61$ V/nm). (C) Note that A, the prefactor of Eq. (2.13) experimentally decreases with the applied field.

maximum of the potential energy $U_i(z)$ of an ion close to the surface, submitted to a high field E [45] (Fig. 2.15). In this approximation, for an ion of charge ne the potential energy is

$$U_i^n(z) \approx \left(\sum I_n - n\phi\right) - \frac{n^2e^2}{16\pi\varepsilon_0 z} - \mathrm{ne}Fz \tag{2.14}$$

when assuming that the attractive force of an ion to the surface is the electrostatic image force (attraction potential $-\frac{n^2e^2}{16\pi\varepsilon_0 z}$ between the ion and the electrostatic image of this ion at the material surface) and the repulsive force is the electrostatic coulomb force in the positive field —neFz. $\sum I_n$ is the sum of the n first ionization energies of the considered element.

The potential energy of the atom is Ua = Λ, the sublimation energy in the material or the bonding energy of a surface atom. The maximum of the ion potential energy is given by the Schottky depression $\sqrt{\frac{n^3e^3F}{4\pi\varepsilon_0}}$ so that the energy barrier $Q_{EV}(F)$ was given by

$$Q_n(F) \sim \left(\Lambda + \sum I_n - n\phi\right) - \left(\frac{n^3e^3F}{4\pi\varepsilon_0}\right)^{1/2} \tag{2.15}$$

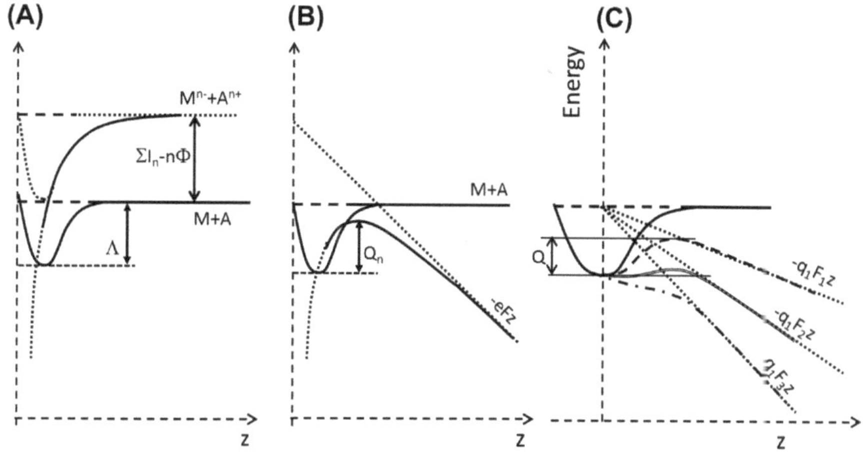

Figure 2.15 Potential energy diagram using the thermoionic cycle as described by Müller in the original description of the evaporation process. (A) Without field, atom is linked to the surface with a cohesive energy Λ, the ion is attracted to the surface by the image force term, and an energy $\Sigma I_n - n\phi$ is required to ionize it. (B) With field, intrinsic repulsion forces from the surface are neglected except the electrostatic force eF on the ion. An energy barrier Q_n must be overcome for an atom to be ionized. (C) McMullen description [46,50], taking into account the electrostatic Maxwell stress $f_C = \frac{1}{2}\varepsilon_0 F^2$. For a sufficient field, an atom can be lifted out from the surface.

Without field, we find the classical energy $Q_{0,n} = (\Lambda + \sum I_n - n\phi)$, which is required to remove an atom from the surface, to ionize n times this atom, the electron returning to the Fermi level of the material of interest. Setting $Q_n = 0$, this equation gives a threshold field or which FEV occurs without any thermal energy that depends on the intrinsic properties of the material and the considered ion. This threshold field F_{EV} is called evaporation field and is expressed by Müller by

$$F_{EV} = \frac{4\pi\varepsilon_0}{n^3 e^3}\left(\Lambda + \sum I_n - n\phi\right)^2 \tag{2.16}$$

This expression is the classical way to predict the low-temperature evaporation field and is used in several textbooks and publications [1–5,7], which for some elements is very close to experimental observations (Table 2.2). In addition, close to FEV, Eq. (2.15) can be simplified to

$$Q_{EV}(F) = C\left(1 - \frac{F}{F_{EV}}\right) \tag{2.17}$$

where $C = Q_0/2$.

Table 2.2 Evaporation Fields From Different Models and Observations

El.	Λ (eV)	I_1 (eV)	I_2 (eV)	I_3 (eV)	ϕ (eV)	C_{11} (GPa)	r_{ws} (nm)	F_{EV}^1 (V/nm)	F_{EV}^2 (V/nm)	F_{EV}^3 (V/nm)	F_{OBS} (V/nm)
Li	1.65	5.39	76.4	122	2.50	14	0.17	14(+)		11	
Be	3.33	9.32	18.2	154	3.90	289	0.12	46(++)		37	34
B	5.92	8.3	25.2	37.9	4.60	526	0.12	64(+)		51	
C	7.4	11.3	24.4	47.9	4.34	1070	0.13	103(++)		63	
Na	1.13	5.14	47.3	71.6	2.30	7.3	0.21	11(+)	20	8	
Mg	1.53	7.65	15.0	80.1	3.70	60	0.17	21(+)		16	
Al	3.34	5.99	18.8	28.4	4.10	114	0.16	19(+)	55	25	
Si	4.67	8.15	16.3	33.5	4.80	190	0.17	33(++)		29	
K	0.941	4.34	31.6	45.7	2.20			7(+)			
Ca	1.83	6.11	11.9	50.9	2.71	27	0.22	18(++)		12	
Sc	3.90	6.56	12.8	24.7	3.5	95	0.18	34(+)		23	
Ti	4.86	6.82	13.6	27.5	4.00	176	0.16	26(++)		30	
V	5.3	6.74	14.7	29.3	4.10	237	0.15	30(++)		35	
Cr	4.1	6.77	16.5	31.0	4.60	314	0.14	27(+)		36	
Mn	2.98	7.44	15.6	33.7	3.80	220	0.14	30(+)	72	30	
Fe	4.29	7.9	16.1	30.7	4.40	278	0.14	33(++)		35	35
Co	4.39	7.86	17.1	33.5	4.40	240	0.14	37(++)		35	36
Ni	4.44	7.64	18.2	35.2	5.00	261	0.14	35(+)		36	35
Cu	3.5	7.73	20.3	36.8	4.60	176	0.14	30(+)		30	30
Zn	1.35	9.39	18.0	39.7	3.80	177	0.15	33(+)		22	
Ga	2.78	6.00	20.5	30.7	4.10	44	0.17	15(+)		18	
Ge	3.98	7.88	15.9	34.2	4.80	129	0.18	29(+)		24	
As	3.0	9.81	18.6	28.4	4.70	26	0.17	42(++)		15	
Rb	0.86	4.18	27.3	40.0	2.10			6(+)	12		

Y	4.4	6.38	12.2	20.5	3.10	75	0.20	24(++)		20	
Zr	6.32	6.84	13.1	23.0	4.20	144	0.18	28(++)		28	
Nb	4.47	6.88	14.3	25.0	4.00	247	0.16	37(++)	81	36	35
Mo	6.81	7.1	16.2	27.2	4.20	450	0.15	41(++)	90	42	46
Ru	6.62	7.37	16.8	28.5	4.50	325	0.15	41(++)		40	
Rh	5.75	7.46	18.1	31.1	4.80	513	0.15	41(++)		43	46
Pd	3.94	8.34	19.4	33.0	5.00	234	0.15	37(+)		32	
Ag	2.96	7.58	21.5	34.8	4.60	132	0.16	24(+)		25	
Cd	1.16	8.99	16.9	37.5	4.10	131	0.17	25(+)		18	
In	2.6	5.79	18.9	28.0	4.10	45	0.18	12(+)		16	
Sn	3.12	7.34	14.6	30.5	4.40	83	0.19	23(++)		26	
Sb	2.7	8.64	16.5	25.3	4.60	70	0.19	30(++)		11	
Cs	0.83	3.89	25.1	35.0	2.10			5(+)			
Ba	1.86	5.21	10.0		2.50			12(++)			
La	4.49	5.58	11.1	19.2	3.30			18(++)			
Hf	6.35	7.00	14.9	23.3	3.50	148	0.17	39(++)		29	
Ta	8.09	5.89	16.0	22.0	4.20	266	0.16	44(+++)		37	
W	8.66	7.98	18.0	24.0	4.50	533	0.16	52(+++)	98	47	57
Re	8.1	7.88	17.0	26.0	5.10	524	0.15	45(++)		46	48
Os	7	8.70	17.0	25.0	4.60			48(++)			
Ir	6.93	9.10	17.0	27.0	5.30	609	0.15	44(++)		47	53
Pt	5.85	9.00	18.6	28.0	5.30	607	0.15	45(++)	86	44	48
Au	3.78	9.23	20.5	30.0	4.30	202	0.17	53(+)	68	29	35
Hg	0.694	10.4	18.8	34.2	4.50			31(+)			
Tl	1.87	6.11	20.4	29.8	3.70			13(+)			
Pb	2.04	7.42	15.0	31.9	4.10	56	0.19	20(+)	36	16	
Bi	2.15	7.29	16.7	25.6	4.30			18(+)			
U	4.00	6.00			3.30			31(+)			

Λ is the sublimation energy; I_n the *n*th ionization energies, Φ the work function of electrons, r_{WS} the Wigner Seitz radius of the considered atom (calculated from the atomic density); C_{11} the elastic constant from the Hook tensor; F_{EV}^1 the Müller model predictions, with the charge state given; F_{EV}^2 the original values of evaporation field from the McMullen model [46]; F_{EV}^3 field evaporation, using Eq. (2.22), scaled to have $F(\mathrm{Ni}) = 36$ V/nm.

The linearity of Q_{EV} with F is experimentally observed in the range 0.8−0.95 F_{EV} for some materials but deviations exist out of this region (Fig. 2.14B).

Nevertheless, the agreement of these equations with experimental measurements is poor. Indeed, it is first necessary to assume a high degree n of ionization in most of the cases to obtain a reasonable F_{EV} value. In addition, the experimental estimation of Q_0 is not consistent with theoretical values (for instance, in Fig. 2.14, Q_0 is estimated to 6 eV for W, far from the values of 12 eV for a W^+, 25 eV for W^{++}, or 45 eV for W^{+++}). Generally, data related to fractional field reduction need to be acquired via experiment. The reasons for these discrepancies are the fundamental flaws existing in the original Müller's formula. Eq. (2.14) neglects a lot of phenomenological parameters that different authors tried to implement to improve the simple energetic approach of Müller. Brandon introduced a criterion to evaluate the best ionization state of the evaporated atom [47,48]. Additional terms could be added to take into account the energetic influence of polarization of surface atoms [49] and ions in the high field region (which is now attributed to vacuum electrostatic-field energy terms [5]). It was also pointed out that at the calculated evaporation field, a Schottky hump maximum is often missing, using Eq. (2.14). Adaptation of the model taking into account the intersection of the ionic energy curve with the atomic energy curve was tempted to solve the problem making the straightforward calculation of F_{EV} more difficult.

But more fundamentally, the fact that the model neglects the real potential to which atom and ion are submitted during FEV, especially within the first fractions of a nanometer of the surface (atom and ion are within the range of strong cohesive forces), and the fact that the real electric field distribution at the atomic scale is not taken into account by these phenomenological models prevent the use of the model in the general purpose of predicting FEV behavior and constants in real materials. As a result, there is a general agreement that the former model has only a pedagogic value (details about the history of the different models existing may be found in Ref. [5]).

(1D) Maxwell Stress Model

It is now found and supported by quantum description of the evaporation process that FEV can be seen as a continuous variation of the charge carried by the escaping atom. As calculated by DFT, the surface atoms are not neutral, but could be considered as partial ion. In particular, field and charge are concentrated on protruding atoms that already have a minimum of remaining atomic bond. Following Gauss's law, at local field of several tens of volts per nanometer, local charge $q(F)$ is of several tenths of e on these atoms (see Figs. 2.2D and 2.5). These partial ions feel the positive field, and are submitted to a Maxwell stress as described in Eq. (2.4). This stress can be sufficient to pull out the atom from the surface, the atom being more exposed to the applied field, and complete ionization being possible at a given point. Once being ionized, the particle leaves the surface freely.

This approach was proposed by McMullen, Perdew, and Rose to interpret the complete FEV of an Al-atom layer modeled with a 1D DFT approach [46,50]. The self-consistent quantum calculation, which models the fundamental state of the electron density when moving a surface ion core slab outward a surface submitted to different applied fields, demonstrates the existence of an energy barrier, very similar to the historical representation of Müller that has to be crossed by the atom to become ionic. Nevertheless, the barrier is formed approximately by the superposition of a classical atomic energy barrier (modeled in the McMullen approach by a universal binding energy curve), and the electrostatic potential of the partial charge $q(F)$, in the field distribution F.

The adhesive energy diagram of an atom of the surface may be described by a universal binding energy curve (as proposed by McMullen et al. [50]),

$$\sigma(z^*) = \alpha\left\{-\left(1+\frac{z^*}{l}\right)\exp\left(-\frac{z^*}{l}\right)\right\} \tag{2.18}$$

Here α is the change in energy required to separate the first layer of atoms from the surface, $z^* = z - z_0$, where z_0 is the equilibrium position of the atomic layer at the surface, and l is a parameter that depends on the steepness of the potential well.

The parameter l can be derived from experimental measurement such as mechanical strength properties. For instance, it is evaluated by

$$l = \left(\frac{\alpha d}{C_{11}}\right)^{1/2} \tag{2.19}$$

where C_{11} is the elastic constant perpendicular to the considered plane (function of the experimental Young modulus and compression modulus), d is the interplanar distance, and α is twice the measured surface energy of the material. As pointed out by Rose et al. α is proportional to the ratio of the cohesive energy of the material to the surface of an atom and may be estimated by $\Lambda/2\pi r_{WS}^2$, (r_{WS} being the Wigner Seitz radius).

The maximal force (rupture force f_r) that can be exerted on the surface is given by the maximum of the derivative of the Eq. (2.18) and is given by

$$f_r = \frac{1}{e}\frac{\alpha_A}{l} \tag{2.20}$$

e being the base of natural logarithms.

The energetic potential of these surface atoms is then bent by the presence of the surface electrostatic strength. If we state that the electrical surface starts at the equilibrium position of the surface atom with no field, this potential can be written as

$$\sigma_F(z^*) = \sigma(z^*) - \frac{1}{2}\varepsilon_0 F^2 z^* \tag{2.21}$$

Evaporation takes place when the stress induced by the field is stronger than the maximal force of adhesion; otherwise a potential barrier exists for FEV (Fig. 2.14C).

This threshold field is given by

$$F_{EV} = \left(\frac{2}{e}\frac{\alpha}{\varepsilon_0 l}\right)^{1/2} \propto \left(\frac{\Lambda \cdot C_{11}}{r_{WS}^3}\right)^{1/4} \tag{2.22}$$

Values found by this model are larger than experimental ones by a factor of about two, the general experimental trends, however, being properly reproduced (see Table 2.2). Note that the critical field depends on thermodynamic properties and on elastics constants. In experiments, the surface is evaporated atom by atom from kink and ledge site positions. The 1D model of McMullen does not include effects such as field enhancement from these protruding positions of area, and cannot reproduce the actual evaporation mechanism from the kink or ledge site positions onto atomic terraces. This could explain the observed ratio of critical field existing between experimental observations and model predictions. However, this approach does not explain how the different stages of ionization occur after desorption. Quantum mechanics calculations are required to fulfil this requirement. Such a model induces a bond-breaking-first prior to ionization, as it was pointed out in Ref. [5].

The high value of the evaporation field calculated in Al was also found numerically when simulating with a 3D DFT model the evaporation of a complete (111) Al layer. In their paper, Sanchez and Lozovoi calculated the field required to lift out a completely full (111) surface of Al and obtained a value of 54 V/nm for the threshold field of evaporation, similar to the value of the semiempirical model of McMullen (55 V/nm) [27]. Reducing the coverage to 1/16 (one atom every four free sites of the (111) atomic plane) enables one to find a much lower value (19 V/nm), showing this impact of local rearrangement of the electronic density at the surface. As a result, it should be possible to rescale the calculated value to take into account this effect, systematic for all the elements. Table 2.2 shows the reasonable agreement of this approach with experimental observations.

(3D) Maxwell Stress Model

Note that using a pair potential approach such as a Morse potential form (as proposed by Kreuzer, for instance [51,52]) rather than a universal energy curve approach, the energy barrier can be estimated analytically.

Indeed the pair potential of the atom to the surface is

$$U(z) = nD\left(\exp\left[-2\left(\frac{z}{l_m}\right)\right] - 2\exp\left[-\left(\frac{z}{l_m}\right)\right]\right) \tag{2.23}$$

with z the distance to the atom position, D the bonding energy for a pair of atoms (proportional to Λ), n the number of bonds to be involved, and l_m a scaling length, similar to l of the universal bonding energy curve. In this case, the maximal rupture force (f_{rm}) that can sustain an atom is given by

$$f_{\mathrm{rm}} = \frac{nD}{2l_m} \tag{2.24}$$

The energetic potential of the atom is then bent by the presence of the surface electrostatic strength, including the field enhancement factor induced by the atom protrusion ($1.2 < \gamma < 3$). This potential can be written as

$$U_{\mathrm{E}}(z) = U(z) - \frac{1}{2} S \times \varepsilon_0 (\gamma F)^2 z \tag{2.25}$$

S is here the atomic surface on which the force is exerted $\left(\sim \pi R_{\mathrm{at}}^2\right)$, and ε_0 the vacuum dielectric permittivity. The evaporation field is then simply found as the field that gives rise to the maximal force f_1 and is given by

$$F_{\mathrm{EV}} = \frac{1}{\gamma}\left(\frac{n}{S}\right)^{1/2} \times \left(\frac{D}{\varepsilon_0 l_m}\right)^{1/2} \tag{2.26}$$

Similar to preceding approaches, an energy barrier $Q_{EV}(E)$ exists with a form depending on the energetic bonding model. The variation of the energetic barrier can be found analytically and is thus given by

$$\frac{Q(F)}{Q(0)}=\sqrt{1-\left(\frac{F}{F_{EV}}\right)^2}+1/2\left(\frac{F}{F_{EV}}\right)^2 \times \ln\left[\left(1-\sqrt{1-\left(\frac{F}{F_{EV}}\right)^2}\right)\Big/\left(1+\sqrt{1-\left(\frac{F}{F_{EV}}\right)^2}\right)\right] \tag{2.27}$$

This equation fits well the curve obtained by DFT [27,53], and in the region $0.8\,F_{EV} < F < 0.95\,F_{EV}$ this expression can be approximated by the classical linear relationship Eq. (2.17).

A more correct complete model should be able to account for the deformation of the tip and the actual forces that exist between the selected atoms involved in FEV and their neighborhood. The 3D nature of these interactions will induce a complex path before the final rupture of bonds. The derivation of present analytical models does not include explicitly the structural properties of metals, missing the link between the instantaneous effects owing to the applied field and the consequent response observed in the metal surface as a result of an extended application of an electric field. Indeed, we need to consider not only the effect of the electrostatic force exerted to the most protruding atom, but also the electrostatic forces applied on all the surface atoms.

These forces pull on the entire surface atoms, and exert a notable mechanical load on the end apex. On a microscopic scale, assuming surface electrostatic forces and classical interatomic potentials between atoms, deformations can be calculated, and the dynamic evolution of the tip could be calculated prior to any FEV. To test this idea, we used in a first approach a Morse model of interatomic potential in an FCC crystal structure, with $D = 0.43$ eV, $1/l = 13.9$ nm^{-1}, and the crystallographic parameters of nickel [54]. The interatomic potential calculation was limited to the 12 first neighbors of Ni atoms (Fig. 2.16). Electric field distribution on a (001) oriented tip composed of about 2 million atoms was used to calculate the electrostatic forces applied on all the surface atoms. Note that the atoms situated at the base of the tip are fixed. By relaxing the atom positions under the application of an average field on the surface, the deformation of the surface of the tip can be observed. In (A) to (C), the local deformation

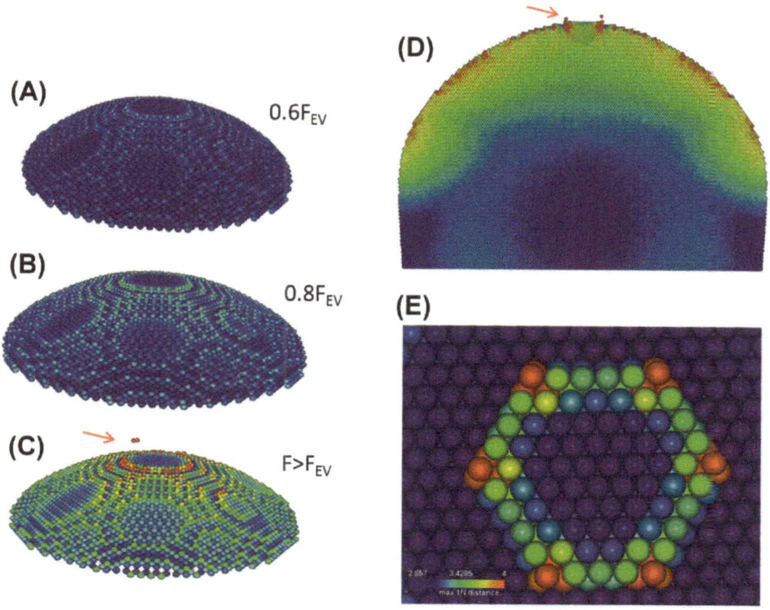

Figure 2.16 Tip shape deformation under the effect of the field-induced surface stress for a Ni specimen. (A)–(C) shows the equilibrium steady state of the tip at, respectively, 0.6 F_{EV}, 0.8 F_{EV}, and $F > F_{EV}$, with F_{EV} defined as the field sufficient to induce single atom evaporation from the surface. Colors represent the amplitude deformation of the nearest neighbor distance of each atom. In (D) a slice showing the internal deformation is shown for $F = F_{EV}$. In (E), the displacement of surface atoms of the (111) pole is evident; terrace edges appear bended under the applied stress.

was represented through a color contrast as a function of the applied field. The local field enhancement induced by atomic surface roughness, is responsible for a nonhomogeneous deformation of the surface at the atomic scale. Atoms in kink-site, edge-site, or ledge-site positions are more protruding and are displaced by the strong pulling action of the electrostatic force. When reaching a field of about 35–40 V/nm, very close to the estimated FEV of nickel, some atoms are found to detach naturally from the surface. We may note that under this constraint, the entire volume is observed to deform significantly. Fig. 2.16D shows a slice inside the tip showing the internal deformations of the tip which are still important a few nanometers deep in the volume. At the tip surface, the nonevaporating atoms present equilibrium positions slightly shifted compared to initial lattice positions.

This model can be improved by inserting temperature and the possibility of surface diffusion [53]. A tentative molecular dynamic (MD) approach based on these principles was, for instance, proposed by Djurabekova et al. [55]; they combined the impact of the electrostatic forces on the surface with classical MD approaches to simulate the evaporation process. Their model accounts concurrently for both electrodynamic effects and interatomic interactions between the atoms. Note that the Laplace equation with the dynamically changing boundary condition on the surface is solved at every MD step, and the following charge redistribution is allowed to depend on the position of surface atoms. The model allows for the dynamic simulation of surface modification under the electric field effect, including the lattice expansion under the field and the migration of charged adatoms. This kind of approach is promising and offers the possibility to input time-varying and spatially varying conditions of temperature or voltage, for a realistic sample. Nevertheless, the primary limitations are the speed at which the simulation can be performed and the accuracy of the specimen model. MD dynamic models are limited to the evolution of the sample at the nanosecond range for a relatively small number of atoms. In a few years, this limitation will, however, certainly be relaxed.

TRAJECTORIES OF IONS AFTER IONIZATION OR FIELD EVAPORATION IN THE ATOM PROBE

Assuming models of FEV, once field evaporated, ions start a trajectory in the vacuum between the surface of the tip to the detector under the action of the strong electric field present in the neighborhood of the sample. These ion trajectories, the velocities of emitted particles, and the timescale of the buildup of the kinetic energy have direct consequences on the imaging process in APT. A brief inspection of these processes is exposed hereafter.

Velocity, Energy, and Time of Flight of Projected Ions

The vacuum in the AP chamber is sufficiently low to neglect interactions of the ion with any particle during the flight. The initial subtle preevaporation movements, induced by diffusion prior to FEV will also be neglected. As a result, the ion will acquire at the end of its flight a kinetic energy equivalent to the difference in potential energy between the tip and the detector. For the considered voltage (1–20 kV) and standard charges of ions ($n = 1$–4), classical mechanical laws also apply (relativistic terms are neglected).

The velocity of an ion is deduced from conservation of energy, and it is assumed that the ion is initially immobile on the tip surface:

$$v = \sqrt{2\frac{ne\Delta V}{m}} \tag{2.28}$$

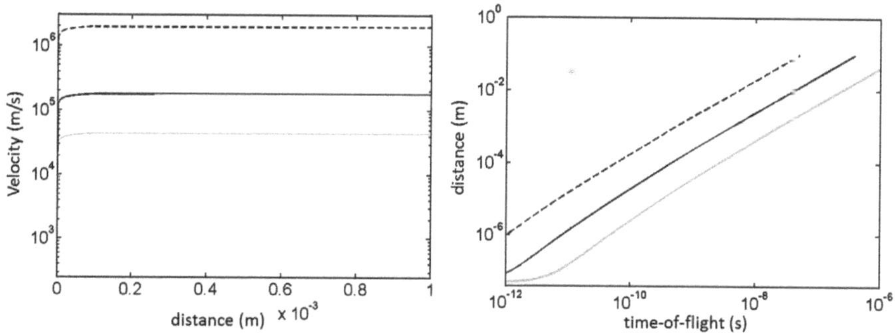

Figure 2.17 Dynamics of ion flights in the AP chamber for H (*black dashed curves*), Fe (*black solid curve*), and Au (*gray curves*). Curves on the left represent the evolution of the velocities as a function of the distance to the tip; curves on the right represent the position as a function of the times of flight.

For ions with $0.25 < m/n < 200$ Da, it gives 4.10^4 m/s $< v < 4.10^6$ m/s. As observed in Fig. 2.17 (right), this velocity is acquired in less than a fraction of a nanosecond, and most of the kinetic energy is acquired in the first 10 μm of the tip to counter electrode distance. The detector is at a distance L of several to tens of centimeters from the tip apex. The ion spends most of its time traveling to the detector in a free flight (constant velocity). It is a fair approximation to assume that the ion acquires quasi instantaneously its kinetic energy, so that the time of flight is given by

$$t = L\sqrt{\frac{m}{2neV}}. \tag{2.29}$$

Experimentally, small systematic deficits in kinetic energy exists (ie, $Q_{0,n} = (\Lambda + \sum I_n - n\phi)$); it demonstrates that a part of the energy required to ionize and lift out atoms from the surface is used during the FEV mechanism. Note that assuming that the thermal vibration energy of the ion can be neglected (Ec $\sim k_B T < 0.05$ eV compared to several kiloelectron volts final energy), the ion paths are independent of the mass, the charge of the ion, and the applied voltage. This result was proved analytically by Smith and Walls in a cylindrical symmetry, but this can be extended to any 3D distribution of the field [56]. This result also neglects second-order dynamic effects in the voltage pulsed mode which are small considering the fast transit of ions in the tip to the counter electrode region (subnanosecond) compared to the pulse duration (2–10 ns).

Field Evaporation in the Atom Probe From Theory to Applications

The theoretical treatment presented previously to interpret and understand FEV was limited to the description of a single atom in a defined unique electric field condition at the surface of the studied material. In practice, the problem is more complex since the tip surface is composed of thousands of atoms submitted to different fields (depending on local roughness) and having different bonding interactions with the surface (depending on the situation of the atom compared to neighbors). In addition, experimentally, only a limited surface is probed by a detector with technical limitations (see Chapter 6: Atom Probe Tomography Detectors for more details). Note also that the experimental temperature is not null and this temperature must be taken into account experimentally. To bridge the gap between theory and experiments, some approximations and considerations are generally used.

From Evaporation Rate to Detection Rate

Let us first consider that under an electric field a surface atom may be field evaporated using the Eq. (2.13). Assuming A being the vibration frequency ν_0 of the atom due to thermal vibrations, or the number of tests to cross the barrier done by the atom per second, the rate of FEV is given by

$$K_a\left(s^{-1}\right) = \upsilon_0 \exp\left(-\frac{Q(F)}{k_B T}\right) \tag{2.30}$$

To simplify, we will assume that Eq. (2.17) $\left(Q_{EV}(F) = C\left(1 - \frac{F}{F_{EV}}\right)\right)$ rules the barrier energy as a function of the field.

At the tip surface, each atom i is submitted to a unique field $F(i) = \gamma_i F$, with F the average field and γ_i the geometrical enhancement factor. As represented in Fig. 2.5, this factor varies between 0.5 and 2, with most of the surface atoms submitted to an average field F. It means that the rate of FEV (K in atoms per second) at the tip surface is given by (Fig. 2.18)

$$\begin{aligned} K\left(s^{-1}\right) &= \sum_{i=1}^{N} \upsilon_0 \exp\left(-\frac{Q(F_i)}{k_B T}\right) \\ &\approx N \int D(F)\upsilon_0 \exp\left(-\frac{C(1 - F/F_{EV})}{k_B T}\right) dF \end{aligned} \tag{2.31}$$

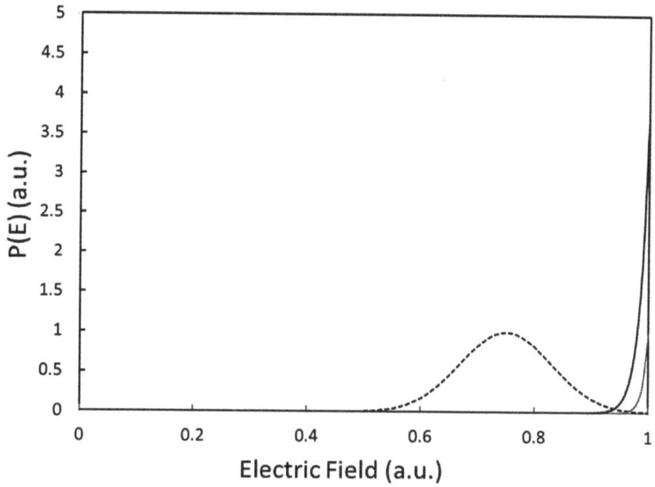

Figure 2.18 Density of surface atoms with applied field of F (*dashed curve*), rate of evaporation of a single atom following [24] (*gray curve*), and rate of evaporation of surface atoms assuming density of surface atoms curve (*solid dark curve*).

where $D(F)$ is the density of atoms submitted to the field F, and N the total number of surface atoms.

This density is different from tip to tip, but we can note that only atoms submitted to the highest electric field contribute to the measured number of evaporated atoms from the surface. To simplify, we replace this complex equation assuming that only a limited number of atoms situated at high-risk places of FEV and submitted to a unique field F should be considered so that

$$K\left(s^{-1}\right) \approx N_{\mathrm{hr}} \upsilon_0 \exp\left(- \frac{C(1 - F/F_{\mathrm{EV}})}{k_{\mathrm{B}} T} \right) \tag{2.32}$$

Experimentally, we must estimate the number of these N_{hr} atoms taking into account both the field of view of the detector, and the detection efficiency $Q_{\mathrm{D}} < 1$ [5,57]. Using this expression, it is straightforward to deduce the evaporation field constant and the activation energy from experimental measurements of this rate, but we may note that a large uncertainty exists in the estimation of N_{hr}. Kellogg, for instance, used a constant value of N_{hr} to deduce curves of Fig. 2.14. Nonetheless, the result of the integral calculation of Eq. (2.24) may differ from one average condition of analysis to another, and may give rise to large variations in N_{hr}. It could explain the strong variations in constant A found at different average applied fields (Fig. 2.14C).

Low field means high temperature to get measurable flux, which reduces the sensitivity of the equation with field, enabling a much higher number of atoms to be involved in the evaporation.

Note that on a standard tip, at low base temperature, N_{hr} can be limited to a few atoms [57].

Field Evaporation at T > *0*

FEV is a theoretical concept that expresses the field required to suppress the energy barrier of an atom at the tip surface. At $T > 0$, atoms may cross the barrier by thermal activation, so that the rate of evaporation is not a pure step function. It is useful to define an evaporation field F_T that gives rise to a constant rate of evaporation K_T at $T > 0$.

Assuming Eq. (2.25), F_T is written as

$$\frac{F_T}{F_{EV}} = 1 - T\left[\frac{k}{C}\ln\left(\frac{N_{hr}\upsilon_0}{K_T}\right)\right] \tag{2.33}$$

This linear variation of the threshold field of evaporation at $T > 0$ is observed experimentally in a large number of materials and is reproducible [58] (Fig. 2.19).

The slope of this law depends on the intrinsic constant C, related to the cohesive energy Λ. Divergences to the linearity exist, in particular at low temperatures or at high temperatures, as the theoretical variation of Q as a function of F is not strictly linear (see Eq. (2.21), for instance). Note that most of the experimental curves were measured in the FIM mode [58]. The imaging gas as well as reshaping effects due to surface diffusion may modify the experimental curves. The tip tends to reshape at high temperature, changing the conditions of field for a constant V and N_{hr} changes with average field and evaporation rate. The presence of imaging gas reduces also the threshold field of evaporation, especially at low temperatures, due to thermal accommodation process [2,3].

Steady State Field Evaporated Shape of the Tip

Let us consider an AP sample of a material composed of a single phase with a single evaporation field at the beginning of the experiment. The tip, prepared by electrochemical or FIB etching, presents an average end radius of curvature with some local imperfections that come from the preparation technique. These protruding areas will be the siege of local enhancement of field induced by the local roughness. These zones will be the first to be field evaporated, and the tip will evolve atom by atom to a smoother surface.

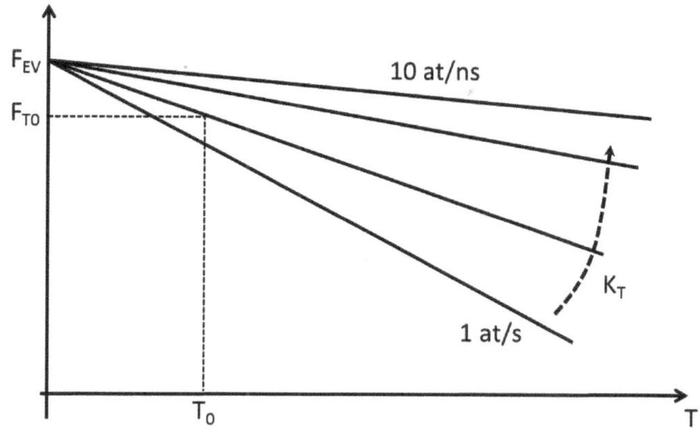

Figure 2.19 Schematic diagram of the evaporation field at $T = T_0$ defined at a fixed rate of evaporation K_T. The evaporation field decreases linearly with the applied field.

The surface will evolve to a self-similar end shape, where the rate of evaporation will be equivalent everywhere. Any roughness at the atomic scale creates a local enhancement of the field that tends to be removed by the induced local rise in evaporation rate.

At a scale sufficient to describe a local radius of curvature, the surface evolves to a smooth surface without any discontinuity. At this scale, the local radius of curvature will evolve to have a constant rate of evaporation. Due to the steep nature of the evaporation rate with field (Fig. 2.18), the local field will be close to F_{EV}, the local evaporation field of the material. If variations in evaporation field exist between or along different crystallographic directions (through variations of the binding energy, for instance), the surface will transform into a quasi-polygonal end shape with regional radii depending on the surface-binding energy (see Chapter 7: 3D Reconstruction in APT: Basics and Advanced Approaches for more details). At the atomic scale, discontinuities still exist, giving a field distribution similar to Fig. 2.18.

When evaporating a multiphase specimen, the same mechanism applies, but the presence of different materials or phases with different thresholds of evaporation must be included. In a first approximation, we may assume that the local radius of curvature will be inversely proportional to the local field at the surface. Assuming a steady state of FEV, each phase at the surface of the sample develops a constant mean curvature surface (eg, the quasi hemispherical cap when the sample is homogeneous) (Fig. 2.20). Indeed, the progressive removal of atoms with high evaporation probability results in a surface with a constant evaporation probability, and thus a constant electric field. The mean radii of curvature of two different phases A and B will compensate

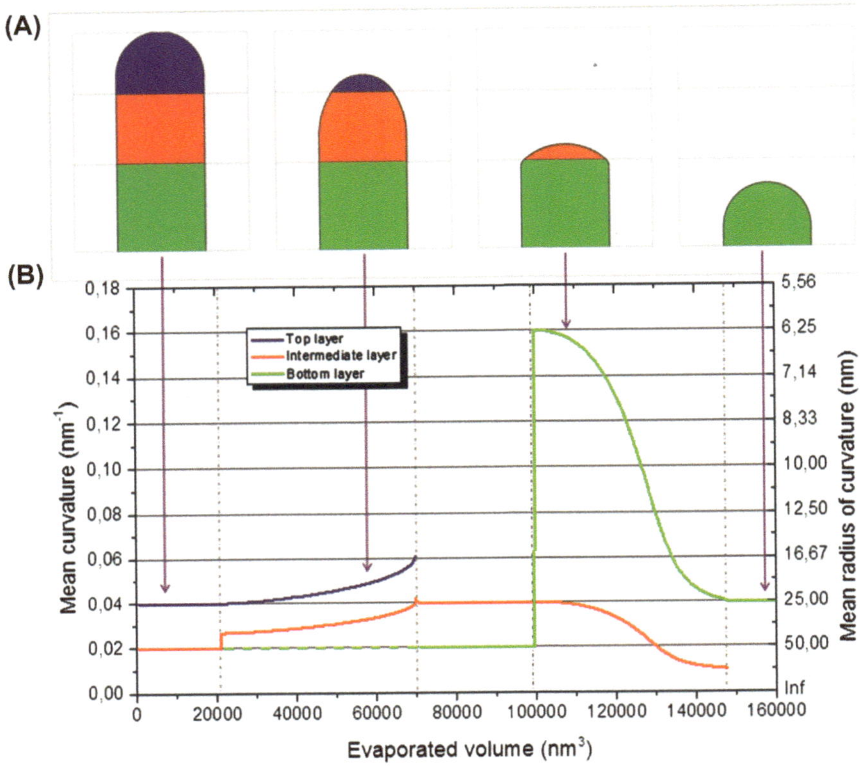

Figure 2.20 (A) Emitter shape during field evaporation. The top blue layer is a high field material, the intermediate red layer is a low-field material, and the bottom green layer is again a high field material. The field ratio of the first interface (between the blue layer and the red layer) is set to $f = 1.5$ while it is set to $f = 0.25$ for the second interface [between the red layer and the green layer]. (B) Mean curvature evolution for each layer of the structure. The color code corresponds to (A).
Reprint from Dynamic evolution and fracture of multilayer field emitters in atom probe tomography: a new interpretation, *Rolland N, Vurpillot F, Duguay S, Blavette D, The European Physical Journal Applied Physics 2015;72(2):2100.* With kind permission of The European Physical Journal (EPJ).

the evaporation field differences as follows if A is defined by its radius R_A and its FEV F_{EV}^{A}, and B by its radius R_B and its FEV F_{EV}^{B}.

$$\frac{R_A}{R_B} = \frac{F_{EV}^{A}}{F_{EV}^{B}} \tag{2.34}$$

At the interface between phases, in the mesoscopic approach, no rupture of the derivative of the surface tangent should exist. Otherwise, local enhancement of field will produce a local fast increase in evaporation rate that will smooth the roughness rapidly. At the atomic scale, this effect can still exist, especially if local atomic bonding is different for these particular atoms.

Under the effect of FEV, the tip will develop a unique shape, depending on the evaporation fields of each species that evolves continuously as phase proportions change during the erosion process of the field evaporated end form. An example is given for a multilayer sample in Fig. 2.20. This process exists even for very small particles or phases, because of the high steepness of the evaporation rate with field.

Pulsed-Field Evaporation

The controlled emission of evaporated atoms is mandatory in AP to perform time-of-flight mass spectrometry. Pulsed FEV can be achieved through two main mechanisms, voltage pulsing and thermal pulsing.

Voltage-Pulsed Evaporation

Let us consider a material with evaporation constants C and F_{EV}, submitted to a base temperature T_0. Under the application of V_{DC}, a standing field F_{DC} exists at the atom surface. We can choose this field to avoid any significant evaporation of atoms ($K_T \ll 1$ atom/s). When applying a voltage pulse to the tip, the evaporation rate can be increased several ranges of magnitude. The amplitude of this voltage peak is chosen to ensure that one atom can be evaporated every N voltage pulses. This rate defines the usual operating evaporation rate (atom/pulse $\ll 1$). Experimentally, a positive variation of voltage can be applied to the sample, or a negative pulse of voltage can be applied to the counter electrode ahead. An electrical drawing of the experimental setup is given in Fig. 2.21. The tip to counter electrode acts as a capacitor of electrical capacitance C_E (10^{-17} F), and the tip presents a resistance R_E, which depends on its electrical properties (resistivity) [59,60]. A difference of voltage $V_0 + V_P$ is transmitted to the tip surface at the top of the voltage pulse that triggers FEV. Note that the bandwidth of the system ($R_E C_E$ product) must be sufficient to ensure good transmission of the evaporation pulse of width t_P. Time-of-flight spectrometry is performed by measuring the difference in time between the maximum of evaporation pulse V_P and the stop time given by the ion impact.

The temporal variation of V gives a temporal variation of the evaporation rate R_T during a cycle containing a unique voltage pulse. Let us assume that the maximum probability of the evaporation pulse $R_{T,max}$ is obtained for $V(t=0) = V_{max} = V_{DC} + V_P$, with V_{DC} the standing voltage on the tip, and V_P the maximum amplitude of the voltage. To quantify the amplitude of the triggering voltage, the voltage contribution is given by $\alpha_P = V_P/(V_P + V_{DC})$, or by the pulse fraction $V_P/V_{DC} = \alpha_P/(1 - \alpha_P)$ (Fig. 2.22).

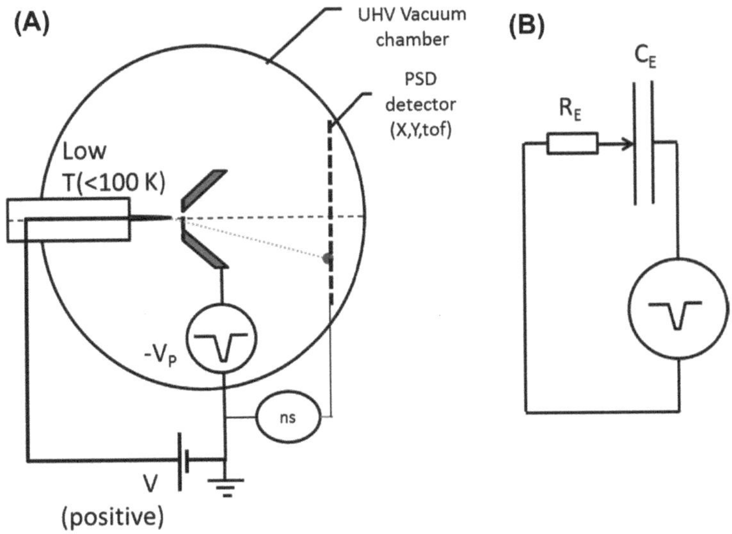

Figure 2.21 Schematic drawing of a voltage pulse AP. (A) The time of flight is measured between the start time (voltage pulse) and the stop time (ion detection). (B) Field is derived from the difference in voltage between the tip V_{DC} and the counter electrode submitted to $-V_P$, ($F = [V + V_P]/k_F R$).

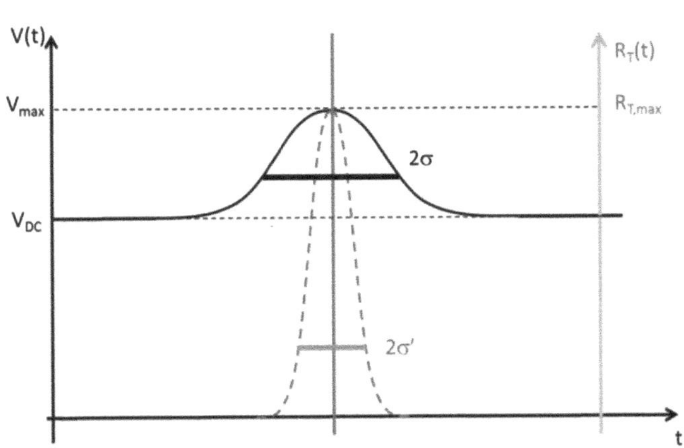

Figure 2.22 Schematic drawing of the voltage pulse (*black solid line*) of maximal voltage V_{max} and the corresponding evaporation pulse of rate $R_T(t)$. The evaporation pulse is about 10 times shorter than the voltage pulse of width 2σ.

Assuming to simplify that the evaporation field is reached at the topmost of the pulse (evaporation rate is very large on the pulse), we can write

$$\frac{V(t)}{V_{max}} = (1 - \alpha_P) + \alpha_P \times f(t) \tag{2.35}$$

where $f(t)$ is the temporal shape of the voltage pulse equal to one for $t = 0$, and null far from $t = 0$. Considering this equation the evaporation rate can be rewritten as

$$\frac{K_T(t)}{K_{T,max}} = \exp\left[-\frac{C}{kT}(1 - V(t)/V_{max})\right] = \exp\left[-\frac{C}{kT}\alpha_P(1 - f(t))\right] \tag{2.36}$$

If we now consider $f(t)$ as a Gaussian pulse of standard deviation $\sigma < 1$ ns, and the fraction α_P not too low (10–30%) it gives

$$\frac{K_T(t)}{K_{T,max}} = \exp\left[-\frac{C}{kT}\left(1 - \exp\left(-\frac{t^2}{2\sigma^2}\right)\right)\right] \approx \exp\left[-\frac{C}{kT}\alpha_P\frac{t^2}{2\sigma^2}\right] \tag{2.37}$$

Indeed, the large value of $\frac{C}{kT}\alpha_P \left(10 < \frac{C}{kT}\alpha_P < 100 \text{ for tungsten values}\right)$ makes the evaporation rate between pulses extremely small, so that it could be neglected. The evaporation pulse is shorter than the voltage pulse $\sigma' = \sigma\sqrt{\frac{kT}{\alpha_P C}}$.

Assuming a voltage pulse ~1 ns in width, an operation evaporation rate of 1 atom/pulse gives an instantaneous evaporation rate on the pulse of about 10 atoms/ns, or 10^{10} atoms/s. At this rate, and considering low working temperature (10–80K), the slope of Eq. (2.33) is divided by 20 (ln (10^9)) compared to DC FEV, and we can consider that the threshold field of evaporation at this temperature is almost equivalent to the evaporation field at 0K (Fig. 2.23).

Temperature-Pulsed Evaporation

The second mechanism that can be used to trigger evaporation is a temperature pulse. Assuming a temperature pulse with a Gaussian shape of standard deviation σ, we can write similarly to the voltage pulse mechanism

$$K_T(t) \propto \exp\left[-\frac{C}{k\left(T_0 + T_P\exp\left(-\frac{t^2}{2\sigma^2}\right)\right)}\alpha_P\right]$$
$$\approx \exp\left[-\frac{C}{k(T_0 + T_P)}\alpha_P\right]\exp\left(-\frac{C\alpha_P}{k(T_0 + T_P)}\frac{t^2}{2\sigma^2} \times \frac{T_P}{(T_0 + T_P)}\right)$$

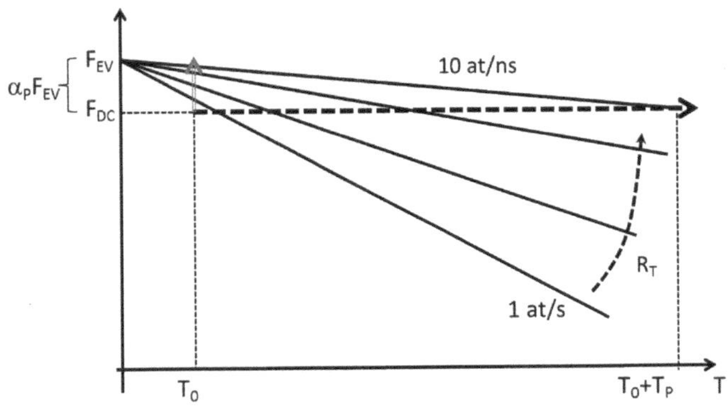

Figure 2.23 Schematic diagram of the pulse evaporation mode in atom probe. At (T_0, F_{DC}), evaporation rate is extremely low. Evaporation can be triggered either by voltage pulse of amplitude $\alpha_P F_{EV}$ or by temperature pulse of amplitude $T_P + T_0$.

and

$$\frac{K_T(t)}{K_{T,\max}} = \exp\left(-\frac{t^2}{2\sigma'^2}\right) \tag{2.38}$$

With σ', the standard deviation, equal to $\sigma' = \sigma\sqrt{k\frac{(T_0+T_P)^2}{\alpha_P C T_P}}$. With standard values ($\alpha_P = 0.2$, $C = 3\text{eV}$, $T_P = 200\text{K}$, and $T_0 = 80\text{K}$), the evaporation pulse width is one range of magnitude shorter than the temperature pulse. However, in contrast to the voltage pulse mode, there must exist a high difference between the peak temperature and the base temperature to avoid evaporation at the DC voltage. Note that the real shape of the temperature pulse is not strictly Gaussian, since the shape of the temperature pulse is governed by the heating and cooling mechanisms in the tip geometry, as it will be discussed in "Chapter 8, Laser Assisted Field Evaporation," and is strongly asymmetric. Nevertheless, we use this simplification in an educational purpose.

Coevaporation Mechanism

Operating detection rate is chosen experimentally to limit multihit impacts onto the detector (more details will be given in Chapter 6: Atom Probe Tomography Detectors). However, due to the statistical nature of the evaporation process, evaporation during the pulse of more than one atom cannot always be avoided.

Considering all the atoms of the surface seen by the detector, the mean time required to detect an ion is

$$\tau_D \propto \exp\left(\frac{C(1 - F/F_{EV})}{k_B T}\right) \tag{2.39}$$

The waiting time τ between two successive events is therefore related to a Poisson process and the probability distribution function $P_w(\tau)$ to get a waiting time τ can be statistically defined by an exponential law:

$$P_w(\tau) = \frac{1}{\tau_D}\exp\left(-\frac{\tau}{\tau_D}\right) \tag{2.40}$$

For low-probability events (which is true at a mean detection rate $K_D < 0.1$ atom/pulse), the probability of detecting n atoms $P_{multi}(n)$ evaporated on the same pulse therefore follows a Poisson law:

$$P_{multi}(n) = \frac{K_D^n}{n!}\exp(-K_D) \tag{2.41}$$

The probability to detect more than one ion would be

$$P_{multi}(n \geq 2) = 1 - \exp(-K_D) - K_D \exp(-K_D) \tag{2.42}$$

An operating evaporation rate of 1% gives a theoretical probability to get more than two hits lower than 5×10^{-5}.

Deviations to this theoretical law are observed experimentally (Fig. 2.24, Ref. [61]). It was shown that even at a low detection rate, the occurrence of multihit events is still important [62]. The atoms that are field evaporated during an APT (or FIM) analysis have been shown to be highly correlated in time and in space. This phenomenon can be explained by a nonuniform distribution of the electric field at the surface of the tip (Fig. 2.24). The atoms that are simultaneously evaporated are situated in regions of the surface where the electric field is similar. When the surface has reached its equilibrium shape, there are several atom positions on a tip surface that are completely equivalent regarding the electric field. Moreover, considering the order of evaporation, the evaporation of one atom can lead to the rapid evaporation of the surrounding neighbors. This correlated evaporation is due to the discrete nature of the geometrical surface. The evaporation of one atomic volume modifies the local roughness of the surface. Surrounding atoms are submitted to a high electric field and are evaporated swiftly. This explains the fact that simultaneous impacts are often situated very close to each other, and that the distribution of the distances between impacts

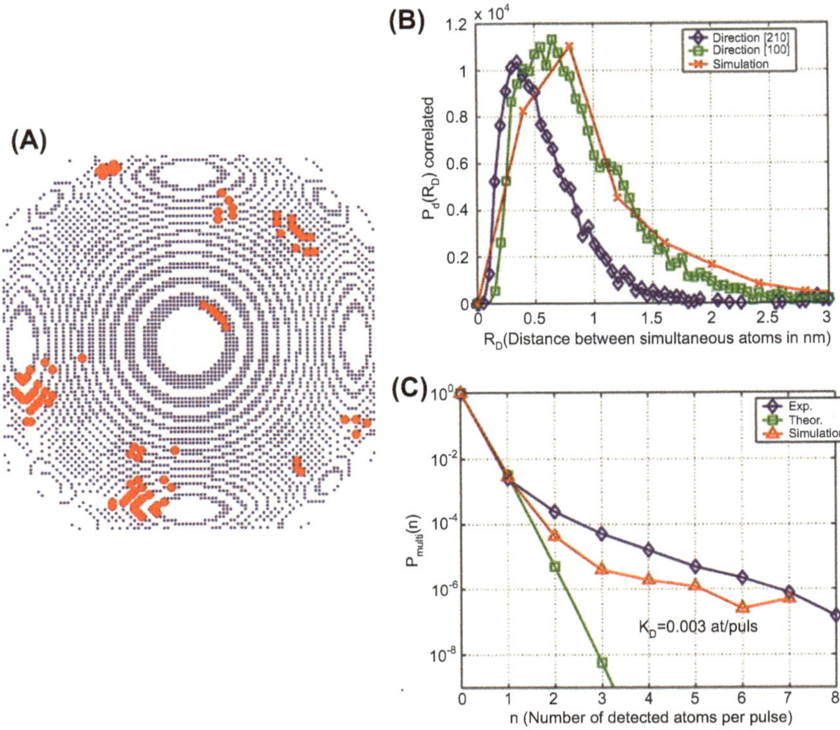

Figure 2.24 Simulated and experimental evidence of coevaporation. (A) Simulated spatial distribution on a tip of the next 100 atoms [in red] in the sequence of field evaporation. (B) Histogram of the distribution of the distance between two impacts onto the detector for the evaporation of an aluminum alloy (two different crystallographic directions involved) and for the simulated example. Correlated evaporation is evident. (C) Distribution of the number of atoms acquired per pulse in an experiment [blue], a simulated case [red], and using Eq. (2.41) [green].

presents a maximum that is very close to the first neighbors distance on a particular crystallographic terrace.

The strong divergence with the Poisson law for the histogram of the number of atoms detected per pulse (Fig. 2.24) can be very well understood by this concept of local dynamic variation of electric field. On a particular pulse, if an atom is likely to evaporate because its electric field is higher than the mean electric field of the surface, its evaporation can increase the probability of evaporation of local neighbors. This will increase strongly the probability detection of more than one atom per pulse as compared to Poisson law.

Preferential Evaporation

In AP, several problems may arise as slow FEV occurs at the DC voltage (between two pulses). For a given material, defined by its constants C and F_{EV}, the number of atoms evaporated between pulses may be evaluated with respect to the number of atoms evaporated on pulses. In voltage mode AP, atoms evaporated on pulses have higher kinetic energy, and are identified in the mass spectrum. Atoms evaporated between pulses and detected without any correlation with pulses, are identified as background noise events. In laser mode AP, background is higher close to the mass peaks of the considered species due to the presence of a slow cooling down tail after the triggering temperature pulse (see Chapter 8: Laser-Assisted Field Evaporation for more details).

In voltage pulse mode, considering a time interval between two pulses given by τ_{cycle} and assuming an evaporation pulse width Δ ($\sim 2\sigma'$ considering Eq. (2.37)), the ratio between the number of atoms evaporated on pulses N_{pulse} to the total number of evaporated atoms N_{total} on a cycle may be written approximately as

$$\frac{N_{pulse}}{N_{total}} \approx \frac{1}{1 + \frac{\tau_{cycle}}{\Delta} \exp\left(\frac{-C\alpha_p}{kT}\right)} \tag{2.43}$$

In laser mode, assuming a pulse temperature T_P, and a maximal temperature $T_M = T_0 + T_P$, this ratio is given by

$$\frac{N_{pulse}}{N_{total}} \approx \frac{1}{1 + \frac{\tau_{cycle}}{\Delta} \exp\left(\frac{-C\alpha_p}{kT_0}\left[\frac{T_P}{T_P + T_0}\right]\right)} \tag{2.44}$$

Increasing the cyclic ratio Δ/τ_{cycle} (increasing repetition rate), decreasing the base temperature, or increasing the pulse contribution α_P decreases the background noise and any loss of atoms between pulses.

Analyzing a material composed of several elements with different evaporation fields and different constants C, makes the evaporation more complex. Atoms with high C (high cohesive energy), also have a strong evaporation field constant. At the top of voltage pulses, these atoms can be more difficult to evaporate than atoms with lower C values. A non—steady state condition can be attained during FEV, in which some atoms (low evaporation field) are evaporated preferentially during pulses. In this case, since on average the tip evaporates continuously, the others are evaporated between pulses, at the DC voltage, to keep the whole evaporation rate constant. Therefore, under such conditions, the measured composition obtained with atoms identified as peaks on the mass spectrum, will not be the composition of the bulk material.

In the simple case of an AB alloy, with A and B defined by their constants ($F_{A,EV}$ the evaporation field of A, C_A the energetic constant of A, $F_{B,EV}$ the evaporation field of B, C_B the energetic constant of B), the apparent composition could be estimated by writing a conservation law that assumes that the total rate of evaporation of atoms is constant over a complete cycle of evaporation (DC + pulse):

$$K_A\Delta + K_{A,DC}\tau_{cycle} = K_B\Delta + K_{B,DC}\tau_{cycle} \tag{2.45}$$

where R_A (respectively B) is the rate of evaporation of A (respectively B) atoms onto the pulse, and $R_{A,DC}$ (respectively B) is the rate of evaporation of A (respectively B) atoms between pulses.

Writing

$$K_A \approx e^{\frac{-C_A\left(1-F_A/F_{A,EV}\right)}{kT}}, \quad K_B \approx e^{\frac{-C_B\left(1-F_B/F_{B,EV}\right)}{kT}},$$

$$K_{A,DC} \approx e^{\frac{-C_A\left(1-F_{A,DC}/F_{A,EV}\right)}{kT}} = K_A \cdot e^{\frac{-C_A\alpha_p F_A/F_{A,EV}}{kT}},$$

and

$$K_{B,DC} \approx e^{\frac{-C_A\left(1-F_{B,DC}/F_{B,EV}\right)}{kT}} = K_B \cdot e^{\frac{-C_B\alpha_p F_B/F_{B,EV}}{kT}} \tag{2.46}$$

It gives

$$\frac{K_A}{K_B} = \frac{1+\frac{\tau_{cycle}}{\Delta}e^{\frac{-C_B\alpha_p F_B/F_{B,EV}}{kT}}}{1+\frac{\tau_{cycle}}{\Delta}e^{\frac{-C_A\alpha_p F_A/F_{A,EV}}{kT}}} \tag{2.47}$$

In this equation, F_A is the electric field above A atoms for the tip biased at the top of evaporation pulses and F_B the electric field above B atoms. It is important to stress that these two fields are not identical, since each phase (A or B) tends to develop its own steady state shape at the evaporation ($R_A/R_B \sim F_B/F_A$, Fig. 2.25). In fact, due to the steep nature of the evaporation rate with the field, at the top of the voltage pulse, the field is very close to the evaporation field of each element, as it was pointed out in the section

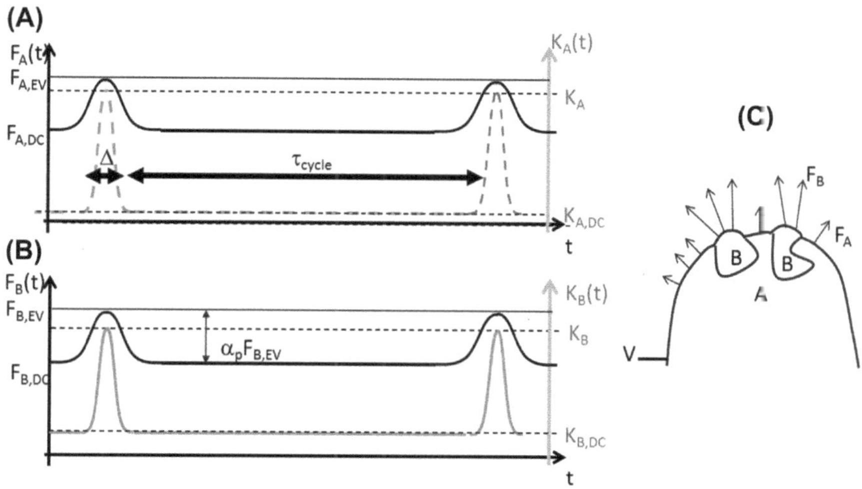

Figure 2.25 Schematic representation of the preferential evaporation bias in the analysis of an alloy AB. In (A) [and, respectively, in (B)] the evaporation rate of atoms A (and B, respectively) is represented during a pulsing cycle (τ_{cycle} the duration between two successive pulses) with a voltage pulse giving rise to an evaporation pulse of duration Δ. Note that the field is different above A and B atoms as represented in (C), and is close to $F_{A,EV}$ for A atoms and $F_{B,EV}$ for B atoms.

Steady State Field Evaporated Shape of the Tip. Introducing $F_A = (1 - \varepsilon_A) F_{A,EV}$ and $F_B = (1 - \varepsilon_B) F_{B,EV}$, with ε a small value, leads to

$$\frac{K_A}{K_B} = \frac{1 + \frac{\tau_{cycle}}{\Delta} e^{\frac{-C_B \alpha_P}{kT}} e^{\frac{C_B \alpha_P}{kT} \varepsilon_A}}{1 + \frac{\tau_{cycle}}{\Delta} e^{\frac{-C_A \alpha_P}{kT}} e^{\frac{C_A \alpha_P}{kT} \varepsilon_B}} \approx \frac{1 + \frac{\tau_{cycle}}{\Delta} e^{\frac{-C_B \alpha_P}{kT}}}{1 + \frac{\tau_{cycle}}{\Delta} e^{\frac{-C_A \alpha_P}{kT}}} \tag{2.48}$$

And the measured concentration of B atoms is (Fig. 2.26)

$$C'_B = \frac{1}{1 + \frac{C_A}{C_B} \frac{K_A}{K_B}} \tag{2.49}$$

In laser mode, equations are slightly different since the electric field is identical during and between pulses, but the temperature is different. If we assume α_P identical for A and B atoms and writing

$$K_A \approx e^{\frac{-C_A \alpha_P}{kT_{max}}}, \quad K_B \approx e^{\frac{-C_B \alpha_P}{kT_{max}}},$$

$$K_{A,DC} \approx e^{\frac{-C_A \alpha_P}{kT_0}}, \text{ and } K_{B,DC} \approx e^{\frac{-C_B \alpha_P}{kT_0}} \tag{2.50}$$

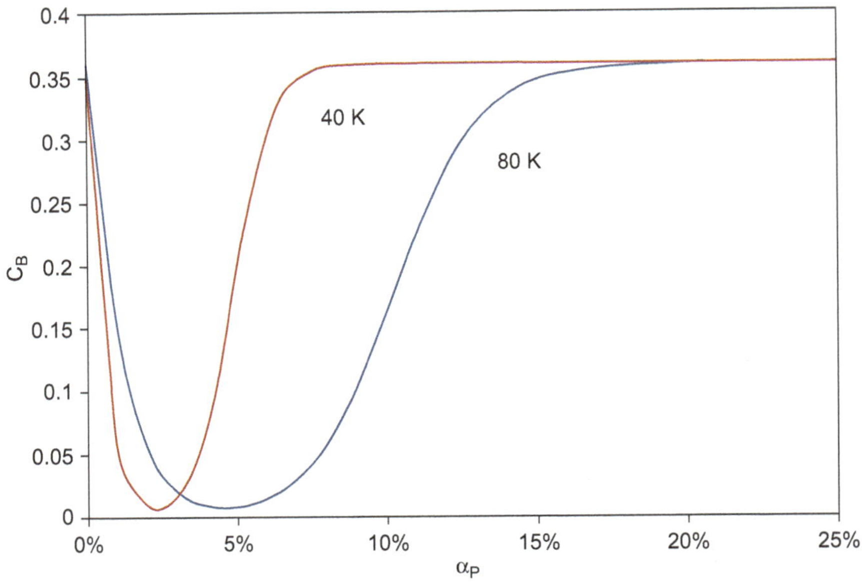

Figure 2.26 Theoretical measured composition in B of an AB alloy in function of the pulse contribution, at two base temperatures of 40K and 80K, taking into account Eqs. (2.41) and (2.14), for $C_A = 1.4$ eV, $C_B = 0.6$ eV, the bulk composition of B being $C_B = 35\%$. Pulse contribution can be decreased at low base temperature.

It gives

$$\frac{K_A}{K_B} \approx \frac{1 + \frac{\tau_{cycle}}{\Delta} e^{\frac{-C_B \alpha_P}{kT_0}\left[\frac{T_P}{T_{max}}\right]}}{1 + \frac{\tau_{cycle}}{\Delta} e^{\frac{-C_A \alpha_P}{kT_0}\left[\frac{T_P}{T_{max}}\right]}} \tag{2.51}$$

To reduce the preferential evaporation of one species compared to the other, it is required to increase the duty cycle by increasing the repetition rate, increasing the pulse contribution, or decreasing the base temperature. This behavior is well known experimentally, and analyzing conditions must be tested to calibrate every new material [63–65].

Surface Diffusion of Atoms Prior to Field Evaporation

The model proposed to explain composition bias in AP neglects a lot of physical effects that can occur before evaporation. In particular, atoms may move before evaporation [66]. This mechanism perturbs both the spatial performance and the composition measurement. The motion of lattice atoms at the surface of the specimen was observed in FIM mode by different authors [67]. It can be shown that for a single atom migrating on a flat crystal surface, the number of jumps between lattice positions of the planar surface is given by

$$N\left(s^{-1}\right) = \upsilon_0 \exp\left\{ -\frac{Q_{\text{jump}}}{k_B T} \right\} \tag{2.52}$$

with υ_0 the vibration frequency related to the diffusion phenomenon and Q_{jump} the activation energy for diffusion. As a result, the mean square displacement at the tip surface (2D problem) is given by

$$\frac{\langle \rho^2 \rangle}{2\tau_0} = D_0 \cdot \exp\left(-\frac{Q_{\text{jump}}}{k_B T} \right) \tag{2.53}$$

with D_0 the diffusivity at the tip surface and τ_0 the observation time. The activation energy may vary from a few tenths of an electron volt to several electron volts in most materials. During the analysis of a sample, an atom stays at the tip surface during a residence time τ_{surface} that depends on the evaporation rate and on the number of atoms on the hemispherical apex. During this interval, the atom is found to move a distance $d \sim \sqrt{2\tau_{\text{surface}} D_0 \cdot \exp\left(-\frac{Q_{\text{jump}}}{k_B T} \right)}$. This phenomenon was, for instance, observed in FIM mode under pulse laser illumination in Fig. 2.27.

Generally, the probability of surface diffusion is very small compared to FEV since in voltage mode activation energy for FEV is only a fraction of an electron volt. Nevertheless, in laser mode, using a high pulse contribution α_P, and under low evaporation rate conditions, the diffusion mechanism can overwhelm the evaporation process.

Surface migration can be a source of bias in the measurement of accurate composition in some materials as it was pointed out by Gault and coworkers [68–70]. Indeed, in AP microanalysis of steels, inconsistencies in the measured compositions of solutes (C, N) have often been reported, as well as their appearance as molecular ions. It was suggested that these issues might arise from surface migration of solute atoms over the specimen surface. Surface migration of solutes was evidenced by FIM observations, and its consequences on AP microanalysis are detailed for a wide range of solutes (P, Si, Mn, B, C, N). Since these segregations follow crystallographic features in the desorption map, the impact of directional diffusion was discussed to understand the artifacts. An example of steel with low content of N and C is presented in Fig. 2.28.

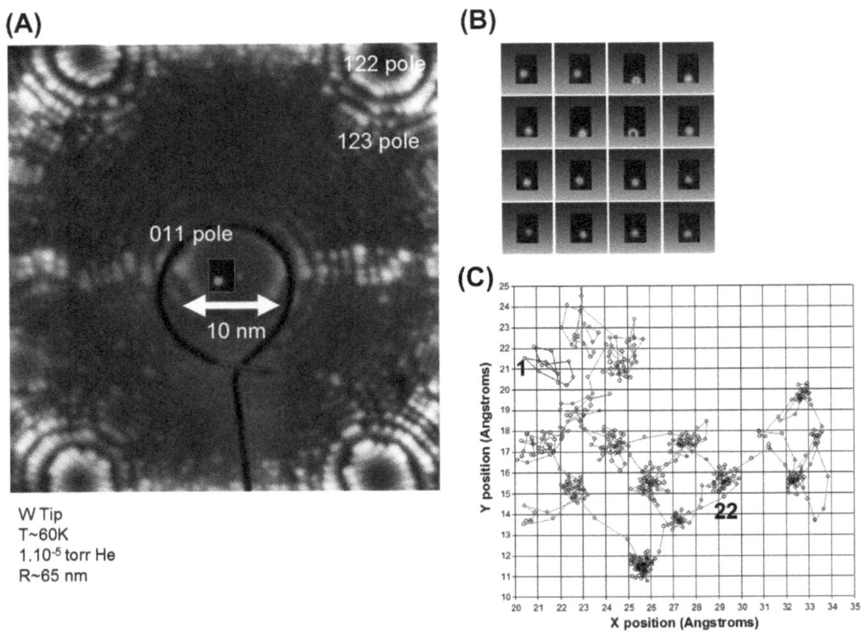

Figure 2.27 Observation of the random walk motion of a surface atom in FIM mode under laser illumination. (A) Zoom on the (011) pole of a tungsten specimen taken at 60K, with the presence of $1 \cdot 10^{-5}$ torr of helium. Image was taken at 40% of the field evaporation threshold with a sensitive CCD camera. (B) The random walk of a single atom is recorded under pulsed laser illumination. (C) Atom position at the tip surface deduced by mass center calculation. Note the presence of defined locations due to the tip crystallography.

This phenomenon primarily affects interstitial elements, and has also been observed for substitutional elements such as Si in steel, impacting the capabilities of composition measurement in APT. The physical explanation of the migration is for the moment neither well documented nor understood, but the impact of the surface field is clear. Solutes are migrating toward high-electric field regions over the specimen surface and the spatial resolving power of the technique is lowered. Other surface processes (ie, surface reconstruction) may also be involved. For each alloy, experimental conditions where surface migration is either avoided or, at least, limited, should exist and be revealed by a systematic investigation. Further investigations are clearly needed to determine the best experimental conditions necessary for avoiding this kind of artifact that is not the general case.

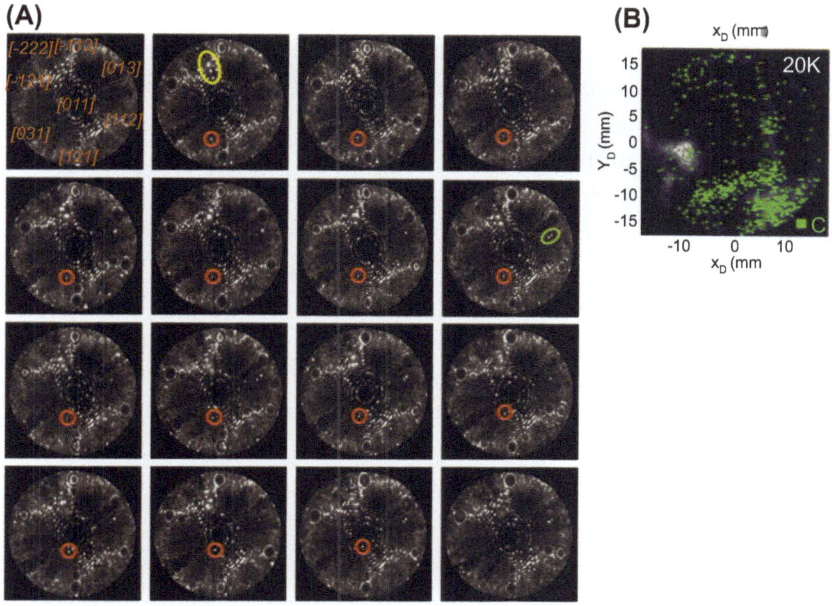

Figure 2.28 (A) Observation of the directional walk motion of a surface atom in focused ion beam mode at 20K. Migration of a bright spot along a zone axis toward the (110) pole is evidenced by the *red circles*, while the *yellow* and *green ellipses* highlight the presence of precipitates within the micrographs. This bright spot most likely corresponds to either N or C *(image taken from Ref. [68])*. (B) In these alloys, high field regions present artificial segregation of C and N atoms as seen on the desorption map.

CONCLUSION

Understanding the basic principles of field ionization, FEV, postionization, surface diffusion, and related phenomena is extremely essential to understand the relationship between the few parameters that can be tuned by the user of an AP and the metrological parameters of this nanoanalyzing instrument. Controlling the electric field is mandatory in this aim. We have introduced in this chapter the way to generate a high static field (or quasi-static in case of voltage pulse) onto a specimen surface whatever its physical nature (metals to insulators). Geometric aspects are important, and must be controlled with care as it will be explained in the specimen preparation chapter.

The way a tip surface is eroded atom by atom and the mechanisms of ion emission have direct consequences on the technological choices used to pulse the tip, and to collect information necessary to reconstruct an accurate image of the probed material. These choices will be presented in the chapters dedicated to mass spectrometry, in the chapter dedicated to laser pulsing, or in the detector chapter, for instance.

The way atoms are removed from the surface will give us the road to backproject ions onto the specimen surface, to build from the raw dataset a correct 3D representation of the atoms before evaporation, and therefore to evaluate the limitations existing in terms of spatial resolution of the technique. This will be discussed in the reconstruction chapter. Finding the limitations of the technique will help us understand the interest to couple or correlate APT with other techniques of characterization such as SEM or TEM. These parameters will also be essential to understand how to extract valuable information from the millions of atoms collected after analysis as it will be explained in the data mining chapter.

REFERENCES

[1] Gomer R. Field emission and field ionisation. Cambridge, MA: Harvard University Press; 1963.
[2] Müller EW, Tsong TT. Field ion microscopy: principles and applications. New York: Elsevier; 1969.
[3] Tsong TT. Atom-probe field ion microscopy. Cambridge, UK: Cambridge University Press; 1990.
[4] Miller MK, Cerezo A, Heatherington MG, Smith GDW. Atom probe field ion microscopy. Oxford: Clarendon Press; 1996.
[5] Miller MK, Forbes RG. Atom-probe tomography: the local electrode atom probe. Springer; 2014.
[6] Larson DJ, Prosa TJ, Ulfig RM, Geiser BP, Kelly TF. Local electrode atom probe tomography: a user's guide. Springer; 2013.
[7] Gault B, Moody MP, Cairney JM, Ringer SP. Atom probe microscopy. Springer; 2012.
[8] http://www.cameca.com/instruments-for-research/atom-probe.aspx.
[9] Kellogg GL. J Appl Phys 1982;53:6383–6.
[10] Kelly TF, Larson DJ. Atom probe tomography 2012. Annu Rev Mater Res 2012;42: 1–31.
[11] Lozovoi AY, Alavi A. Phys Rev B 2003;68:245416.
[12] Durand E. Electrostatique et Magnétostatique. Masson et Cie; 1953.
[13] Tsong TT. Field ion formation. Surf Sci 1978;70:211–33.
[14] Gomer R. Field emission and field ionisation. Cambridge: Harvard University; 1961. p. 45–6.
[15] Gipson GS. An improved empirical formula for the electric field near the surface of field emitters. J Appl Phys 1980;51(7):3884–9.
[16] Bajikar SS, Kelly TF, Camus PP. Electrostatic analysis of local-electrode atom probes. Appl Surf Sci 1996;94(5):464–71.
[17] Gipson GS, Eaton HC. The electric-field distribution in the field-ion microscope as a function of specimen shank. J Appl Phys 1980;51:5537–9.
[18] Loi ST, Gault B, Ringer SP, Larson DJ, Geiser BP. Ultramicroscopy September 2013; 132:107–13. http://dx.doi.org/10.1016/j.ultramic.2012.12.012.

[19] Greiwe G-H, Balogh Z, Schmitz G. Ultramicroscopy 2014;141:51.
[20] Arnoldi L, Silaeva EP, Gaillard A, Vurpillot F, Blum I, Rigutti L, Deconihout B. J Appl Phys 2014;115(20):203705.
[21] Sévelin-Radiguet N, Arnoldi L, Vurpillot F, Normand A, Deconihout B. EPL Europhys Lett 2015;109(3):37009.
[22] Peralta J, Broderick SR, Rajan K. Ultramicroscopy September 2013;132:143–51.
[23] Tsukada M, Tamura H, McKenna KP, Shluger AL, Chen YM, Ohkubo T, et al. Ultramicroscopy May 2011;111(6):567–70.
[24] Karahka M, Kreuzer HJ. Ultramicroscopy September 2013;132:54–9.
[25] Ono T, Sasaki T, Otsuka J, Hirose K. Surf Sci 2005;577(1):42–6.
[26] Ono T, Hirose K. J Appl Phys 2004;95:1568.
[27] Sanchez CG, Lozovoi AY, Alavi A. Mol Phys 2004;102:1045–55.
[28] Kreuzer HJ, Nath K. Surface Science 1987;183(3):A183.
[29] Oberdorfer C, Eich SM, Schmitz G. Ultramicroscopy 2013;128:55–67.
[30] Rolland N, Vurpillot F, Duguay S, Blavette D. Microsc Microanal 2015;21(06): 1649–56.
[31] (a) Jeffreys H. On certain approximate solutions of linear differential equations of the second order. Proc Lond Math Soc 1924;23:428–36.
(b) Brillouin L. La mécanique ondulatoire de Schrödinger: une méthode générale de resolution par approximations successives. C R Acad Sci 1926;183:24–6.
(c) Kramers HA. Wellenmechanik und halbzählige Quantisierung. Z für Phys 1926; 39(10–11):828–40. Bibcode:1926ZPhy...39..828K.
(d) Wentzel G. Eine Verallgemeinerung der Quantenbedingungen für die Zwecke der Wellenmechanik. Z für Phys 1926;38(6–7):518–29. Bibcode:1926ZPhy.. 38..518W.
[32] Fowler RH, Dr Nordheim L. Electron emission in intense electric fields. Proc R Soc A May 01, 1928;119(781):173–81.
[33] Van Eekelen HAM. Surf Sci 1970;21:21–44.
[34] (a) Müller EW. Z Phys 1951;131:136.
(b) Inghram MG, Gomer R. J Chem Phys 1954;22:1279.
(c) Müller EW, Bahadur K. Phys Rev 1956;102:624.
[35] Suchorski Y, Medvedev VK, Block JH. Appl Surf Sci 1996;94–95:217–23.
[36] Devaraj A, Colby R, Hess WP, Perea DE, Thevuthasan S. J Phys Chem Lett 2013;4: 993–8.
[37] Diercks DR, Gorman BP, Kirchhofer R, Sanford N, Bertness K, Brubaker M. J Appl Phys 2013;114:184903.
[38] Riley JR, Detchprohm T, Wetzel C, Lauhon L. J Appl Phys Lett 2014;104:152102.
[39] De Castilho CMC, Kingham DR. Surf Sci 1986;173:75–96.
[40] Haydock R, Kingham DR. Surf Sci 1981;103:239–47.
[41] Ernst N. Surf Sci 1979;87:469–82.
[42] Kingham DR. Surf Sci 1982;116:273–301.
[43] Müller EW. Naturwiss 1941;29:533.
[44] Kellogg GL. Phys Rev B 1984;29:4304.
[45] Müller EW. Phys Rev 1956;102:618–24.
[46] McMullen ER, Perdew JP. Phys Rev B 1987;36(5).
[47] Brandon DG. Brit J App Phys 1963;14:474.
[48] Brandon DG. Surf Sci 1964;3:1–18.
[49] Tsong TT. Surf Sci 1978;70:211.
[50] McMullen ER, Perdew JP, Rose JH. Solid State Commun 1982;44(6):945–9.
[51] Kreuzer HJ, Nath K. Surf Sci Lett 1987;183(3):A183.
[52] Wang LC, Kreuzer HJ. Surf Sci 1990;237(1–3):337–46.
[53] Vurpillot F, Oberdorfer C. Ultramicroscopy 2015;159:202–16.
[54] Flahive PG, Graham WR. Surf Sci 1980;91:449–62.

[55] Djurabekova F, Parviainen S, Pohjonen A, Nordlund K. Phys Rev E 2011;83:026704.
[56] Smith R, Walls JM. J Phys D Appl Phys 1978;11:409.
[57] Gruber M, Vurpillot F, Bostel A, Deconihout B. Surf Sci 2011;605(23–24):2025–31.
[58] Wada M. Surf Sci October 1, 1984;145(2–3):451–65.
[59] Melmed AJ, Martinka M, Girvin SM. Appl Phys Lett 1981;39:416.
[60] Gilbert M, Vurpillot F, Vella A, Bernas H, Deconihout B. Ultramicroscopy 2007; 107(9):767–72.
[61] De Geuser F, Gault B, Bostel A, Vurpillot F. Surf Sci January 25, 2007;601(2):536–43.
[62] Renaud L, [Ph.D. thesis], Rouen; 2001.
[63] Miller MK, Smith GDW. J Vac Sci Techno 1981;19:57.
[64] Miller MK, Russell K. Surf Interface Anal 2007;39:262–7.
[65] Takahashi J, Kawakami K. Surf Interface Analysis 2014;46:535–43.
[66] Ehrlich G. Surf Sci 1977;63:422–47.
[67] Kellogg GL. J Appl Phys 1981;52:5320.
[68] Gault B, Danoix F, Hoummada K, Mangelinck D, Leitner H. Ultramicroscopy 2012; 113:182–91.
[69] Yamaguchi Y, Takahashi J, Kawakami K. Ultramicroscopy 2009;109(5):541–4.
[70] Yao L, Gault B, Cairney JM, Ringer SP. Philos Mag Lett 2010;90:121–9.

CHAPTER THREE

Basics of Field Ion Microscopy

F. Danoix, F. Vurpillot
Groupe de Physique des Matériaux, University and INSA of Rouen, Normandie University, France

INTRODUCTION

Field ion microscopy (FIM) is one of the early steps leading to the development of atom probe tomography (APT). It shares with APT the necessity of an extremely intense electric field at the specimen surface, and most of the projection laws necessary to interpret collected data. FIM development started in the mid-1930s from the early work on field emission of electron by Müller, and was officially successfully achieved in October 1956 [1]. It was, and is still, almost only applied to conductive materials, even if some recent attempts have been conducted on laser-assisted FIM. This chapter will therefore mostly concentrate on FIM of metals and alloys. But some examples of application to nonmetallic sample will be presented.

The invention of the different kinds of microscopes represented each time a step forward for science, particularly in materials science. As better and better microscopes were built, it was discovered that there exists a limit that cannot be exceeded. This is connected with the wave characteristics of the used imaging radiation. Using light waves, it is impossible to distinguish details smaller than the wavelength of the light. For a conventional microscope using visible light, the resolution is some hundreds of nanometers. A great breakthrough in microscopy came when it was found possible to produce an image of an object using electrons, as used in transmission electron microscopy (TEM) or scanning electron microscopy (SEM), intrinsic wavelength of electron used in these instruments being a fraction of Angstrom. However, this requires the use of complex and expensive electron optics to reach atomic scale performances. This requirement was overcome in scanning tunneling microscopy, where this image is formed by the punctual measurement of electronic current emission from atoms itself. By scanning point by point the specimen surface with a mechanical device (a tip) enables to build an image of the topology of atomic surface of materials. This makes it possible to depict individual atoms, that is, to study in the greatest possible detail the atomic structure being examined.

Atom Probe Tomography
ISBN: 9780128046470
http://dx.doi.org/10.1016/B978-0-12-804647-0.00003-6

The bottleneck is obviously the need to remove any vibration from the surface during the measurement, and the fact that the surface can evolve during the scan process, which is not instantaneous.

In the landscape of microscopy techniques, FIM is a really particular technique. First, FIM is based on the field ionization of rare gas atoms near the surface of a sharp tip [2]. These gas ions are the vector for the formation of an image of the atomic positions at the specimen surface. The intrinsic resolution of the field ion microscope is thus theoretically given by the wavelength associated to the emitted gas ions. For usual experimental conditions (Ne as imaging gas, specimen voltage 5 kV), this wavelength is 10^{-4} nm, far below the resolution necessary to resolve atomic positions. Second, as it will be described later, a field ion microscope is an angular projection microscope, and therefore insensitive to micron vibration of the sample. The deduced spatial magnification is proportional to the ratio between the specimen-to-screen distance over the specimen radius of curvature, and the projection is only function of the shape of the sample. No complex ion optic device is necessary to produce a magnified image of the surface. Commonly, the magnification is in the million range. Third, all the surface atoms are imaged at the same time. Combined with its intrinsic resolution, the magnification of the microscope makes it possible to resolve individual atomic locations at the specimen surface. It explains why FIM was the first instrument to achieve true atomic resolution in the real space, only 5 years after its birth, and decades before scanning tunneling microscopy or TEM.

However, the field ion microscope does not directly map atomic positions at the specimen surface, but the local variations of the electric field. As shown in the previous chapter, the locus of the highest electric field is right on top of protruding atoms in kink position, the field ion micrograph contrast is thus directly related to atomic positions. For this reason the field ion microscope is viewed as the only microscope able to image individual atoms. It should, however, be kept in mind that not all atoms at the specimen surface will be imaged, but only the most protruding ones at the surface.

BASIC PRINCIPLES

Electric Field at the Specimen Surface

Based on the simple previous description, only a limited number of ingredients are necessary to control and understand the field ion image formation. As explained, the sample needs to be sharp enough to generate the required high electric field. Simple electrostatic arguments can be applied to describe the intensity of the electric field produced by such a polarized specimen. As the physical treatment of field generation has already been covered in chapter "Field Ion Emission Mechanisms," only orders of magnitude will be presented here.

Let us consider a single metal sphere of radius R, bearing a charge Q distributed at the surface. Electrostatics says that the voltage V and the electric field F at the specimen surface are given, respectively, by

$$V = \frac{1}{4\pi\varepsilon_0}\frac{Q}{R}, \quad F = \frac{1}{4\pi\varepsilon_0}\frac{Q}{R^2} \tag{3.1}$$

It directly gives a simple relationship between voltage and electric field:

$$F = \frac{V}{R} \tag{3.2}$$

It follows that a voltage V applied to a metal sphere of radius R will generate an electric field at the surface equal to V/R. Considering that electric fields required for both field ionization and field evaporation, are in the range 10^{10} V/m, and that voltages in the kilovolts range are easily achievable, it appears that radii as small as a few tens of nanometers are required. Of course, FIM (and APT) specimens are not isolated spheres, but needles, which, for the sake of simplicity, are usually described as a truncated cone ended by a quasi-hemispherical end cap. In this configuration, the same basic description still applies to the hemispherical cap, provided a correction factor in the voltage/field relationship is added, leading to

$$F = \frac{V}{k_F R} \tag{3.3}$$

where k_F is a constant called the field factor. Analytical or numerical calculations (see, for instance, the early work by Gomer [3], and more details are given in chapter "Field Ion Emission Mechanisms") lead to the conclusion that the field factor is primarily depending on the shank angle of the truncated cone. Other parameters, such as the shape of the specimen holder, or the electrostatic configuration of the field ion microscope vessel (including the detector) will also influence the field factor, and exact values can only be obtained through numerical simulations, with the exact knowledge of all these geometrical parameters [4–7]. It is commonly accepted that k_F varies from three to eight, but it should be kept in mind that this value is specimen-instrument dependent.

Image Formation

Once a sharp specimen is obtained (see chapter "Atom Probe Sample Preparation" for specimen preparation), an image of its surface can be obtained on a detector after gas atoms are ionized and repelled. The characteristics of the microscope will thus be determined by the associated ion optics. The projection laws are depending on the ion trajectories from the specimen to the detector. As the trajectories are determined by the ion paths in an electrostatic field, they are related to the electric field distribution all along the way to the detector (Fig. 3.1). Again, analytical or numerical calculations can be used to determine the electric field configuration, and express the projection laws.

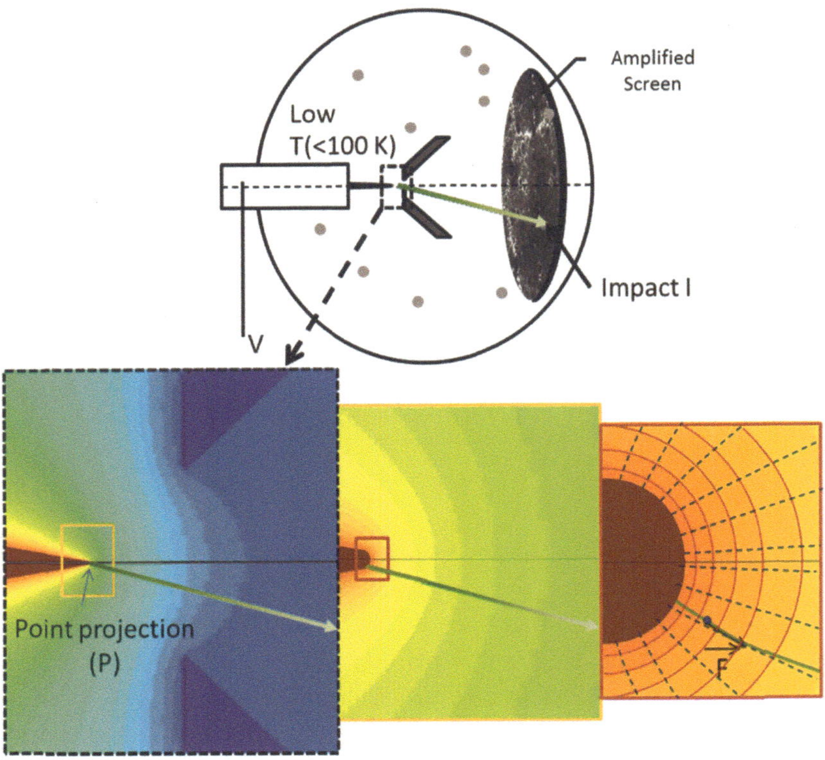

Figure 3.1 Schematics of the ion trajectories in a field ion microscope: in the inset the equipotential lines from the polarized specimen to the grounded detector. In dotted lines on the last inset, the corresponding electric field lines. An ion leaving the specimen surface with no initial velocity will be accelerated according to the local electric field direction, and hit the detector at position I. As seen from the detector, the ion from its starting position seems to have followed a direct trajectory originating from P.

The classical model assumes that, right after ionization, the ions, with no initial velocity, are repelled normally from the specimen surface. During their path to the detector, the ions are influenced by the electrostatic field lines, bent by the presence of the shank of the specimen. The amplitude of the electric field, and thus its capability of significantly affecting the ion trajectories, is significantly reduced with the distance to the specimen. In the final section, the ion trajectories become asymptotically linear. As for the field factor, the ion trajectories depend on the shape of the sample, in particular to its shank angle. Experimentally, the field ion microscope is often considered as a point projection microscope, the center of projection, P, being no longer the center of the end hemispherical cap of the specimen, C, but moved further away into the shank but still along the tip axis. This shift is characterized by the distance CP, expressed in radius of curvature units as $CP = mR$ (see Fig 3.2) or $OP = (m + 1)R$ if O is the tip apex [8,9].

m is a compression factor, taking values smaller than one, usually close to 0.6. Similar to the field factor, m depends on the specimen and electrostatic configuration of the microscope. It reflects the ion trajectories bending due to the presence of the shank of the specimen (in the case of an isolated sphere, m would be null, all trajectories radial, and the projection purely stereographic).

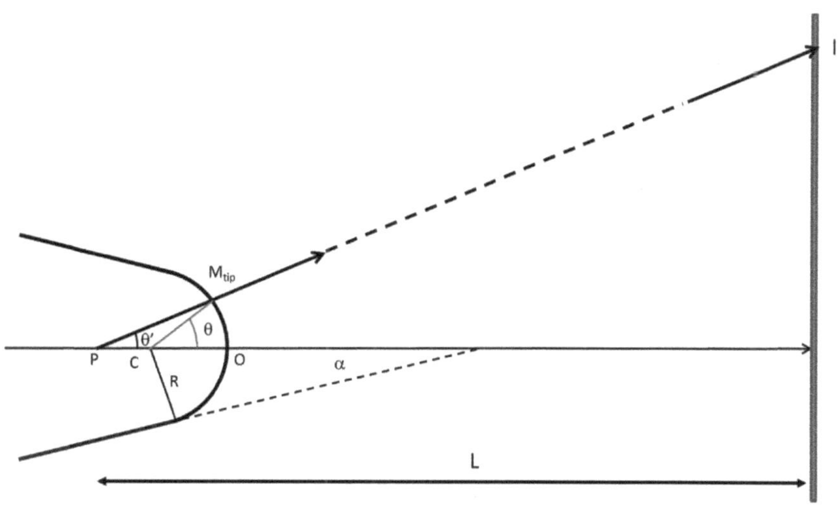

Figure 3.2 Definitions of angles in the projection of a point M_{tip} on tip to an impact I. Note that *L* is the macroscopic distance between the tip and the screen orders of magnitude larger than OP.

As a result, in a first approach, the observed impact angle (θ') of an ion hitting the detector in position A is directly connected to the crystallographic angle (θ) corresponding to its initial position at the specimen surface (Fig. 3.2), I, by the following relation:

$$\theta \approx (m+1)\theta'$$

The $(m + 1)$ factor is generally named the image compression factor (ICF or ξ).

Considering that the radius of curvature of the specimen is negligible with respect to the tip to detector distance, the magnification M of the field ion microscope is simply given by

$$M = \frac{L}{(m+1)R}$$

where L is the distance between the specimen and the detector.

In existing microscopes, L is usually in the range 2–10 cm, leading to a magnification in the order of a million times. With such a magnification, and the intrinsic resolution due to gas ions, single atoms may be resolved, making the strength of FIM.

The geometrical parameters k_F and m will be shown to be important parameters for accurate APT reconstructions, and can be derived from FIM experiments. Indeed, as will be demonstrated in chapter "3D Reconstruction in APT: Basics and Advanced Approaches," the reconstruction accuracy depends on the precision of the two parameters m and k. With the ICF potentially varying between 1.3 and 2, and the field factor k_F in the range 2–8, APT reconstruction can be severely affected. FIM can be used to measure both m and k_F. In pure metals, the ICF can be closely determined using the crystallographic angles modification, and the field factor deduced from the specimen end radius. The presence of visible atomic planes (or poles) is required for such measurements (see chapter "Three-Dimensional Reconstruction in Atom Probe Tomography: Basics and Advanced Approaches" for more details).

FIELD ION MICROSCOPY IN MATERIALS SCIENCE

Introduction

As previously discussed, the field ion microscope is basically a projection microscope, the magnification of which is, under standard operating conditions, of several millions, and a resolution of a fraction of a nanometer. With such performances, true atomic resolution is potentially achievable.

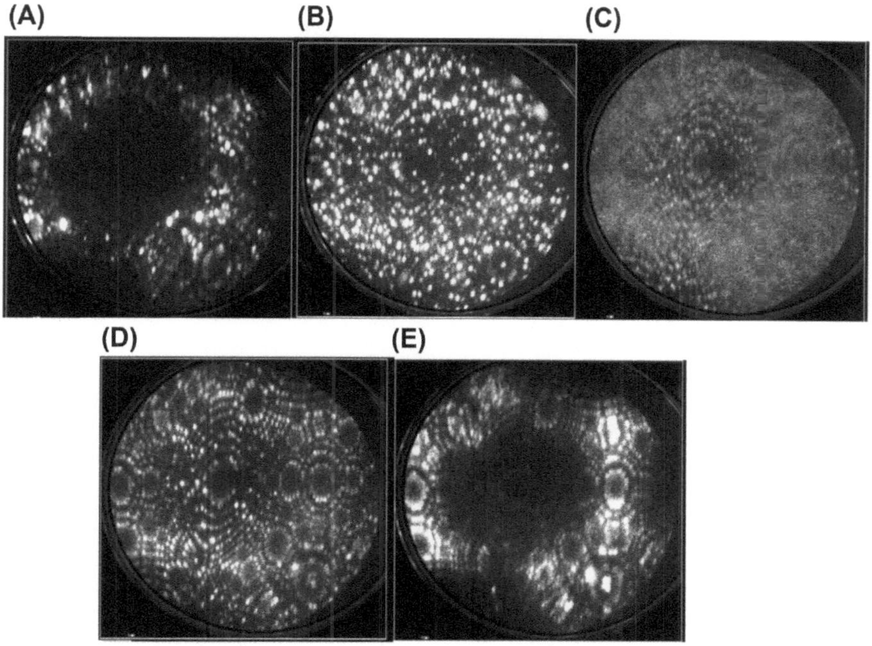

Figure 3.3 Specimen development as observed in field ion microscopy for a tungsten specimen, imaging gas: He, imaging gas pressure = 2×10^{-5} torr, $T = 80$ K. Field of view about 50 nm. (A) Voltage 5 kV, the contaminated surface is first imaged giving rise to a disorganized atomic structure; (B) field evaporation of the contaminated surface is observed at about 6 kV; (C) tip at 8 kV, the metallic *W* surface appears, and evaporates but electric field is too high (above the critical field of ionization in the free space, or best image field); (D) back to 7 kV, the atomic surface of *W* is clearly resolved, but no evaporation is observed, field being lower than the evaporation field of *W*; (E) 6 kV only highly curved regions give rise to ionization.

Indeed, FIM was the first ever method that allowed individual atoms to be observed. The first observation was made in 1956 by Bahadur and Müller at Penn State University on W [1]. As shown in Figs. 3.3 and 3.4, on a field ion micrograph, individual bright spots are clearly resolved. Each bright spot corresponds to a unique atom, indicating that true atomic resolution is achieved. Under simple assumptions, which have been partly discussed previously, invaluable information regarding the atomic scale structure of investigated materials can be derived from FIM.

Specimen Development

A clean field ion micrograph of the specimen surface (ideally free of contamination and atomically resolved) can only be obtained if the upper

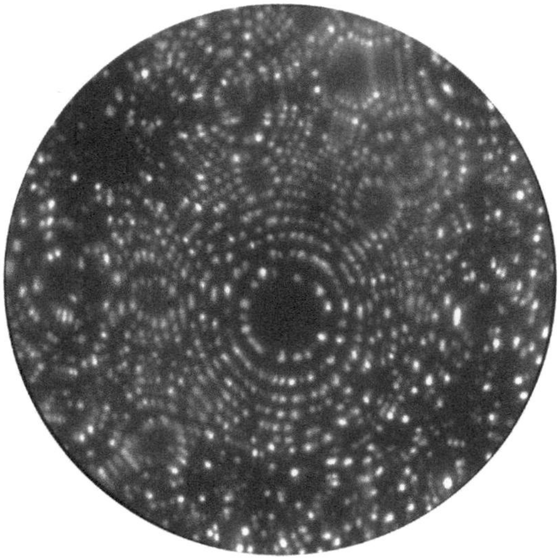

Figure 3.4 Field ion micrograph of a pure iron specimen. Concentric rings are clearly observed, and individual Fe atoms at the specimen surface resolved. Imaging conditions: $T = 20K$, specimen voltage = 4.5 kV, imaging gas: Ne, imaging gas pressure = 5×10^{-5} torr. Field of view about 50 nm. The central pole is oriented along a <111> direction of the crystalline lattice.

contamination layers formed during the specimen preparation (either electrochemically or by ion milling) are removed. This removal can be achieved in situ by field evaporation. Again, physical description of field evaporation process is detailed in chapter "Field Ion Emission Mechanisms," only experimental aspects will be covered here. The controlled erosion of the tip is obtained through field evaporation of the upper contamination layer atoms. This evaporation is tuned by gently increasing the specimen voltage. When the evaporation field of surface atoms is reached, they are field evaporated. This evaporation is continued until specimen atoms are observed. This transition is easy to observe experimentally, as usual; "contamination" species give rise to large and blurred spots, whereas specimen atoms are much well defined and visible throughout the detector (Fig. 3.3). In addition, the evaporation sequence is more regular when the specimen itself field evaporates.

The extreme sensitivity of the erosion process to the local amplitude of the surface electric field (see chapter "Field Ion Emission Mechanisms"), and the small field penetration in conductive materials, makes the field evaporation of specimen atoms easy to control. As a result, the tip surface of the specimen develops as a relatively smooth and regular surface, even down to the nanometer scale. In the case of pure materials, the extremity of the tip end form can be described, as a first approximation, by a unique radius of curvature R (at least over a field of view of ±40°). Only second-order local curvatures induced by lattice binding energy variations, work function variations, or field variations are generally observed (see Fig. 3.3E). At the atomic scale, the atoms that are likely to be the next to field evaporate are the atoms situated at the border of lattice terraces of the studied materials (kink sites or ledge sites). These protrusions give rise to field enhancement of several percent, as compared to the field present at the center of terraces, explaining the highest evaporation probability.

Image Interpretation

In most materials, atoms are periodically arranged in three dimensions (3D) to form a crystalline structure. Symmetries arising from the interception of this rigid lattice by the hemispherical end shape of the specimen generate a series of concentric rings, named poles, each of them being the trace of the emergence of the corresponding crystallographic direction out of the specimen. Between, or in some cases inside, these concentric rings, parts of atomic planes can be observed. The micrograph presented in Fig. 3.3 is characteristic of a crystalline specimen. Depending on the respective orientation of the crystal and the specimen axis, the observed symmetries can be rotated. In the case of very simple (but common) crystal structures (eg, simple, face or body center cubic, hexagonal), different symmetries will be observed. The analysis of these symmetries can be interpreted in the simple framework of stereographic projection, and the structure of the analyzed crystal structure derived, even if FIM should not be regarded as a crystallography instrument.

It is clear that amorphous materials, because of the lack of crystal periodicity, will not show such concentric rings, as illustrated in Fig. 3.5. This micrograph is obtained in an amorphous AlSm alloy, which has been partially devitrified: the microstructure consists pure Al nanocrystallites (each individual particle having its own crystal orientation) embedded in an amorphous AlSm matrix [10]. The amorphous matrix is characterized by the absence of any poles: as the symmetries observed on field ion micrographs arise from the regular arrangement of atoms in the crystal lattice, these symmetries cannot be observed in the amorphous phase.

This micrograph also highlights the importance of symmetries to index field ion micrographs, and derive the orientation of the observed crystal. Indeed, all the individual crystallites, because of their nanometric size,

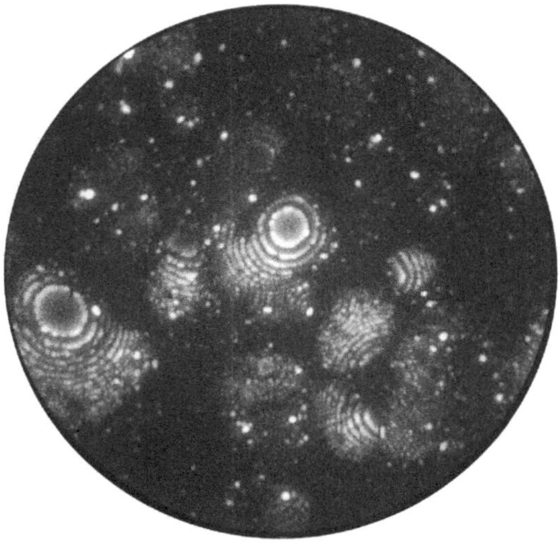

Figure 3.5 Al8at%Sm amorphous alloy partially devitrified. Individual pure Al nanocrystallite, appearing in bright contrast, shows typical ring structures, whereas no such features appear in the amorphous matrix (darkly imaged). Imaging conditions: $T = 20K$, specimen voltage = 5.5 kV, imaging gas: He, imaging gas pressure = 1×10^{-5} torr. Field of view about 100 nm.

only reveal a limited part of a pole figure, making for most of them, impossible to derive their specific orientation. Symmetries are shown to be a key element in the FIM indexation process.

Point Defects

As atoms can be individually imaged, one could assume that vacancies would be easy to identify, and therefore numbered, as they would appear as a "lack of atom" in the field ion image (Fig. 3.6). Unfortunately, there are many artifacts that can result in the absence of a bright spot corresponding to an atomic position in micrographs. First of all, as field evaporation is a statistical process, even if unlikely, evaporation may take place from locations where it is not expected. In worst cases, atoms can be evaporated from inside fully resolved atomic planes. In addition, in alloys, solute atoms may show with a dark contrast, or preferentially evaporate (which may in some cases be assisted by field corrosion), leading to the biased conclusion that a given atomic site is not occupied. Therefore, any attempt for determining the vacancy number density requires carefully controlled experimental conditions, often based on the examination of evaporation sequences. Such an approach has been successfully used by Berger [11] who could determine the vacancy concentration in quenched refractory metals.

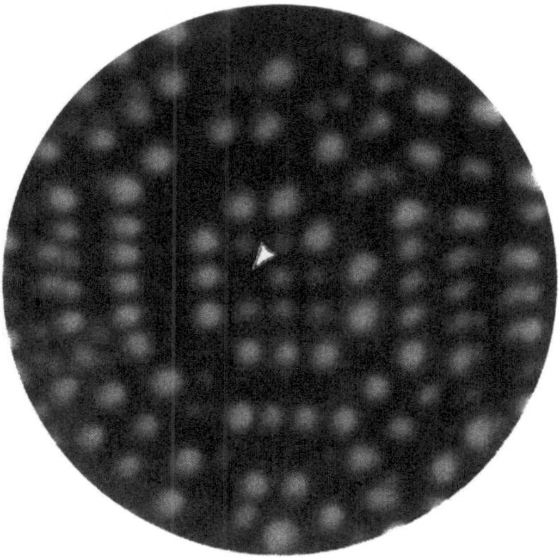

Figure 3.6 Observation of a vacancy, indicated by the *white arrow head*, in a (112) plane in Pt (C. Martin, University of Rouen).

This approach was further developed to study the distribution of vacancies produced by the impact of a high-energy Kr^{2+} ion. This distribution was reconstructed in 3D, where the atom-by-atom evaporation sequence is computed and treated with the aid of an early image analysis system [12]. Going further in the study of aggregate of point defect, tentative studies of voids have been conducted, but, due to severe distortions in the electric field distribution at the specimen surface affecting the observed contrast, only qualitative description of such voids could be obtained by FIM [13,14].

In contrast to vacancies, autointerstitials are another type of defects in pure metals that FIM should be able to resolve. Indeed, these defects should be characterized by the presence of additional brightly imaging spots in micrographs. As in the case of vacancies, other sources of bright spots, such as preferential retention of solute atoms, field adsorption and corrosion, exist, and make unambiguous identification of autointerstitial very tedious. An elegant, but indirect, proof of the presence of autointerstitials was given by Seidman et al. [15], by studying the contract modification at the specimen surface, with the autointerstitial atom lying a few atomic planes under the surface.

Line Defects

The second major type of lattice defects that could be studied by FIM is the linear defects, namely dislocations. Numerous studies of dislocations have been conducted in the 50′ and 60′, and it is beyond the scope of this contribution to provide a complete description of these studies. The reader interested should refer to the monograph by Bowkett and Smith [16]. The contrast observed by FIM originates from the atomic displacements generated by the presence of a dislocation. The only possibly observed displacements have to have a component normal to the specimen axis. For example, for a dislocation having a Burger's vector component normal to the specimen surface, a characteristic spiral will replace the set of concentric rings (Fig. 3.7). If the Burger's vector is parallel to the specimen surface, then no modification of the local symmetry will be generated.

Because only components normal to the specimen surface generate atom stacking modifications in the plane of observation (ie, specimen surface), the

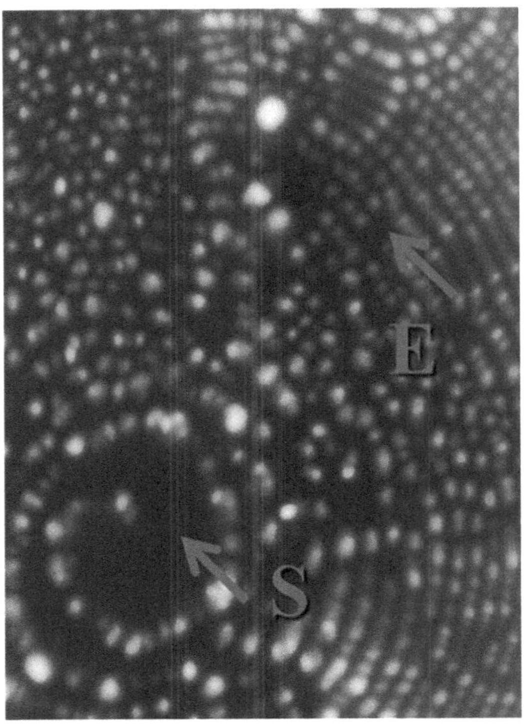

Figure 3.7 Field ion micrographs of a mixed dislocation emerging at the surface of a pure iron specimen. The screw component (S) is indicated by the *lower arrow* and the edge component (E) by the *upper arrow*. Imaging conditions: $T = 20$K, specimen voltage = 8.4 kV, imaging gas: Ne, imaging gas pressure = 5×10^{-5} torr. Field of view about 20×30 nm^2.

complete determination of a Burger's vector is not always possible. Only if the dislocation is observed to emerge from two different poles this can be possible.

One of the major problems related to dislocation study by FIM is the potential influence of the mechanical stress resulting from the externally applied electric field. TEM observations of FIM specimen prior and after FIM imaging have unambiguously shown dislocation glide [17]. It should be noticed that no direct observation of dislocation glide has been reported so far. In addition, it is not clear how the emergence of a dislocation core will be modified when emerging at the free surface of the specimen.

Planar Defects

As FIM allows the crystal orientation to be determined, any perturbation of the symmetric arrangement can be detected. In the situation where an internal interface is present in the observed region, any misorientation across this interface will generate a symmetry breaking, as shown in Fig. 3.8. Only in the unlikely event of a misorientation corresponding exactly to a crystal symmetry, this breaking will clearly appear on the field ion micrograph. Simple geometrical consideration are generally enough to deduce the

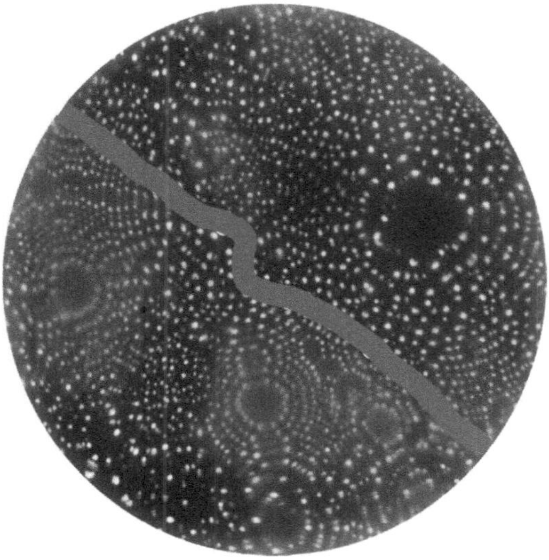

Figure 3.8 General grain boundary in pure iron. The position of the grain boundary is revealed by the global symmetry changes, highlighted here by the *thick gray line*. Both crystal orientations can be independently estimated. Imaging conditions: $T = 20$K, specimen voltage = 6.3 kV, imaging gas: Ne, imaging gas pressure = 5×10^{-5} torr. Field of view about 80 nm.

specific orientations on each part of the interface. Depending on the sophistication of the projection model used, the precision of the orientation can be reduced down to less than 2° [16].

Specific case of internal boundaries is twin boundaries. Because all atom planes are continuous across twin boundaries, very little contrast is visible in FIM, explaining the limited application of FIM to twin boundaries.

Phase Contrast

In the case of multiphase materials, the evaporation fields (closely related to the cohesive energies) are unlikely to be identical in each phase. As a consequence, the overall shape of the specimen surface will differ from the assumed smooth hemisphere, different local radii of curvature will develop on top of each phase emerging at the specimen surface: the local radii of curvature will be inversely proportional to the cohesive energy of the related phases. The different local radii of curvature can be regarded as the response of the specimen during field evaporation to accommodate the differences in evaporation fields, in such a way that the electric field at the specimen surface is always equal to the local evaporation fields of the different phases [18]. This local evolution of the specimen surface has several consequences for FIM. First, as the ionization rate of imaging gas ions is also inversely proportional to the local radius of curvature, it is clear that the most (respectively less) cohesive phases will generate more (respectively less) ionization, and thus will appear brighter (respectively darker) than the surrounding matrix. In addition, with a higher local field, the ionization rate of imaging gas atoms will be enhanced, resulting in an enhancement of the local brightness. Of course, the opposite effect will apply to low evaporation field phases (Figs. 3.9–3.11). This is the basic mechanism responsible for phase (in its metallurgical sense) contrast in the FIM.

In addition, as explained previously, the local variation of the radius of curvature will locally affect the image magnification. Indeed, a local radius of curvature lower than that of the overall specimen will result in a higher local magnification. As a consequence, high cohesive phase will appear brighter and more strongly magnified than the rest of the specimen. This effect, referred to as the local magnification effect, is also responsible for reconstruction artifacts that will be treated in chapter "3D Reconstruction in APT: Basics and Advanced Approaches."

FIM is a very efficient technique to study materials with nanometer-scale inclusions, as it is most likely that both phases have different cohesive energy.

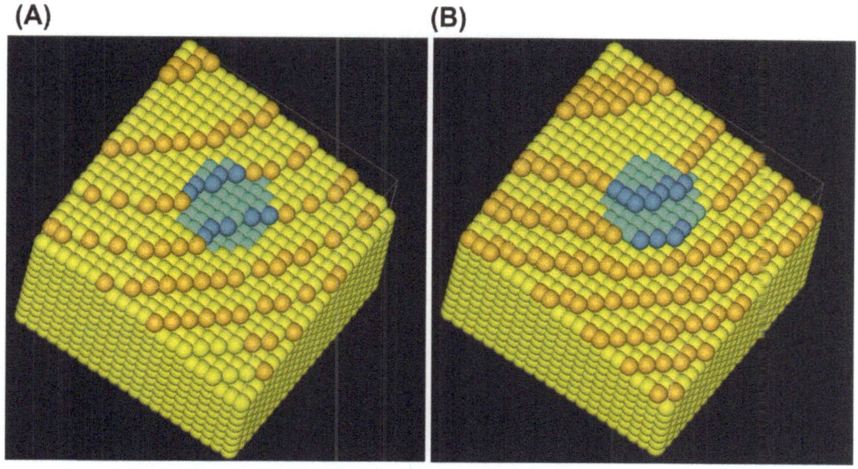

Figure 3.9 Simulation of a specimen surface with a low-field (A) and high-field (B) phase developing different local radii of curvature than the matrix one.

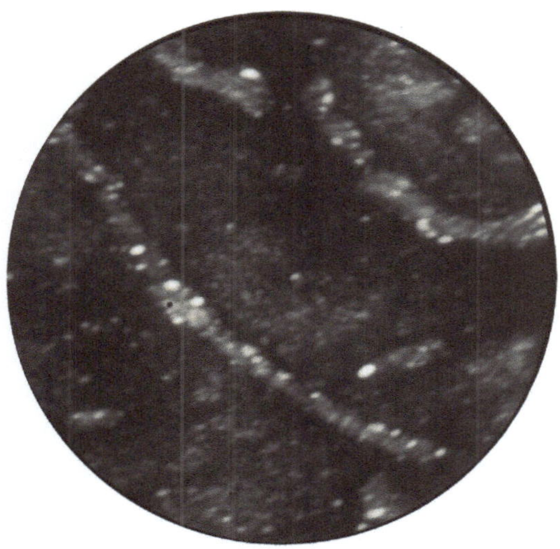

Figure 3.10 Field ion micrograph of chromium nitride Cr_2N platelets in iron matrix. The nitrides have a higher evaporation field, and thus appear in bright contrast, with respect to the Fe matrix. Imaging conditions: $T = 80K$, specimen voltage $= 9.0$ kV, imaging gas: Ne, imaging gas pressure $= 5 \times 10^{-5}$ torr. Field of view about 125 nm.

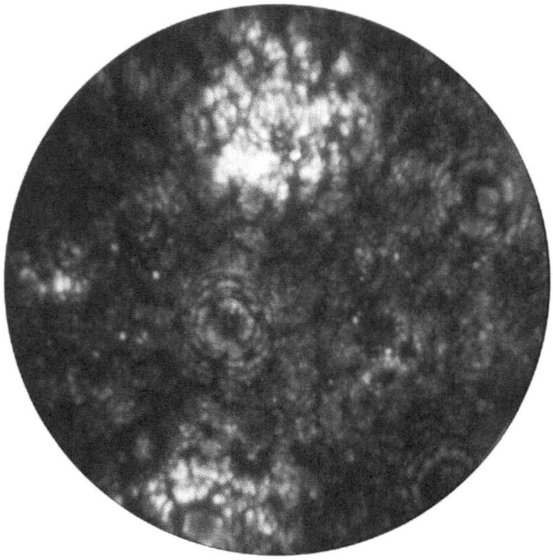

Figure 3.11 Field ion micrograph of a Fe-49Cr alloy spinodally decomposed at 500°C. The Cr-rich domains appear as darkly imaging veins all over the specimen surface. Imaging conditions: $T = 80\text{K}$, specimen voltage = 10.5 kV, imaging gas: Ne + H, imaging gas pressure = 5×10^{-5} torr Ne + 10^{-5} torr H. Field of view 150 nm.

Because the origin of the contrast is different from the ones used in the various TEM operation modes, both techniques are complementary.

However, it should be kept in mind that, because magnification is phase dependent, measuring, for example, precipitate size directly by FIM will always lead to biased values, the bias increasing with the ease of observation of both phases, ie, with the difference in cohesive energies.

Example of a brightly imaging phase is shown in Fig. 3.10. The high-field phase, which develops as nanometer-thick platelets, is clearly seen in a brighter contrast. Some evidence of atomic alignments can be seen on the right-hand side of the micrograph. A low index pole is also barely seen in the ferritic matrix. The curved shape of the upper platelet can be explained as it corresponds to the interception of the planar platelet with the hemispherical shape of the specimen apex. Its actual form is portion of an ellipse, the shape of which depends on the respective orientation of the platelet and the specimen axis.

Fig. 3.11 shows another situation, where two phases are interconnected, as in spinodally decomposed iron–chromium alloys. In this case, the low-field phase is the Cr-rich α' phase, which appears as dark contrast "veins." FIM can help measuring the mean distance between Cr (respectively Fe)-rich domains, ie, the wavelength of the spinodal structure. Note that because this alloy is located in the middle of the miscibility gap, the actual volume fraction of both phases is 50%. However, the volume fraction of the observed veins is clearly lower. This is a consequence of the local magnification effect, as the trajectories of ions originating from the low-field region are focused, resulting in an apparent lower lateral extent.

Provided the magnification of the micrograph is carefully estimated, FIM is of great help in estimating some characteristic of a contrasted second phase, such as particle size, number density, and volume fraction.

Even if successfully applied to a number of metallurgical applications, these estimates need to be considered with care. Two main aspects must be taken into account. First, a field ion micrograph is only a two-dimensional section of the investigated material. The collected data must therefore be corrected to account for this geometrical constrain. In addition, as illustrated in the case of spinodally decomposed Fe—Cr alloy presented in Fig. 3.11, the local radius of curvature of the second phase is different from the average one of the specimen. The projection law is locally affected, and so is the magnification. Brightly imaging phases will be overmagnified, and the size of brightly imaging particles, as measured on the field ion micrographs, will be overestimated. The size of particles exhibiting a dark contrast will conversely be underestimated.

Three-Dimensional Field Ion Microscopy

Exactly as when developing the specimen, field evaporation can be used to investigate the materials in depth. When the voltage applied is raised, even in the FIM mode, the specimen will need to develop a larger radius of curvature to return to equilibrium. This increase in radius of curvature can only be achieved by evaporating the specimen. When carefully controlled, individual atoms can be field evaporated from the specimen surface, atomic layer by atomic layer. Examination of evaporation sequences offer the possibility of investigating, not only the surface, but also the depth of the specimen, at the ultimate atomic resolution. The analysis then turns from 2D to 3D. As briefly mentioned in the case of point defect, the investigation of evaporation sequences overcomes some of the limitations of the field ion microscope, as true 3D information becomes available. Following the pioneering work of Seidman [12], who processed semimanually individual field ion micrographs, progress in computer power and data analysis makes automatic processing of evaporation sequences possible. With modern computer, thousands of high-resolution field ion micrographs can be processed, in the aim of rendering a true 3D atomic 3D reconstruction. This method, alternatively called three-dimensional, or computerized, FIM (3DFIM [19] or cFIIT [20]) has recently been established as a characterization method providing 3D reconstruction at the atomic scale (Fig. 3.12). It has successfully been used to determine characteristic features of decomposed alloys (precipitate size, number density, volume fraction). Because this method is still in its early years, it is expected that it can also be used to study lattice defects in materials [21].

Even if based on individual field ion micrographs, it appeared that more information is contained in 3DFIM reconstruction. Indeed, as "classical" field ion micrographs are recorded at equilibrium, evaporation sequences are dynamic. Surprisingly, the contrasts are slightly different, and may bring complementary information. As shown in Fig. 3.13, a complex microstructure can be reconstructed in 3D. Among the advantages of 3DFIM, illustrated in Fig. 3.13, the number density of particles can be directly measured. Indeed, as this method allows atomic resolution, calibration of the reconstructed volume is exact. Determining the number of particles in the volume leads directly to the number density [22]. The precision of the method only depends on the relative characteristic length scales (volume and antiparticle distances), but not on any assumption made to deduce 3D information from projected volumes (like in standard FIM and TEM). Another advantage of 3DFIM is that particle size can be measured along the analysis direction, where the bias introduced by local magnification is not active. Particle size along the specimen axis can, for example, be measured counting the number of atomic planes evaporated

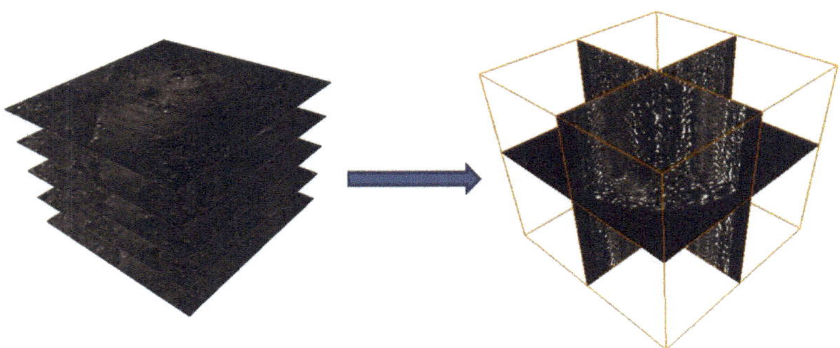

Figure 3.12 Principle of the three-dimensional field ion microscope. Series of individual field ion micrographs (on the left) are recorded during field evaporation sequences, and digitally compiled to generate a three-dimensional reconstruction using an image processing software.

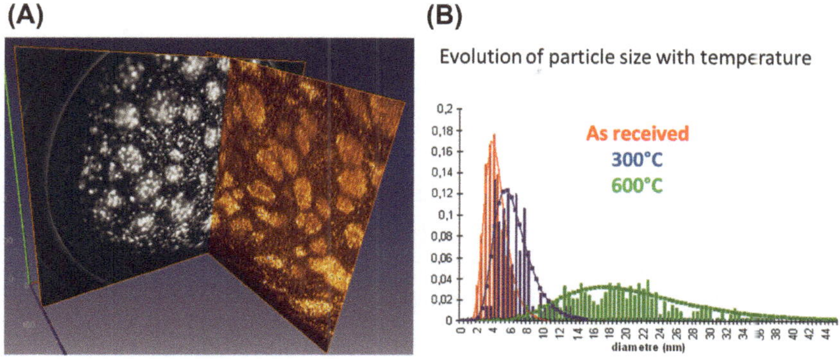

Figure 3.13 (A) 3DFIM reconstruction (in color) based on individual micrographs (black and white). Precipitates appear in bright contrast. (B) Estimate of the particle mean diameters estimated from 3DFIM reconstruction after different aging treatments. *From Cazottes S. [Ph.D. thesis]. France: University of Rouen; 2008 (in French) [24].*

during the evaporation of a given particle. This protocol can be applied to many particles, and precipitate size distribution histograms established [23,24].

Laser-Assisted FIM and Applications to Nonmetals

Application of FIM to insulators or to semiconductors (SCs) is generally more difficult than applications to metals. Several studies have been conducted in the past to image these kinds of samples, and it was proved that it was possible for a large number of SCs (see Refs. [25,26] for more details). Modern instruments and specimen preparation techniques improved considerably the facility of application of this microscope on these kinds of instrument. The difficulty comes principally from the strong electrical resistivity of the materials when analyzing SCs and insulators (especially at low temperature, free carriers density is then close to zero, and extrinsic carriers are frozen). When imaging an SC tip, the ion current during the imaging process is in the picoampere range (each spot emits about 10^3–10^4 ions/s, and the image is composed of thousands of spots). Electronic charges emitted during the ionization process must be drained inside the resistive materials. The electrical resistance R of the material is proportional to its length, and to the resistivity ρ, but inversely proportional to the tip section (about 100×100 nm^2). The simple Ohm law predicts that a picoampere current drained by a millimeter length nanowire of doped Si ($\rho \sim 10^4\ \Omega$ m) should generate an ohmic drop in the kilovoltage range. It should therefore be difficult to obtain the high electric field at the end apex necessary for ionizing the imaging gas; this also changes strongly the electric field distribution around the specimen and the ion trajectories. It should be noted that this drop is, however, limited by the maximal field that can exist inside the materials (limited by the internal breakdown field).

This shortcoming is overcome by first the use of samples with lengths in the micron range (see specimen preparation chapter for more details) and second the possibility in atom probe (AP) to assist the imaging process with laser pulses directly applied to the end apex. As described in Ref. [25], the images are experimentally considerably improved by illuminating the sample with light of appropriate wavelength. Under the application of laser pulses focused at the end apex, the resistance of the material can be reduced by photoconductivity effects (detailed calculations and measurements may be found in Ref. [26]). Examples of FIM obtained for Si and GaN are presented in Fig. 3.14 and examples in the literature exist for other materials such as ZnO, or MgO, GaAs, CdSe, or SiC [24,25].

Concentric rings, due to crystallographic directions can be observed in experimental images, showing quasiatomic resolution. Nevertheless, the regularity of images is not as perfect as it could be observed in metals. This is probably because when an atom is evaporated, a local reconstruction of the surface takes place immediately, to minimize the number of dangling bonds (see chapter "Field Ion Emission Mechanisms" for more details). The

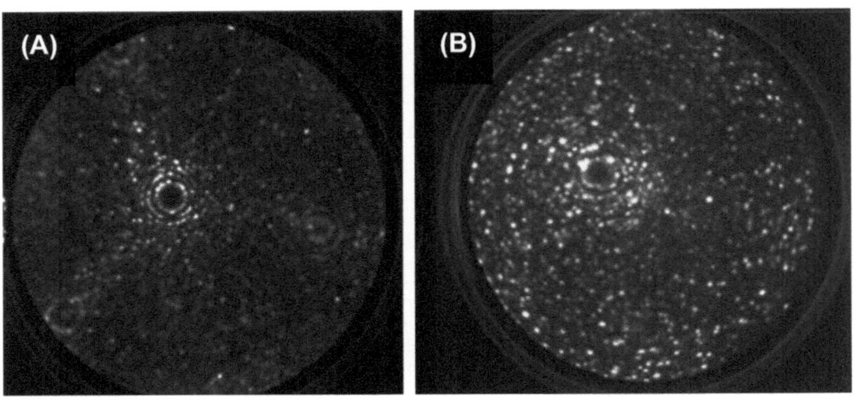

Figure 3.14 Field ion images images at 20K with a mixture of Ne and H (5×10^{-5} torr). (A) *p*-Doped silicon *(reproduced with permission from Arnoldi L, Silaeva EP, Gaillard A, Vurpillot F, Blum I, Rigutti L. Energy deficit of pulsed-laser field-ionized and field-emitted ions from non-metallic nano-tips. J Appl Phys 2014;115(20):203705 [27])* oriented along the <111> direction; (B) Wurtzite GaN material, *(Courtesy of L. Rigutti.)* the *c*-axis of the structure is evident.

use of hydrogen in the imaging gas mixture generally improves the quality of images, probably because hydrogen would inhibit the surface reconstruction process.

FIM can thus be used in material science applications of SCs or insulators. As for metals, differences in evaporation fields are used to produce a significant contrast between phases or objects observed with this microscope. For instance, in Fig. 3.15A, silicon carbide particles, 2–3 nm in diameter, are

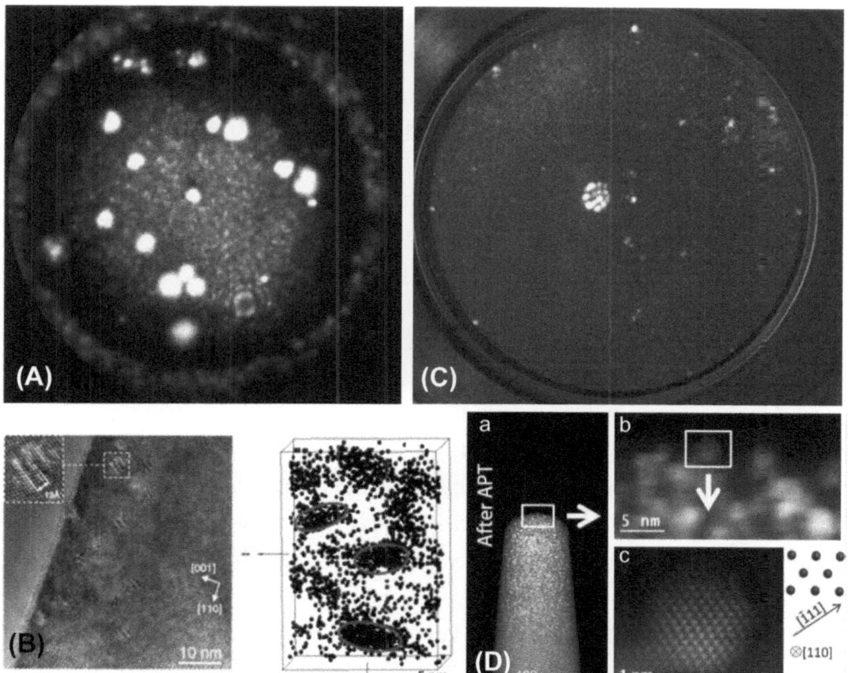

Figure 3.15 Field ion microscopy images at 20K with a mixture of Ne and H (5×10^{-5} torr). (A) SiGe with small SiC particles *(From Estivill R, Grenier A, Duguay S, Vurpillot F, Terlier T, Barnes JP. Quantitative investigation of SiGeC layers using atom probe tomography. Ultramicroscopy 2015;150:23–9 [30])* oriented along <001> direction; (B) transmission electron microscopy image of the surface and atom probe tomography (APT) reconstruction showing the same presence of particles; (C) Au particles in a MgO matrix in bright contrast. Crystalline structure of the Au particle is evident. *(Courtesy of D. Shinde.)*. (D) These particles were imaged using transmission electron microscopy *(Devaraj A, Colby R, Vurpillot F, Thevuthasan S. Understanding atom probe tomography of oxide-supported metal nanoparticles by correlation with atomic-resolution electron microscopy and field evaporation simulation. J Phys Chem Lett 2014;5(8):1361–7. [28] and Lefebvre W, Hernandez-Maldonado D, Moyon F, Cuvilly F, Vaudolon C, Shinde D, et al. Ultramicroscopy 2015;159:403–12. [29]).*

easily revealed in a SiGe matrix. SiC is a high evaporation field material compared to the SiGe matrix, giving rise to a strong difference in contrast. Note the presence of (111) poles that can be used to calibrate the size of the particles. This result is confirmed by TEM observation (Fig. 3.15B) or APT measurement. In Fig. 3.15C, crystalline gold particles are imaged in a matrix of MgO, as observed in TEM [28,29].

CONCLUSION

FIM is an essential auxiliary instrument of APT. This instrument can be used before or after AP experiments to provide a static single image of the sample surface, providing crystallographic or size information of object embedded in the studied material. In a lot of cases, the spatial resolution of images are better than atom probe tomograms, and the visualization of important crystallographic axes or features more evident. This information is highly valuable for extracting reconstruction parameters (see chapter "Three-Dimensional Reconstruction in Atom Probe Tomography: Basics and Advanced Approaches" for more details). It should also be noted that the image of one point is the result of the ionization of thousands of image gas atoms. The theoretical imaging efficiency can approach 100% and reinforces the interest to compare FIM images to AP images. It should also be noted that image distortions are much less severe in FIM, because the image gas is ionized from static positions on the tip, at a distance of a few Angstroms from the surface. In Fig. 3.15C, gold particles are observed as well-defined crystalline clusters, whereas APT measurements show nonexpected blurred clusters [27].

At present, FIM images are rarely used as part of current LEAP investigations. The reason is principally the absence of a real dedicated FIM instrument inside modern LEAP machines. In the local electrode atom probe (LEAP), FIM images may be digitally captured by measuring ion gas impact positions with the delay-line detector. The detector microchannel plates (MCP) gain must be very high to ensure capturing significant position signals, and therefore the pressure of image gas in the chamber must be reduced to 10^{-7}–10^{-8} torr in this mode. Image is built, after posttreatment, from several tens of thousands of impacts per second This imaging process is slightly different in old FIM instruments where MCPs, polarized at low gain, are coupled to a phosphor screen directly recorded by a CCD camera. In this mode, images are produced by millions of ion gas impacts per second on the screen, which enables one to adapt more easily the imaging conditions to the sample.

REFERENCES

[1] Müller E, Bahadur K. Field ionization of gases at a metal surface and the resolution of the field ion microscope. Phys Rev 1956;102(3):624.

[2] Müller EW. Z Fur Phys 1951;131:136.

[3] Gomer R. Field emission and field ionisation. Cambridge: Harvard University; 1961. p. 45−6.

[4] Gipson J. J Appl Phys 1980;51:3884.

[5] Bajikar SS, Kelly TF, Camus PP. Appl Surf Sci 1996;94−95:464.

[6] Schlesiger R, Schmitz G. Ultramicroscopy 2009;109:497.

[7] Nishikawa O, Ohtani Y, Maeda K, Watanabe M, Tanaka K. Mater Charac 2000;44:29.

[8] Walls JM, Southworth HN. J Phys D-Appl Phys 1979;12:657.

[9] Southworth HN, Walls J. Surf Sci 1978;75:129.

[10] Gloriant T, Danoix F, Lefebvre W, Greer A. Adv Eng Mater 2007;9:151.

[11] Berger AS, Seidman DN, Balluffi RW. Acta Metall 1973;21:123.

[12] Wei CY, Current MI, Seidman DN. Phil Mag 1981;44:459.

[13] Brenner SS, Seidman DN. Radiat Effs 1975;24:73.

[14] Godfrey TF, Lewis RJ, Smith DA, Smith GDW. J Less-Comm Met 1976;44:319.

[15] Seidman DN, Lie KH. Acta Metall 1972;20:1045.

[16] Bowkett KM, Smith DA. Field ion microscopy. 1970. North Holland (Amsterdam).

[17] Loberg B, Norden H. Ark Foer Fys 1970;40:413.

[18] Vurpillot F, Bostel A, Blavette D. Ultramicroscopy 2001;89:137.

[19] Vurpillot F, Gilbert M, Deconihout B. Surf Interface Anal 2007;39:273.

[20] Semboshi S, Al-Kassab T, Gemma R, Kirchheim R. Ultramicroscopy 2009;109:593.

[21] Dagan M, Hanna LR, Xu A, Roberts SG, Smith GDW, Gault B, et al. Imaging of radiation damage using complementary field ion microscopy and atom probe tomography. Ultramicroscopy 2015;159:387−94.

[22] Jessner P. [Ph.D. thesis]. France: University of Rouen; 2010 (in English).

[23] Akré J, Danoix F, Leitner H, Auger P. Ultramicroscopy 2009;109:518.

[24] Cazottes S. [Ph.D. thesis]. France: University of Rouen; 2008 (in French)

[25] Tsong TT. Atom-probe field ion microscopy. Cambridge, UK: Cambridge University Press; 1990.

[26] Miller MK, Cerezo A, Heatherington MG, Smith GDW. Atom probe field ion microscopy. Clarendon Press, Oxford; 1996.

[27] Arnoldi L, Silaeva EP, Gaillard A, Vurpillot F, Blum I, Rigutti L. Energy deficit of pulsed-laser field-ionized and field-emitted ions from non-metallic nano-tips. J Appl Phys 2014;115(20):203705.

[28] Devaraj A, Colby R, Vurpillot F, Thevuthasan S. Understanding atom probe tomography of oxide-supported metal nanoparticles by correlation with atomic-resolution electron microscopy and field evaporation simulation. J Phys Chem Lett 2014;5(8): 1361−7.

[29] Lefebvre W, Hernandez-Maldonado D, Moyon F, Cuvilly F, Vaudolon C, Shinde D, et al. Ultramicroscopy 2015;159:403−12.

[30] Estivill R, Grenier A, Duguay S, Vurpillot F, Terlier T, Barnes JP. Quantitative investigation of SiGeC layers using atom probe tomography. Ultramicroscopy 2015;150: 23−9.

CHAPTER FOUR

Atom Probe Sample Preparation

I. Blum, F. Cuvilly, W. Lefebvre-Ulrikson
Groupe de Physique des Matériaux, University and INSA of Rouen, Normandie University, France

Since the first experiments, the number of ions collected in atom probe [now atom probe tomography (APT)] datasets has increased from several hundreds to hundreds of millions. This remains still 15 orders of magnitude less than the amount of atoms contained in a mole of material. A simple calculation shows that, even for the most optimistic case (ie, 100 millions of ions collected per day on each APT worldwide), it would take nothing less than 10^{12} days to collect a gram of matter. This is roughly the time necessary for the sun to become a red giant and to absorb the earth and all atom probes at its surface.

The art of specimen preparation for APT hence consists in selecting an infinitely small piece of matter and to confer it the specific shape necessary to enable field evaporation. This means that the specimen must be a needle of sharp shank angle (typically less than 20°), with a radius of curvature in the range of 20−100 nm at its apex. An even more drastic requirement might be the selection of a feature of interest in a piece of material. Fulfilling this requirement may sometimes be equivalent to looking for a needle in a haystack before manipulating this needle, orienting it, and sharpening it for an analysis in APT.

The aim of this chapter is to describe the standard methods of specimen preparation for APT. The first part briefly introduces the principles of sample preparation by electrochemical polishing and its limitations. The second and main part of this chapter introduces applications of focused ion beam (FIB) milling to APT sample preparation. Fundamentals of FIB technology and principles of scanning electron microscopy (SEM) imaging are briefly introduced. The effect of FIB milling on the specimen structure is discussed based on simulations. Ways to preserve the specimen structure (by means of capping layer, annular milling, and low beam energy) are described. The most widely used procedures for sample preparation by FIB are then introduced (ie, the lift-out method, cross-section, and back-side preparation) and attention is paid to site-specific analysis, which is enabled by the use of SEM-FIB.

Atom Probe Tomography
ISBN: 9780128046470
http://dx.doi.org/10.1016/B978-0-12-804647-0.00004-8

SPECIMEN PREPARATION BY ELECTROPOLISHING

Specimen preparation by electropolishing is the standard method that can be applied to bulk metals and alloys. The several steps leading to a needle [1–4] from a blank (or a wire) are illustrated in Fig. 4.1. The basic principle consists in forming a neck in the central region of a specimen blank and to later control the breaking of the specimen into two pieces by applying a low voltage (or low current) in the final step. This way, two specimens can be prepared simultaneously as long as the manipulator is able to retrieve the part of the specimen fallen in the beaker. The choice of electrolytes depends on the specimen and a list of electrolytes established from the literature [1–3,5–12] has been summarized in appendix B of the following reference [13].

If the first steps of electropolishing do not lead to a sharp and clean needle, it is still possible to make a new specimen from a bad one. This can be done by applying a micropolishing of the specimen, as illustrated in Fig. 4.2. The micropolishing method is often referred to as the "microloop" method [1], insofar as it relies on the use of a cathode consisting of a small loop made of noble metal, the role of which is to maintain a droplet of electrolyte across which the specimen can go. As for the standard electropolishing, the micropolishing consists in forming a neck before breaking it to obtain

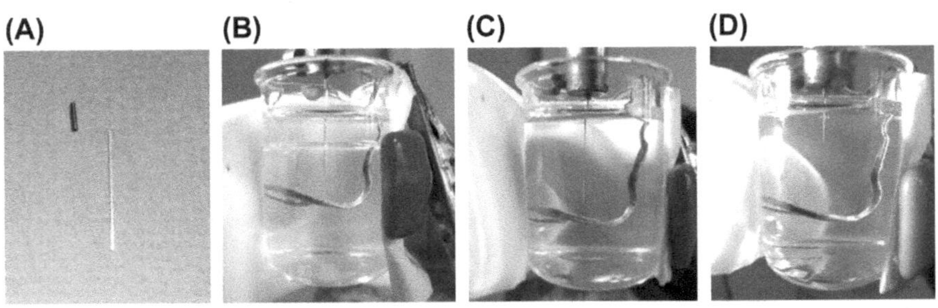

Figure 4.1 (A) Typical specimen blank (here, an aluminum-based alloy) used to prepare an atom probe tomography needle. The dimension of the specimen blank is $0.3 \times 0.3 \times 20$ mm^3. The blank is later fixed in a nickel tube of 0.5 mm diameter, which is mounted on a steel rod to ease its manipulation in electrolytes. (B) Illustration of the double layer method. An electrolyte (here, 25% perchloric acid in 75% acetic acid) is deposited above an inert liquid (Galden perfluoropolyether). The gold cathode is visible in the beaker. (C) In a second step, once a neck has been formed in the central part of the blank, the specimen is immersed in a less concentrated electrolyte (here 5% perchloric acid in 95% acetic acid) and no inert liquid is used. (D) At the end of the second step a sharp needle is formed and the specimen must be rinsed with alcohol.

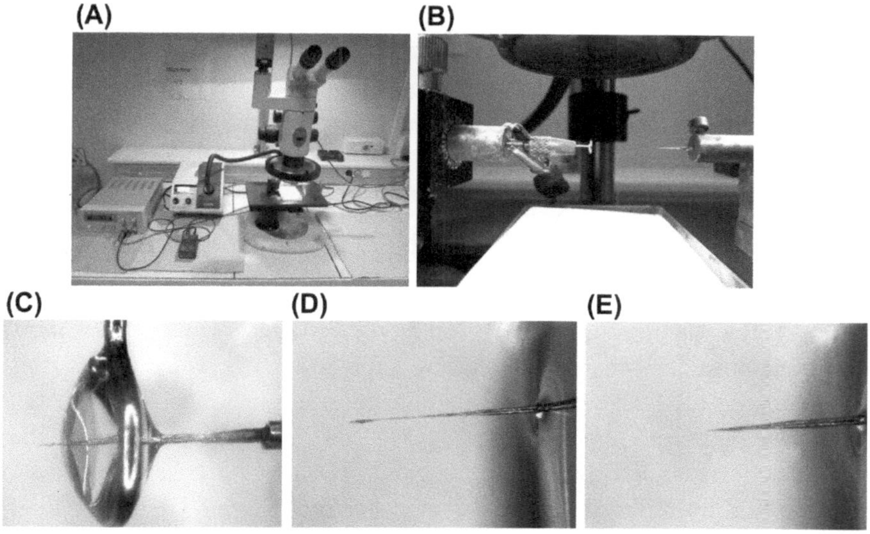

Figure 4.2 Specimen preparation using the microloop method. (A) Overview of the assembly composed of a direct current voltage power supply, an optical microscope operating in reflection, a specimen stage, and a platinum microloop of 5 mm diameter. (B) Closer view of the specimen facing the microloop under the objective lens. (C) A drop of low concentrated solution (here 2% perchloric acid and 98% 2-butoxyethanol) is put in the microloop. In the first step, a neck is formed at short distance from the apex by leaving the specimen apex outside from the electrolyte. (D) Image of a neck obtained by following the method illustrated in (C). (E) A needle is obtained by breaking the neck with short applications of voltage.

a sharp needle. Once a needle is formed (Fig. 4.2E), the specimen must then be quickly withdrawn from the droplet and rinsed with alcohol. Since the optical microscope does not allow inspecting the apex of an APT needle, observation of the specimen in a transmission electron microscope (TEM) prior to insertion in APT is helpful.

Although relatively easy to apply, the electropolishing method is not suitable for every material or every research topic. The selection of interfaces in metals and alloys by electropolishing is possible, although time-consuming and not easily reproducible. Applying short-voltage pulsing to a needle in an electrolyte (eg, 1-ms pulses under 50 V in a mixture of 98% 2-butoxyethanol–2%perchloric acid for Al-based materials) allows drastically thinning needles when applying this to submicrometer thick apex. This protocol may allow selecting grain boundaries or approaching a phase from the tip apex. However, this methodology has become obsolete with the development of SEM-FIB. Moreover, if various rates of polishing of phases are encountered in an alloy, it might be relevant to avoid electropolishing to prevent the formation of complex morphologies at the needle apex. The next section, devoted to the introduction of specimen preparation by FIB, presents how features of interest can be selected in inorganic materials.

SAMPLE PREPARATION USING FOCUSED ION BEAM MILLING

Unlike electropolishing, FIB can be used to prepare atom probe samples from almost any material, including semiconductors [14] and insulators [15]. Because FIB can be used to manipulate only small pieces of material, APT specimens that are prepared by FIB are generally made using the lift-out process, which consists in extracting a piece of material from a given substrate and mounting it on the end of a macroscopic tip. Then the chunk of material itself is milled using the ion beam to give it its final tip shape suitable for APT analysis. In the following, a standard way of performing the lift-out and mounting will be presented. Thanks to the versatility inherent to the FIB technique, the lift-out and mounting can be performed using many different methods, which are each adapted to specific applications. In some cases, other methods can be developed to simply bypass the lift-out and/or mounting steps, usually to reduce the sample preparation time and increase the sample throughput. A nonexhaustive selection of such methods is presented at the end of this chapter.

The selection of a region of interest (ROI) in the SEM can be done because of the existence of specific material—beam interactions and the use of related detection systems. Furthermore, SEM offers reasonable resolution to allow imaging of nanometer size features. Before further considering the specimen preparation, the main capabilities of SEM and associated techniques are summarized.

SEM-FIB Layout

Introduction to SEM Imaging Modes

SEMs are extremely relevant instruments for the preparation of APT needles. Large specimens can be introduced in their chamber of analysis in which specific regions of interest can be identified, therefore allowing site-specific analysis in APT. In addition, a variety of imaging modes are available in SEM. Their potentialities (resolution, depth of focus, and chemical sensitivity) are intrinsically related to the physics of electron—matter interaction. The volume of electron–matter interaction in SEM is often described as a pear of interaction. Indeed, the incident electrons follow trajectories contained in a three-dimensional volume with a pear shape, the size of which depends on the target and on the kinetic energy of incident electrons, as illustrated in Fig. 4.3.

Various signals can be detected in SEM, which are related to the interactions between incident electrons and charges in the specimen. The main particles collected by SEM detectors are the following:

1. Auger electrons (AEs), originating from the extreme surface of the specimen (0.5- to 2-nm-thick layer).
2. Secondary electrons (SEs), originating from a depth up to 5 nm in average. Although they are generated in the whole volume of

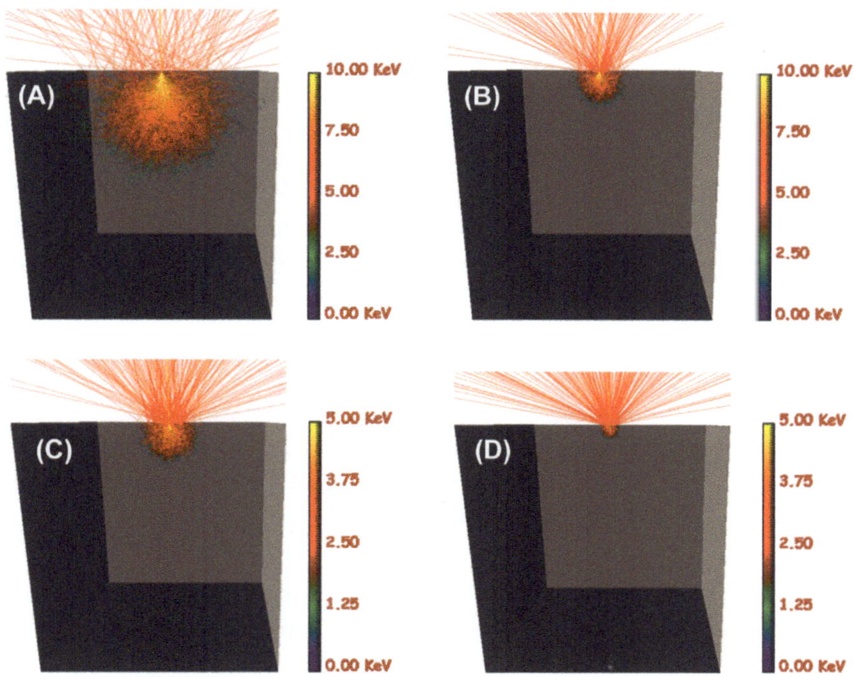

Figure 4.3 Monte Carlo simulation of electron trajectories in a SEM sample performed using the CASINO program developed at the University of Sherbrooke [16] A narrow electron beam (1 nm in diameter) is incident of the top surface of each volume. All simulation boxes have a volume of $2 \times 2 \times 2\ \mu m^3$. An aluminum specimen is considered in (A) and (C), whereas an iron specimen was used in (B) and (D). The energy of the incident electron beam is 10 keV in (A,B) and 5 keV in (C,D). The electron trajectories are displayed as a function of the kinetic energy of each electron. High-energy backscattered electrons (BSE) are mostly emerging at the top surface of the specimen.

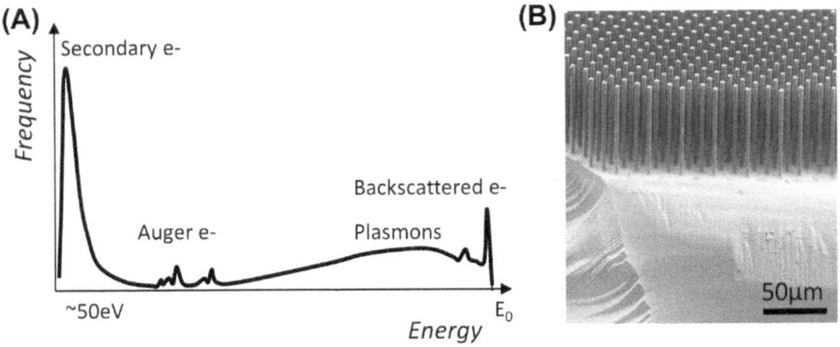

Figure 4.4 (A) Schematic representation of the distribution of electron energy. (B) Topographic contrast by secondary electron imaging of silicon pillars.

interaction (cf. Fig. 4.3), their low energy (around several tens of electron volts) only allows those generated close to the surface to be detected.

3. BSEs, having the largest energy (cf. Fig. 4.4A). They come from a much larger volume, which is about half of the volume of interaction. They are due to Rutherford scattering by atomic nuclei in the sample [17]. BSEs can be scattered elastically by atoms from the specimen and diffraction patterns can hence be formed by these electrons. The technique associated with this phenomenon is electron backscattered diffraction (EBSD) or transmission Kikuchi diffraction (TKD) [18–20] if diffraction patterns are observed in transmission mode. These two techniques allow determining and mapping the orientation of crystals, which might be of great interest for the characterization of interface prior to analysis in APT.
4. X-rays, which originate from transitions of atomic orbitals of electrons in the specimen. They are generated by the whole volume of interaction and can travel through the specimen because of their significantly smaller cross-section of interaction (and absorption coefficient) compared to electrons.

Based on the above considerations, it appears that the spatial resolution is the largest for Auger and secondary electrons, intermediate for BSEs, and worst for X-ray elemental mapping. Nevertheless, the rate of particle emission strongly varies. Emission of AEs is, for instance, a phenomenon of weak efficiency. In addition, it must be kept in mind that the resolution depends on the volume of interaction. This volume of interaction decreases as the density of the specimen increases and is drastically reduced for thin specimens, such as APT needles, for which most of the incident electrons are transmitted through the specimen.

The (elastic) cross-section of interaction for Rutherford scattering varying with the square of the atomic number Z [17], BSEs will provide "Z-contrast." The inelastic scattering cross-section also varies as a function of Z, which explains why SE images also contain some Z-contrast [17]. However, the most interesting feature of SE imaging relies on the topographic contrast. Because of their low energy and of the possibility to place the SE detector laterally, the proportion of SEs emerging at a point of the specimen surface which will be detected will strongly vary as a function of the orientation of the surface with respect to the detector (see Fig. 4.4B). Fig. 4.5 provides a schematic view of a typical SEM chamber layout. The Everhart-Thornley detector [21] used for SE imaging is typically polarized so that only low-energy electrons are attracted to it, and can be placed at a remote location in the chamber. BSEs are mostly emitted in the direction of the incident beam and are too energetic to be deflected by the electric field of the Everhart-Thornley detector [21]. BSEs are sometimes detected by positioning a detector at the end of the electron column. Different technologies exist to discriminate between different types of electrons to observe different contrasts, including detectors that can be placed inside the electron column, as it is the case for an "in-lens" detector. Such a detector is placed in the axis of the pole pieces and is very useful

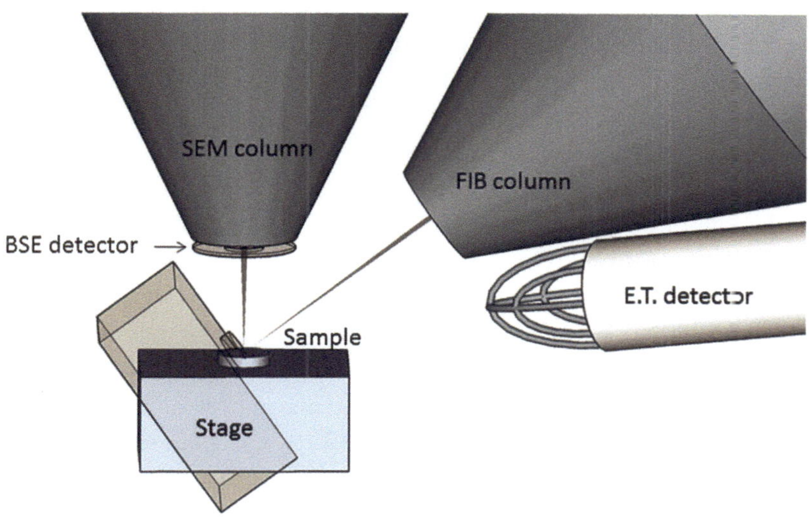

Figure 4.5 Schematic of a scanning electron microscope/focused ion beam (dual beam) chamber with an example of a backscattered electrons detector (BSE) and an Everhart-Thornley (E.T.) detector for the collection of secondary electrons. The stage can be tilted to face the focused ion beam column.

when one needs to operate at small working distances. Both SEs and BSEs can be collected by an in-lens detector. However, the topographic contrast is weaker because the shadowing effect at the specimen surface is less (the incident beam and the direction of detection are similar). By polarizing a grid in front of the in-lens detector, most of the SEs can be removed. This corresponds to an SE energy filtering mode. Finally, it must be mentioned that the exact configuration of in-lens detector may vary as a function of the supplier, although the advantage of short working distance is preserved.

Introduction to Focused Ion Beam Milling

FIB systems use positively charged gallium ions as the primary ion source. Gallium is a suitable element because it has a very low melting temperature ($\sim 30°C$) and low steam pressure and ionizes mostly as Ga^+ (the ratio $[Ga^+]/[Ga^{++}]$ is about 10^{-4}), therefore allowing a very narrow energy distribution of the ion beam [22]. Gallium ions are commonly accelerated at 30 keV. This energy is of the same order of magnitude of the electron energy in the electron beam, but because Ga ions are much heavier than electrons, they carry more momentum and thus can affect the structure of the sample. Low energies are usually used for the final stages of milling (typically 2 keV). It is worth mentioning that SEs can be generated by low-energy FIB milling. This gives imaging capabilities from an additional perspective, which is particularly useful for manipulating objects in three dimensions. This imaging mode is commonly designated as "FIB imaging."

To protect features of interest prior to investigation in APT, capping layers are usually deposited. In addition, micromanipulation of specimens in SEM requires welding them to a micromanipulator. Both welding and deposition are realized via a process of ion beam–induced deposition (IBID). This process relies on the use of a precursor gas (eg, $W(CO)_6$ for the deposition of W and $(CH_3)_3Pt(C_pCH_3)$ for the deposition of Pt) containing volatile elements and a nonvolatile element. This gas is introduced with a gas injection system. The chemisorption of the gas on the specimen surface allows the FIB-assisted IBID of the element used for the welding or capping layer.

In FIB mode, the ion beam is scanned over the sample surface and the beam current can be adjusted from 1 pA for imaging mode to several nanoamperes for milling mode. A fraction of the Ga^+ ions hitting the sample's surface exchange energy with the material by creating cascades of collisions ion solid interactions produce SE emission, sputtered atoms and ions, implantation of Ga, and also amorphization or recrystallization. The sputtering yield depends on [22]:

1. the incident angle. The maximum sputtering yield is reached near 75° due to the higher probability of collision cascades near the surface at higher angle.
2. the atomic number (Z) of bulk material. Elements such as gold and zinc are easier to sputter than platinum, for instance. The complete list of Ga ion sputter yields can be calculated from SRIM simulations [23].
3. the crystallographic orientation. When atom columns are aligned with the gallium ions trajectory, a higher penetration of Ga ions is observed. Then less sputtering and less SEs occur.

The beam current, or Ga^+ flux, can be adjusted depending on the applications. For imaging application, beam currents below 100 pA are generally used, whereas for milling, currents between 100 pA and tens of nanoamperes can be used. Low beam currents will give a smaller probe size, usually down to a few nanometers, which give a better imaging or milling resolution, whereas large currents offer a faster milling rate at the cost of a large probe size, usually up to more than 100 nm. It has already been mentioned that the collision cascades caused by the ions impacts with the sample can cause the formation of many point defects, to such a point that it can lead to the amorphization of the material surface [24]. In some cases, other effects have also been observed like the formation of Ga-rich phases [22] or the dissolution of nanoprecipitates [25].

How deep the material will be affected by implantation depends on the nature of both the ion and the substrate and the energy of the incident ion. The concentration profile after implantation can be estimated using Monte Carlo simulations [23]. Fig. 4.6A–C shows simulated Ga implantation profiles in Si for Ga energies of 30, 10, and 2 kV. As shown in these results, the implantation depth of Ga in Si can be estimated to be equal to 1 nm/kV. However, depending on the material, significant damages can still exist below the implantation peak. If the ROI is located on the surface, a protective layer can be deposited on the sample so that Ga implantation damages only occur within this layer or to avoid unintentional milling of the surface. Fig. 4.6D shows that 30 nm of Ni is sufficient to stop most Ga ions from damaging the underlying substrate. This beneficial effect of a Ni protective layer has been demonstrated in APT by Thompson et al. [26]. Complex phenomena and quantitative evaluations of experimental conditions can be retrieved in some cases but require a deep understanding of the technique

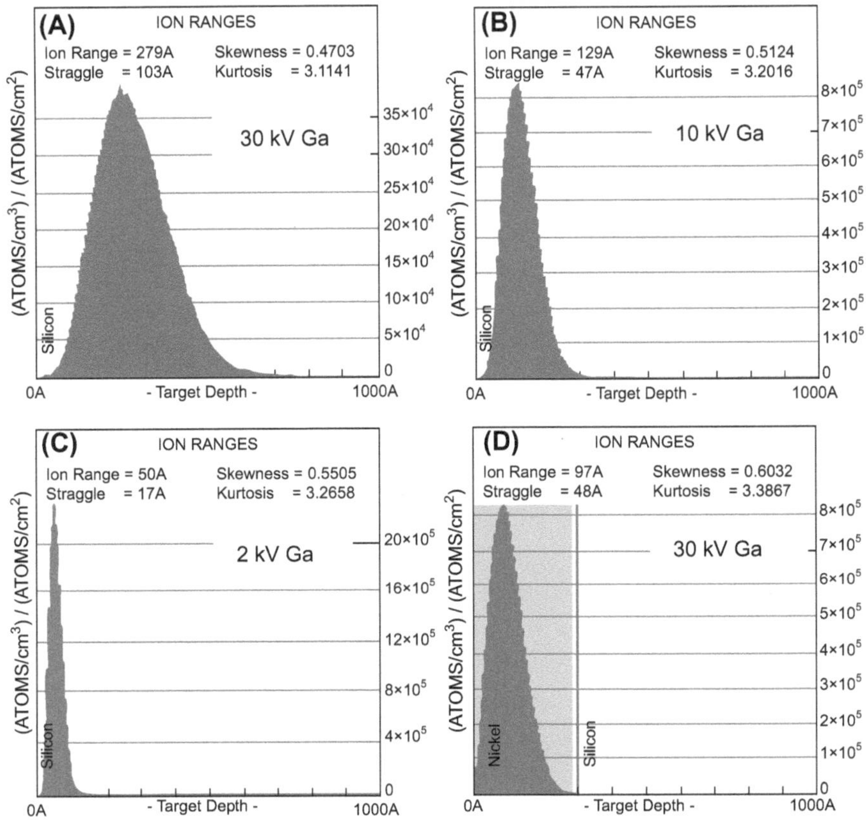

Figure 4.6 Simulated Ga implantation profiles in Si with an energy of (A) 30, (B) 10, and (C) 2 kV, and (D) simulated implantation profile in the same conditions as (A) with a 30-nm Ni layer on a Si substrate (simulations performed with the software SRIM [23]).

and a careful evaluation of both the material and the experimental conditions.

FIB is a very useful technique as its fields of application are far beyond the specimen preparation for APT. For deeper insight into the FIB procedures, the reader can, for instance, refer to the following references [22,27]. The methodologies enabling the fabrication of APT needles by FIB are discussed in the following.

The Lift-Out Process

The Lift-Out Method Step by Step

Fig. 4.7 presents the successive steps of the lift-out method. In the first step of the lift-out (Fig. 4.7A), a capping layer is deposited on the ROI using IBID.

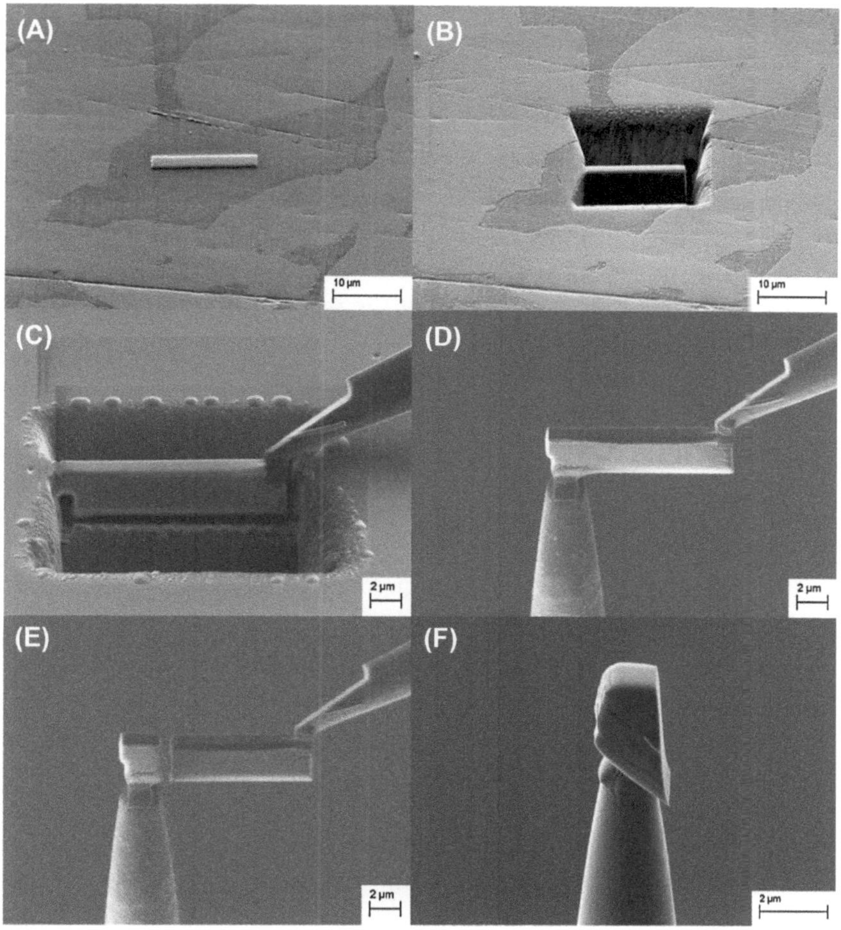

Figure 4.7 Focused ion beam images showing the different steps of the lift-out and mounting method performed on a duplex stainless steel with (A) the deposition of a capping layer, (B) the milling of three trenches around the capped region, (C) the milling of a horizontal cut through the wall and welding of the micromanipulator tip, (D) mounting of the wall on a support tip, (E) cutting of the end of the wall, and (F) side view of the sample mounted on the support tip.

The capping is about 2−3 μm wide and ~100 nm thick. The main purpose of this capping layer is to protect the ROI from the damages caused by Ga implantation during the subsequent steps of the sample preparation. If the capping layer is deposited using IBID, the sample is tilted perpendicular to the ion beam, to obtain a film with a homogeneous thickness.

In the second step (Fig. 4.7B), trenches are cut on each side of the capping layer, forming a wall that is attached on one side. The trenches should be at least 4 and 6 μm wide, the wider trench being on the side of the ion beam when the sample is tilted back to 0°. A wide trench allows the ion beam to reach the bottom of the wall to cut it in step (c). The two trenches can be milled using a single pass of the beam starting away from the protective cap and ending near the edge of the protective cap. This naturally produces trenches that are the deepest closest to the wall. They can be milled using a beam current of several nanoamperes. Using parallelogram-shaped trenches, as in this case, prevents the side walls and redeposited material from hindering the view of the sides of the wall after the different milling steps.

In Fig. 4.7C, the sample is tilted back to 0° tilt angle. The ion beam is then used to cut a horizontal trench, at least 1 μm wide, through the base of the wall and on one side, forming a cantilever. Imaging by SEM during milling then allows observing when the wall has been cut through. A micromanipulator is then inserted and welded to the end of the cantilever using IBID. The wall is then cut free and extracted. The cut can be performed using a current of 1 nA or more, and should be at least 1 μm wide. Wide cuts allow more space to avoid collisions during the actual lift-out of the wall. Also too small cuts can be filled with redeposited material during subsequent milling steps, which is not always clearly visible, making the wall impossible to lift-out. For these reasons, wide cuts are usually preferred even if they can be time-consuming for materials with a low sputtering yield.

Until this point, this lift-out method is very similar to what is commonly done to lift-out TEM lamellas, with the only difference being the morphology of the wall that is extracted from the substrate [28]. Then the free end of the wall is welded to a support tip (Fig. 4.7D). Depending on the microscope's layout, gas molecules can sometimes be shadowed by the chunk and not reach far enough under it, which yields a reduced deposition rate. If this occurs, a minimum amount of material can be deposited for the chunk to stick to the support tip, and more material can be added later at the end of the mounting, when the support tip can be tilted to an angle that reduces the shadowing effect.

The wall is then cut to leave a small chunk of material on the support tip (Fig. 4.7E). Steps (D) and (E) are repeated with the rest of the wall to produce as many mounts as necessary. Usually a wall that is 20 μm long can be used to produce about five mounts. Fig. 4.7F shows a side view of a final chunk of material mounted on a W tip.

An alternative method that is widely used and that is almost equivalent to this one consists in producing not a wall but a bar with a V-shaped cross-section [4,13,29–34]. This is done by cutting one first trench with a tilt angle of 30°, rotating the stage 180°, and making another trench from the opposite side, the two trenches meeting under the ROI. One advantage of that method is that it involves less milling. However, it requires welding the mount on the support tip from both sides. On the other hand, a significant advantage of the wall method is that both the deposition of the capping layer and the milling of the complete wall do not require any stage rotation and sample realignment. Hence, if the instrument allows it, these two steps can easily be automated and multiple walls can be prepared overnight, which often optimizes the instrument's usage time.

Annular Milling

The next step consists in using the Ga ion beam to mill the mount and give it a tip shape with the required dimensions. Fig. 4.8A shows a chunk of $TbCo_2$ thin film on a Fe substrate as mounted on a W tip. By using an annular milling pattern (an example is given in Fig. 4.8F), the mounted sample is milled down to a cylinder of 1–2 μm diameter (Fig. 4.8B). Then, thanks to a step-by-step reduction of the inner diameter of the milling pattern, a conical shape can be obtained (Fig. 4.8B–E) with a ~50 nm curvature radius at the apex. During these different steps, the milling parameters often need to be adjusted. High beam currents produce large probe sizes, which cause the beam to mill the sample out of the defined milling pattern. The milled region is not sharply defined, resulting in large cone angles. For this reason, the beam current is reduced as the annulus inner diameter is reduced, so as to not damage the ROI. In turn, this causes the milling rate to reduce, and thus the outer diameter of the milling pattern also needs to be gradually reduced to increase the Ga flux per unit area. Different specific annular milling parameters can be found in the literature [4,34,35]. In general, the outer diameter of the first milling pattern is ~5 μm and the inner diameter is 1–2 μm, and the inner diameter of the last milling pattern is ~0.2 μm. The outer diameter may be adjusted to the shape of the support tip. Indeed, the first milling steps usually use a large dose of Ga, which can significantly mill the support tip and create microtips at the outer limit of the pattern. Such tips can also evaporate during the analysis and interfere with it. Thus, the outer diameter should also be large enough so that these tips are several micrometers below the analyzed tip.

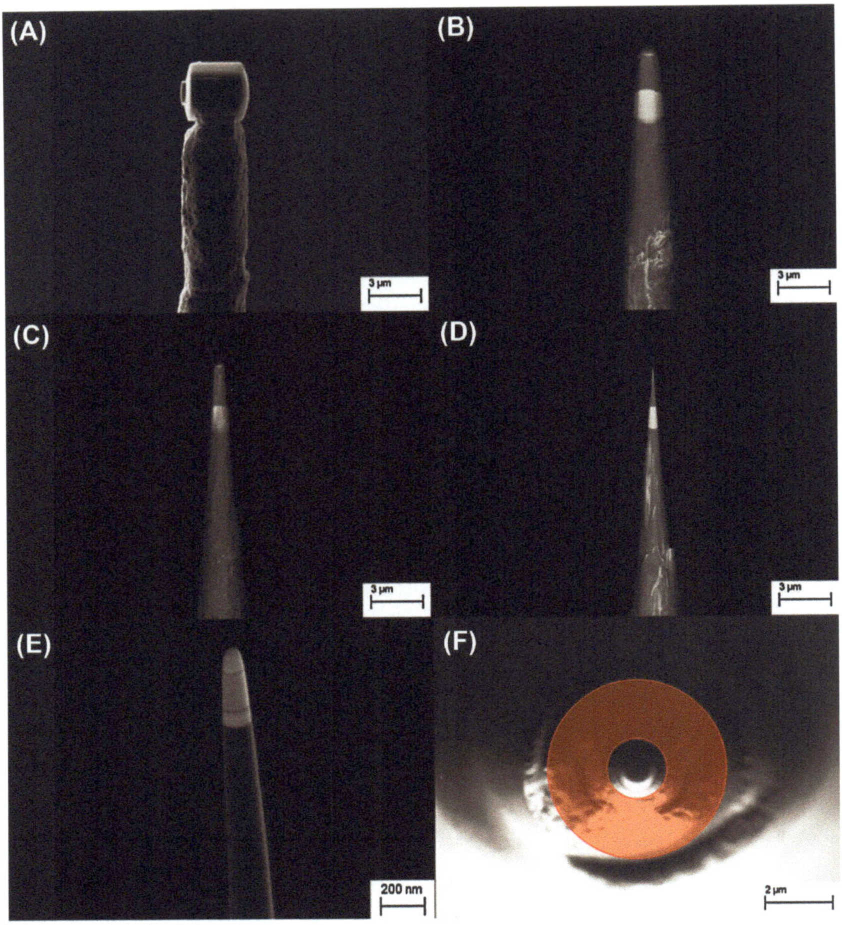

Figure 4.8 Scanning electron microscopy images showing the different steps of annular milling with (A) the initial sample mounted on the support tip, (B) the cylinder that is obtained after the first milling step (C), (D) and (E) the tip after the different steps of annular milling with gradual reduction in the pattern diameter, and (F) an example of a milling pattern superimposed with the focused ion beam image of the tip. The sample consists of a $TbCo_2$ thin film on a Fe substrate.

The annular milling parameters can also be adjusted to control the tip morphology to a certain extent. This can be useful for different reasons. In laser-assisted APT, large shank angles provide a larger heat sink, and thus a faster cooling rate for the tip apex and reduce the extent of the thermal tails in the mass spectrum (cf. Chapter 8: Laser-Assisted Field Evaporation). Conversely, a too large shank angle will require the voltage to rise faster during the analysis, which can reduce the length of the analysis by causing tip fracture because of the too high stress that is put on the tip or simply because the maximum voltage is reached too early. The tip shape can be controlled using different methods. First, increasing the Ga dose (ie, the milling time) in each step will produce tips with a smaller shank angle. On the other hand, large shank angles can be obtained by using more milling steps and a small Ga dose in each step. Also, if the capping layer has a lower sputtering yield than the underlying sample, it will act as a mask for the sample and produce small shank angles, all the more that the capping layer is thick. Other methods exist, like using milling patterns with nonuniform scanning speed, which allow the use of a specific function [36] and tailor the shape of the tip.

Low-Energy Cleaning

The annular milling is performed at a high Ga beam energy, usually 30 kV, to achieve a good spatial resolution and sputtering rates. Such high energy causes ions to be implanted in the sample, which can severely damage the surface region of the sample as mentioned earlier in this chapter. The depth of the damaged region depends on the energy of the ion beam and the material to be implanted. The cascades of collisions in the material can amorphize the material over a certain depth [24], dissolve precipitates originally present in the tip [25], induce grain growth [37], or can form Ga-containing phases in some cases [22]. It is thus important to not include this part of the material in the analysis. It can be removed by milling the tip with a low-energy Ga beam, 2–5 keV. The beam is rastered uniformly on the tip. This process also leaves damages but over a smaller depth, typically a few nanometers, depending on the sample material.

Capping Layer

A protective capping layer is generally deposited on the sample surface prior to lift-out. The main purpose of this layer is to protect the sample from damages that can be caused by the 30-kV Ga beam. This is particularly important when the ROI is close to the surface and extensive milling of the sample to remove any damaged region is not possible. In this case, Ga is implanted only in the capping layer and the ROI is left undamaged. Different parameters need to be taken into account when choosing the material that is used for the capping layer [4]. First, the material needs to be chosen so as to facilitate sample preparation by FIB. Materials that form grains that are too small can cause several grain boundaries to be present in the capping covering the sample during the different stages of annular milling, and contrasts in milling rates between the different grains can result in a final inhomogeneous tip morphology [4]. Thus, the grain size needs to be either larger than the tip size or much smaller, so that, in average, the layer behaves like a homogeneous material. Then the capping material needs to have a milling rate that is ideally equal to the one of the underlying sample. A too high milling rate can cause the capping to be completely sputtered away before the end of the annular milling process. If no material with a low enough sputtering rate can be deposited, it is, however, also possible to counter this effect by depositing a larger thickness of capping and performing the different milling steps with a lower beam current. The low current gives a smaller probe size, which increases the milling accuracy and protects further the center of the annulus during milling. Conversely, a sputtering yield that is lower than one of the samples reduces the ability to control the tip morphology. Milling the remaining capping layer requires large Ga doses, which also mill a lot of material from the sample below, which eventually leads to tips with very small shank angles. This effect can, however, be desired, for materials that have good thermal conductivity or high evaporation field. In these respects, materials like tungsten or platinum that are widely used in FIB systems generally form good capping layers. They also have high evaporation field, however, which is partly due to the fact that FIB deposited metals contain a lot of carbon. This high evaporation field can cause early tip fracture by increasing the stress on the tip, or because of the fast increase in evaporation rate when the interface is reached. For this reason, FIB deposited capping layers are generally completely removed during the low-voltage cleaning, unless the sample material also has a high evaporation field. This cannot be done, however, if the ROI is very close to the surface. In such a case, some of the capping needs to be retained so that the ROI is not damaged and does not risk to be removed during milling. If this is done, additional considerations need to be taken into account when choosing the capping layer. In such a case, the material needs to have an evaporation field that is similar to the sample material, so as to reduce the risk of fracture (in the case of high field on low field), and to reduce local magnification and reconstruction artifacts [38]. Another important criterion is that it needs to exhibit good adhesion to reduce the risk of interfacial fracture under the stress caused by the analysis. In addition, if possible, it should not have common peaks with elements present in the ROI, to facilitate data treatment and data interpretation.

Mounting Techniques

The standard lift-out method that was presented earlier is limited to samples having a flat surface and where the ROI can easily be targeted using plane view SEM, if needed, and where a tip can be prepared and analyzed in the direction normal to the surface. Although these requirements are met in most analyses, various cases exist that require other sample preparation methods.

Site-Specific Lift-Out for Interface Analysis

The spatial resolution of the atom probe is better along the direction of analysis than in the perpendicular directions [39–41], thus a concentration profile across a heterophase interface or grain boundary is most accurate when the interface is oriented perpendicular to the direction of analysis. Grain boundaries where orientations are not optimal can sometimes appear broader by several nanometers [40,41]. With the conventional sample preparation method, however, interfaces cannot be placed perpendicular to the

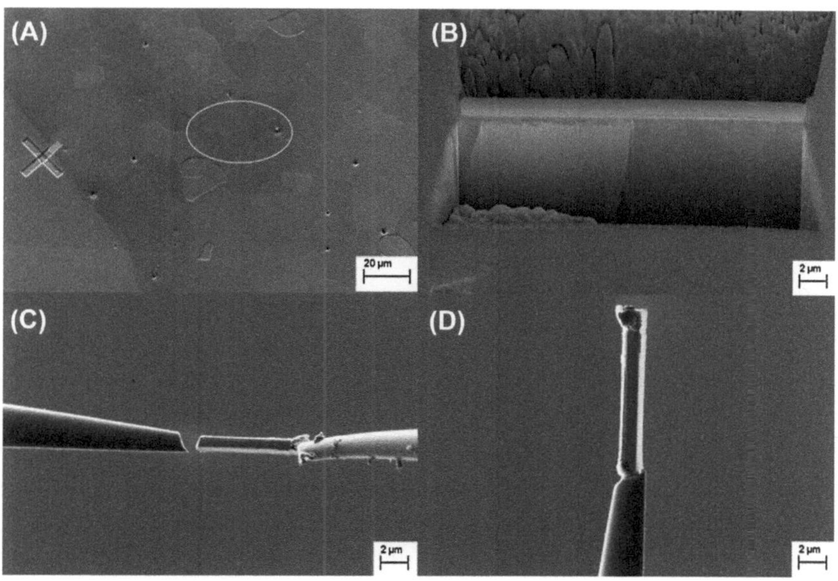

Figure 4.9 Scanning electron microscopy images describing the lift-out method for the analysis of grain boundaries on an austenitic steel, with (A) the selection of the region of interest, (B) milling of a wall prior to lift-out, (C) mounting of the wall parallel to a support tip, and (D) specimen mounted on the support tip, ready for cutting and annular milling.

direction of analysis because they need to intersect the surface to be observed by SEM during sample preparation.

Fig. 4.9 presents a sample preparation method that can be used to perform such a lift-out with the interface having an optimal orientation. First, an interface must be identified on the surface (Fig. 4.9A). This can be done by taking advantage of the channeling contrast, which is sometimes mostly visible using ion imaging. At this stage, EBSD can also be used to identify the interface of interest, and can be used to characterize the crystallographic orientations of the different grains. Then a wall is milled perpendicular to the interface. This cross-section allows one to measure the grain boundary orientation and to check if it is perpendicular to the sample surface (Fig. 4.9B). Then a standard lift-out is performed, except in this case the resulting bar is mounted on a horizontal support tip (Fig. 4.9C,D). The excess material above the ROI can be removed and a regular annular milling can be performed, placing the interface in the vicinity of the tip apex.

Cross-Sectional Analysis

Cross-sectional analysis is particularly interesting for the analysis of thin films. Using the standard lift-out method, the direction of analysis is perpendicular to the thin films. This means that the spatial resolution across the interfaces will be optimized. However, the analysis volume will be limited by the diameter of the tip. If a good resolution across the interfaces is not required, much larger volumes of interface can be analyzed by preparing the tip parallel to the interfaces [42,43]. In addition, samples that contain thick insulating layers can be difficult to analyze because of poor thermal and electrical conductivity to the support tip. In such a case, preparing the tip parallel to the substrate can provide direct contact between the ROI and the support tip [43]. To do so, a capping layer is first deposited on the sample surface. This layer should be at least 500 nm thick. Then, a regular lift-out is performed. If the wall method is used, the undercut must be done at the top edge of the wall to be extracted (Fig. 4.10A). Then the extracted wall is rotated 90° along its main axis, placing one edge on the bottom (Fig. 4.10B). Such large rotations are generally not possible in the microscope, but can be performed by mounting the wall on a horizontal support tip that is rotated ex situ. Then different sections of the wall are mounted on support tips using the regular method (Fig. 4.10C) and the tip is rotated so that the interface is visible by SEM. A capping layer is deposited on the mount and annular milling is performed around the interface area. For this method, it is particularly important that the first thick capping layer

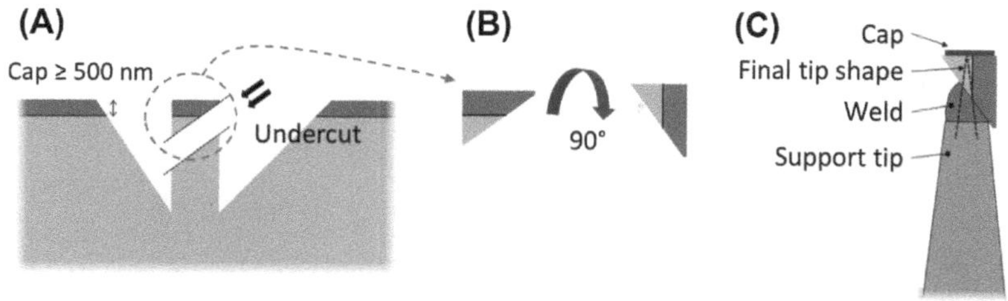

Figure 4.10 Schematic of the lift-out process for the analysis of a material in a direction parallel to the surface with (A) milling of a horizontal trench through the wall and under the capping layer, (B) 90° rotation of the extracted piece of material, and (C) mounting on a support tip and deposition of a capping layer prior to annular milling

has similar sputtering rate and evaporation field as the substrate and that it exhibits good SEM contrast with the substrate if the thin films are not clearly visible. Other effective methods to produce APT samples in the direction of cross-section can be found elsewhere [4,43,44]. This method is also interesting for the analysis of nanoparticles. In this case, the nanoparticles are dispersed on a substrate before depositing the first coating. Cross-sectional analysis of the nanoparticles at the interface can then provide larger analyzed volumes and more statistically significant results [44].

Backside Analysis

In some cases, it can be interesting to analyze a sample from the backside to check for the presence of artifacts. For example, surface retention of high field elements during analysis can have an effect on the shape of concentration profiles. Segregation profiles around grain boundaries that are actually symmetric can appear asymmetric because of surface retention [45]. The analysis of multilayers having various evaporation fields can also produce distortions in the final reconstruction that depend on the direction of analysis [38]. Backside analysis can be used to verify the presence of such artifacts [38,46].

For backside sample preparation using the wall method, the undercut must be done at least 1 μm below the ROI in the wall. An example is presented in Fig. 4.11, where the ROI is an interface between a thin film and the substrate (Fig. 4.11A). After lift-out, the sample is cut and rotated twice (Fig. 4.11B,C). The rotation procedure can be performed using the same method as for the cross-sectional analysis. The chunk can be mounted on a support tip, and a capping layer is deposited on top of it before annular milling (Fig. 4.11D). An alternative to this method, using the V-shape lift-out method, can be found elsewhere [4].

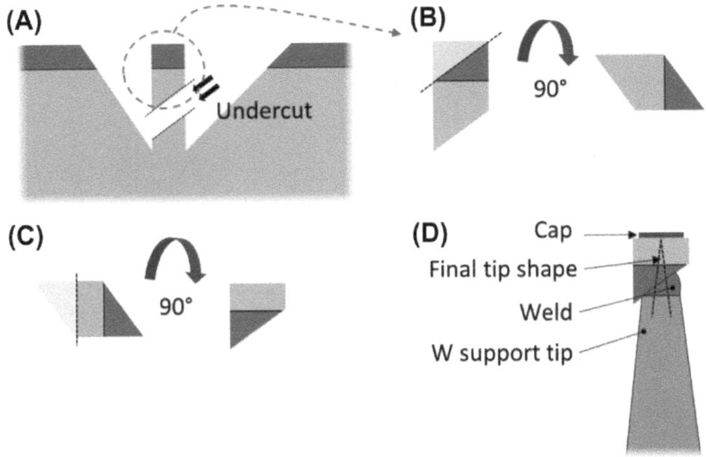

Figure 4.11 Schematic of the lift-out process for backside analysis with (A) milling of a horizontal trench through the wall and under the region of interest, (B) and (C) series of cutting and rotating by 90° to reverse the orientation of the sample, and (D) mounting on a support tip and deposition of a capping layer prior to annular milling.

Transmission Kikuchi Diffraction

In some cases, the placement of an interface close to the tip apex can be challenging. First, the contrast might not be strong enough both in SEM and FIB to see the interface. Second, during the last stages of annular milling, redeposited material can also hide the interface. In such a case, TKD [18–20] can be used to visualize the interface. One advantage of TKD over EBSD is that electron diffraction patterns can be analyzed independent of the surface orientation with respect to the electron beam, whereas it needs to be tilted by 70° in the case of EBSD. For this reason, TKD is particularly adapted for the analysis of tip-shaped samples, provided they are thin enough. An example of an annular milling procedure taking advantage of TKD is presented in Fig. 4.12. TKD can be performed on a tip while it is still aligned with the ion column (Fig. 4.12A). This makes it particularly handy for annular milling, as it is possible to alternate TKD analysis and annular milling without realigning the sample. Although no interface is visible using conventional SEM imaging (Fig. 4.12B), interfaces are visible in TKD (Fig. 4.12C). The distance between the ROI and the tip apex can be measured and the correct amount of material can be gradually removed to place the interface in the vicinity of the tip apex.

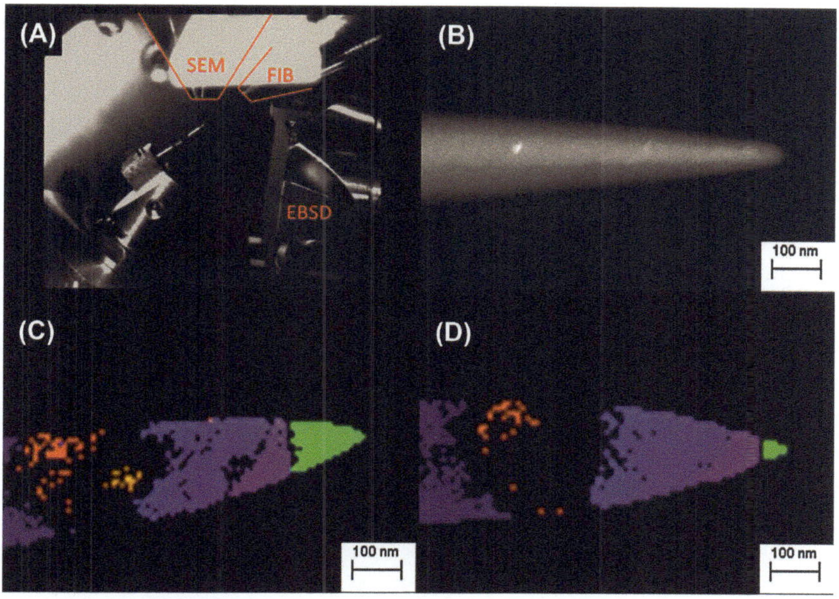

Figure 4.12 Example of transmission Kikuchi diffraction experiments on a martensitic steel sample between steps of annular milling with (A) a photograph of the setup in the scanning electron microscope (SEM) chamber, (B) SEM image of the tip, (C) transmission Kikuchi diffraction pattern where different crystal orientations are visible, and (D) the same tip after an additional milling step to bring an interface close to the tip apex for atom probe analysis. The chosen interface corresponds to a former austenite grain boundary. *EBSD*, electron backscattered diffraction; *FIB*, focused ion beam. *Courtesy of Claire Debreux—PhD thesis.*

Alternatives to the Lift-Out

One disadvantage of sample preparation using FIB is that the lift-out method is time consuming. Performing one lift-out and mounting five chunks on support tips can often require 3 h of FIB time or more. This aspect is all the more important in laboratories where FIB availability is the bottleneck limiting the number of atom probe analyses that can be performed. Thus, any sample preparation method that allows circumventing part of the FIB processing can be very useful. One technique to analyze stacks of thin films consists in elaborating them not on a bulk substrate but on top of micropillars [47].

Several efficient specimen preparation methods have taken advantage of the pillar morphology. For instance, D. Perea et al. could grow InAs nanowires on GaAs (111) substrate [48], therefore reducing the requirement of

FIB milling and allowing the analysis of the catalyst/nanowire interface. The possibility to grow nanowires on microposts has also been revealed to be a very efficient way to select a single nanobject for APT analysis [49]. Fig. 4.13B shows an example of Si micropillars, made through the Bosch process, or deep reactive ion etching [50]. These pillars are then dispersed on a glass substrate. Then, using an optical microscope equipped with a micromanipulator, a support tip is dipped in conductive epoxy glue and then put in contact with one of the micropillars (Fig. 4.13A,C). This can be done with a simple manual hydraulic micromanipulator. Then FIB is eventually used only to perform annular milling of the micropillar (Fig. 4.13D). Numerous examples of the use of this technique can be found in the literature [51–56]. For small particles, the top of the support tip must be cut using FIB to give a flat surface to the particle (Fig. 4.14A), or a micropillar can be first glued to the support tip, and then the particle is glued to the

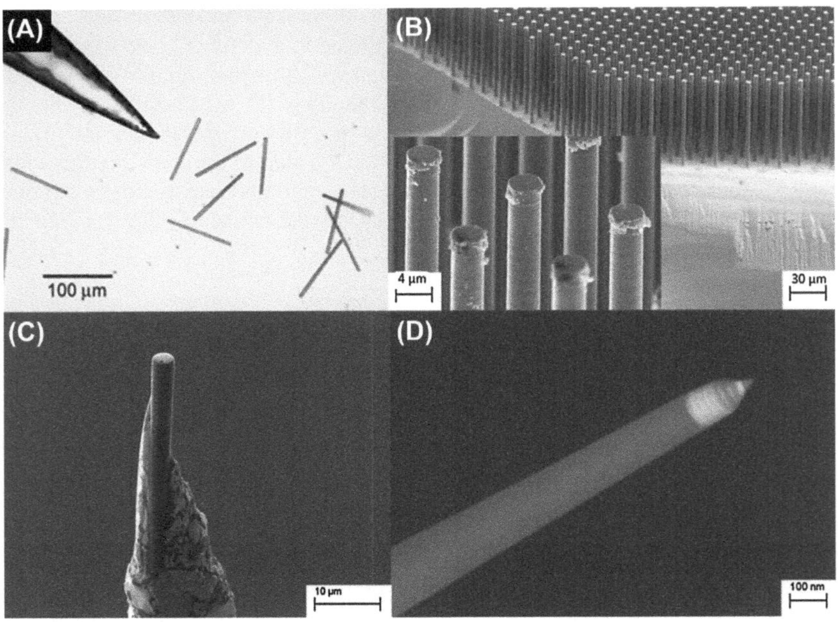

Figure 4.13 Description of the sample preparation process for the analysis of thin films deposited on Si microposts with (A) the gluing of Si microposts to a tungsten needle that was mounted on a micromanipulateur and dipped in epoxy glue (optical image), (B) an example of such Si microposts fabricated by the Bosch process, on their original substrate, (C) SEM image of a Si micropost after gluing to a tungsten needle, (D) SEM image of the Si tip after annular milling, showing the region of interest at the end of the tip, in this case a multilayer structure.

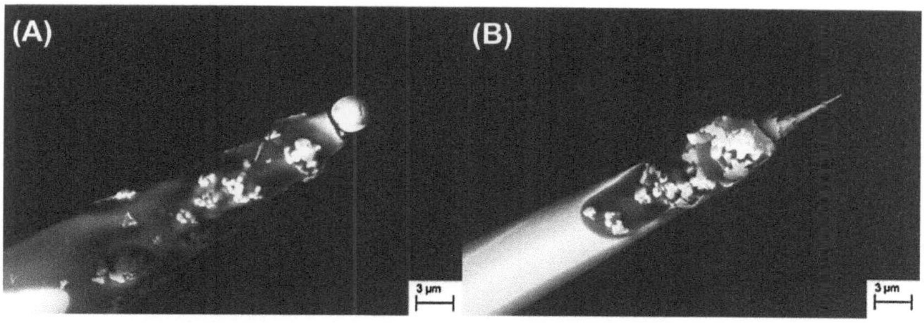

Figure 4.14 Scanning electron microscopy images of the preparation of a tip out of a microparticle with (A) the microparticle glued to a support tip. The microparticle was lifted off from a substrate using a tungsten tip with end cut flat by FIB and dipped in epoxy glue and (B) the same sample after annular milling and ready to be analyzed by atom probe tomography.

top of the micropillar, and again, FIB has to be used only for the annular milling part (Fig. 4.14B). Bulk samples can also be ground to a size that is small enough to be lifted off, prior to FIB milling.

CONCLUSIONS

The drastic improvement of micromanipulation, microwelding, and ion beam milling in SEM has accompanied the development of APT since 2000. The use of subpicosecond laser, the large field of views, and the rapid acquisition rates of APT have since opened new fields of application of this technique (cf. chapters in this book related to these topics). Looking for a needle in a haystack has now become a routine experiment for APT users thanks to the combination of SEM imaging techniques (eg, TKD, EBSD, and SE and BSE imaging), IBID, micromanipulation, and FIB milling. Nevertheless, some challenges still remain. The analysis of soft matter or biological materials by APT, if relevant, will require foreseeing new dedicated protocols. As for the analysis of liquid phases, convincing proof of concept has been demonstrated via the application of cryotransfer [57] that may open new fields of application of the technique.

REFERENCES

[1] Miller MK, Cerezo A, Hetherington MG, Smith GDW. Atom probe field ion microscopy. Oxford Science Publications - Clarendon Press; 1996.
[2] Miller MK. Atom probe tomography: analysis at the atomic level. New York: Springer; 2000.
[3] Miller MK, Smith GDW. Atom probe microanalysis: principles and applications to materials problems. Pittsburg, PA: Materials Research Society; 1989.

[4] Larson D, Prosa T, Kelly T. Local electrode atom probe tomography - a user's guide. Springer; 2013.
[5] Marquis EA, Seidman DN. Acta Mater 2005;53:4259–68.
[6] Knipling KE, Karnesky RA, Lee CP, Dunand DC, Seidman DN. Acta Mater 2010;58: 5184–95.
[7] Mousa M, Wanderka N, Timpel M, Singh S, Krüger M, Heilmaier M, et al. Ultramicroscopy 2011;111:706–10.
[8] Sebastian JT, Rusing J, Hellman OC, Seidman DN, Vriesendorp W, Kooi BJ, De Hosson JTM. Ultramicroscopy 2001;89:203–13.
[9] Sebastian JT, Seidman DN, Yoon KE, Bauer P, Reid T, Boffo C, et al. Phys C Supercond 2006;441:70–4.
[10] Harzl M, Leisch M. Appl Surf Sci 1999;144–145:41–4.
[11] Wanderka N, Glatzel U. Mater Sci Eng A 1995;203:69–74.
[12] Bagot PAJ, Cerezo A, Smith GDW. Surf Sci 2008;602:1381–91.
[13] Gault B, Moody MP, Cairney JM, Ringer SP. Atom probe microscopy. New York: Springer; 2012.
[14] Kelly TF, Larson DJ. Annu Rev Mater Res 2012;42:1–31.
[15] Kuhlman KR, Martens RL, Kelly TF, Evans ND, Miller MK. Ultramicroscopy 2001; 89:169–76.
[16] Drouin D, Réal Couture A, Joly D, Poirier-Demers N. H.D. Monte Carlo simulation of electron trajectory in solids. Casino version 3.2.0.1, Casino Copyright ©. 2011.
[17] Newton RG. Scattering theory of waves and particles. Berlin, Heidelberg: Springer Berlin Heidelberg; 1982.
[18] Keller RR, Geiss RH. J Microsc 2012;245:245–51.
[19] Trimby PW. Ultramicroscopy 2012;120:16–24.
[20] Babinsky K, De Kloe R, Clemens H, Primig S. Ultramicroscopy 2014;144:9–18.
[21] Everhart TE, Thornley RFM. J Sci Instrum 1960;37:246–8.
[22] Giannuzzi LA, Stevie FA. Introduction to focused ion beams: instrumentation, theory, techniques and practice. US: Springer; 2005.
[23] Ziegler JF, Ziegler MD, Biersack JP. Nucl Instrum Methods Phys Res Sect B Beam Interact Mater Atoms 2010;268:1818–23.
[24] Giannuzzi LA, Geurts R, Ringnalda J. Microsc Microanal 2005;11:828–9.
[25] Miller MK, Russell KF, Thompson K, Alvis R, Larson DJ. Microsc Microanal 2007; 13:428–36.
[26] Thompson K, Bunton JH, Kelly TF, Larson DJ. J Vac Sci Technol B Microelectron Nanom Struct 2006;24:421.
[27] Fang F, Xu Z. In: State-of-the-art nanomanufacturing using ion-beam technology. Springer; 2015. p. 1279–315.
[28] Giannuzzi LA, Stevie FA. Micron 1999;30:197–204.
[29] Thompson K, Lawrence D, Larson DJ, Olson JD, Kelly TF, Gorman B. Ultramicroscopy 2007;107:131–9.
[30] Cairney JM, Saxey DW, McGrouther D, Ringer SP. Phys B 2007;394:267.
[31] Saxey DW, Cairney JM, McGrouther D, Honma T, Ringer SP. Ultramicroscopy 2007;107:756.
[32] Saramat A, Svensson G, Palmqvist AEC, Stiewe C, Mueller E, Platzek D, et al. J Appl Phys 2006;99.
[33] Miller MK, Forbes RG. Atom-probe tomography: the local electrode atom probe. New York: Springer; 2014.
[34] Miller MK, Russell KF. Ultramicroscopy 2007;107:761.
[35] Larson DJ, Petford-Long AK, Ma YQ, Cerezo A. Acta Mater 2004;52:2847–62.
[36] Miller MK, Russell KF, Thompson GB. Ultramicroscopy 2005;102:287.
[37] Park C-M, Bain JA. J Appl Phys 2002;91:6830.

[38] Marquis EA, Geiser BP, Prosa TJ, Larson DJ. J Microsc 2011;241:225–33.
[39] Gault B, Moody MP, de Geuser F, Haley D, Stephenson LT, Ringer SP. Appl Phys Lett 2009;95:34103.
[40] Blavette D, Duval P, Letellier L, Guttmann M. Acta Mater 1996;44:4995–5005.
[41] Portavoce A, Blum I, Hoummada K, Mangelinck D, Chow L, Bernardini J. Defect Diffus Forum 2012;322:129–50.
[42] Mangelinck D, Hoummada K, Portavoce A, Perrin C, Daineche R, Descoins M, et al. Scr Mater 2010;62:568–71.
[43] Martin AJ, Weng W, Zhu Z, Loesing R, Shaffer J, Katnani A. Ultramicroscopy 2015; 161:105–9.
[44] Heck PR, Stadermann FJ, Isheim D, Auciello O, Daulton TL, Davis AM, et al. Meteorit Planet Sci 2014;49:453–67.
[45] Felfer PJ, Gault B, Sha G, Stephenson L, Ringer SP, Cairney JM. Microsc Microanal 2012;18:359–64.
[46] Larson DJ, Prosa TJ, Geiser BP, Egelhoff WF. Ultramicroscopy 2011;111:506–11.
[47] Larson DJ, Martens RL, Kelly TF, Miller MK, Tabat N. J Appl Phys 2000;87:5989.
[48] Perea DE, Allen JE, May SJ, Wessels BW, Seidman DN, Lauhon LJ. Nano Lett 2006;6: 181–5.
[49] Perea DE, Lensch JL, May SJ, Wessels BW, Lauhon LJ. Appl Phys A Mater Sci Process 2006;85:271–5.
[50] Ayon A, Braff R, Lin C, Sawin H, Schmidt M. J Electrochem Soc 1999;146:339–49.
[51] Miller MK. Microsc Microanal 2005;11:862–3.
[52] Larde R, Lechevallier L, Zarefy A, Bostel A, Juraszek J, Le Breton JM, et al. J Appl Phys 2009;105.
[53] Ngamo M, Duguay S, Cristiano F, Daoud-Ketata K, Pareige P. J Appl Phys 2009;105.
[54] Cojocaru-Mirédin O, Cadel E, Vurpillot F, Mangelinck D, Blavette D. Scr Mater 2009;60:285–8.
[55] Duguay S, Vurpillot F, Philippe T, Cadel E, Larde R, Deconihout B, et al. J Appl Phys 2009;106:106102.
[56] Blavette D, Duguay S. Eur Phys J Appl Phys 2014;68:10101.
[57] Gerstl SSA, Wepf R. Microsc Microanal 2015;21:517–8.

CHAPTER FIVE

Time-of-Flight Mass Spectrometry and Composition Measurements

C. Pareige, W. Lefebvre-Ulrikson, F. Vurpillot, X. Sauvage
Groupe de Physique des Matériaux, University and INSA of Rouen, Normandie University, France

INTRODUCTION

Since 1967 and the development of the first atom probe (AP) by Müller, Panitz, and McLane at Pennsylvania State University [1], time-of-flight mass spectrometry has been the core part of the instrument. The simple assumptions and basic equations established at that time are still routinely used in modern APs to determine the very nature of field evaporated atoms. The final results, ie, both composition measurements and three-dimensional reconstructions, rely on mass spectrum quality and processing. The quality of the mass spectrum depends not only on the physics of field evaporation but also on AP design (flight length) and hardware devices eventually used to reduce energy deficits (reflectrons). It also depends on the operator and analysis conditions including voltage, detection rate, specimen temperature, pulse fraction, or laser energy. The final result is also related to the processing carried out by the operator who typically needs to perform calibration, mass resolution optimization, background noise subtraction, peak identification, mass range definition, and peak deconvolutions. The aim of this chapter is to move progressively from the general concepts of mass spectrometry introduced earlier by Müller and Panitz to the physical mechanisms involved in the construction of real mass spectra and finally to practical and accurate data processing.

GENERAL CONCEPTS AND DEFINITIONS

Mass Spectrum Computation

In APs, atoms are field evaporated, which means that they are ionized and accelerated in the electric field surrounding the sample surface (see Chapter 2: Field Ion Emission Mechanisms). Some of these ions are collected on a detector used to determine the time and the position of

Atom Probe Tomography
ISBN: 9780128046470
http://dx.doi.org/10.1016/B978-0-12-804647-0.00005-X

the impact (see Chapter 6: Atom Probe Tomography: Detector Issues and Technology). The time of flight of an ion is the time required to fly from the sample surface to the detector, which is called the flight length L (Fig. 5.1). The time of flight can be measured only if the evaporation is triggered by a voltage or laser pulse.

Then, assuming that ions have no initial velocity and a constant velocity during the flight (ie, assuming that the acceleration path can be neglected comparing to the flight length) and that their whole energy is converted into kinetic energy, then the mass over charge ratio M of ions is written as Ref. [1]:

$$M = \frac{m}{n} = 2eV\left(\frac{t_f}{L}\right)^2 = kV\left(\frac{t_f}{L}\right)^2 \tag{5.1}$$

where m is the mass of the ion, n the charge, V the total voltage applied to the specimen, e the elementary charge of the electron, L the flight length, and t_f the time of flight. M is often given in arbitrary mass units (amu) or daltons; if L is in [m], t_f in [μs], V in [kV], then $k = 0.192973$.

In the case of laser pulsing, V is the direct current (DC) voltage (V_{DC}) applied to the sample. In the case of electric pulsing, V is the sum of the DC voltage (V_{DC}) and the pulse voltage amplitude (V_P). V_p is typically in the range of 10−30% of (V_{DC}).

In modern APs, typical flight lengths are between 10 and 50 cm, which gives typical time of flights in a range of about 1−5 μs for an evaporation voltage in a range of 1−10 kV.

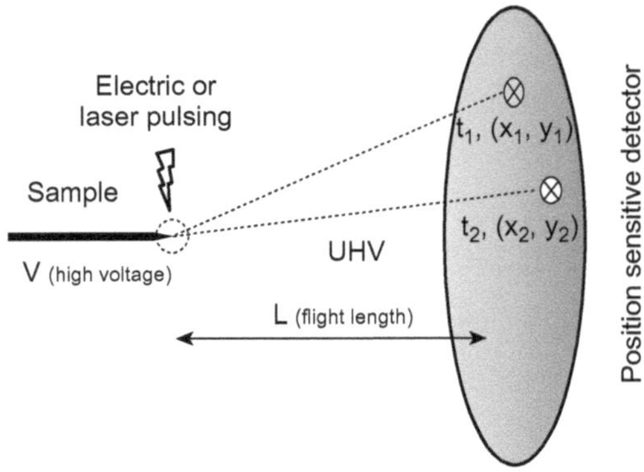

Figure 5.1 Schematic representation of the simplest configuration of an atom probe.

Mass Spectrum Presentation and Interpretation

A mass spectrum is a histogram that gives the distribution of the mass over charge ratio of collected ions. As shown in Fig. 5.2, there are typically peaks that are exhibited and that correspond to specific ions.

The resolution of the mass spectrum is defined as the mass range of each class used to display the mass spectrum. It is important to note that increasing this resolution, ie, decreasing the range of classes, does not necessarily lead to better visualization since for an infinitely small value, each ion basically stands alone and no peak will be exhibited. On the other hand, a rough resolution, with a large range of classes, leads to artificial peak broadening and overlaps.

The identification of peaks on mass spectra rely both on prior knowledge of the analyzed material chemistry and on the natural abundance of isotopes. For example, in Fig. 5.2, the different isotopes of Mo have been identified. Such identification is important because it allows the precise calibration of the mass spectrometer. Indeed, one of the specificities of the AP technique is that the mass spectrometer is usually calibrated for each experiment. It is

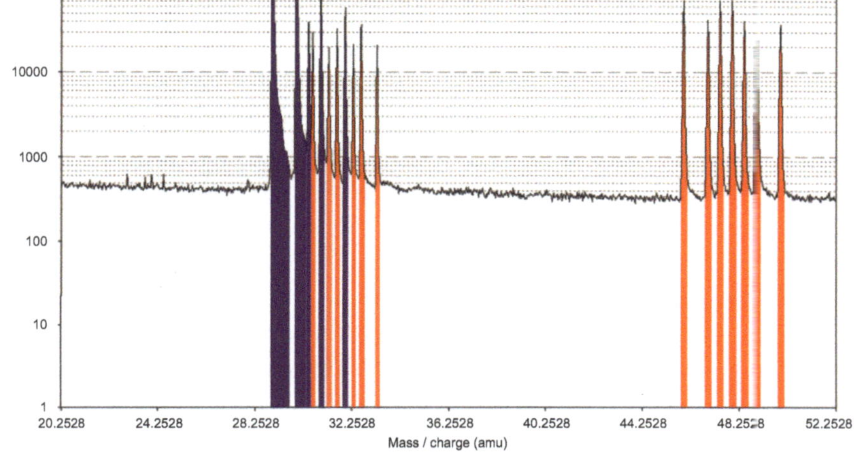

Figure 5.2 Mass spectrum of a Ni–Mo alloy.

necessary because small variations in the flight length from one analysis to another can hardly be avoided since it depends on the specimen length. The important point is that this calibration is important for the reconstruction since the magnification also directly depends on the flight length.

During the field evaporation process, atoms from the sample surface are ionized and field evaporated. The charge of collected ions depends on the occurrence of postionization, the probability of which is linked to the mean electric field surrounding the sample (see Chapter 2: Field Ion Emission Mechanisms). Then, as shown in Fig. 5.2, a given species could be detected as ions with different charges. In some cases, molecular ions could also be collected. They might result from chemical reactions under high electric field on the specimen surface with residual molecules from the vacuum chamber (like hydrogen). They might also originate from strong atom bonds that cannot always be broken under high electric field in the analyzed material.

To determine compositions, the ratio of ions is simply computed but it means that each peak should be identified and the corresponding mass ranges defined. In case of overlaps between ions (for example, in Fig. 5.2, ${Ni_{64}}^{2+}$ and ${Mo_{96}}^{3+}$ fall on the same position at 32 amu), peak deconvolution based on the natural abundance of isotopes should be carried out. The composition accuracy is also affected by the background noise and the mass resolution.

Background Noise, Mass Resolution, and Detection Limit

Background Noise

The background noise is clearly exhibited in Fig. 5.2 where the base line does not stand at the origin. It is relatively constant over the full mass range of the mass spectrum because it originates from the random detection of ionized gaseous atoms/molecules coming from the analysis chamber or atoms/molecules adsorbed at the surface of the specimen. These atoms/molecules being ionized at the DC voltage, are detected at any time during the opening of the detection window giving rise to a random contribution. There is also another small contribution of the detector itself. The microchannel plates (MCPs) used as the amplification system (Chapter 6: Atom Probe Tomography: Detector Issues and Technology) sometimes spontaneously create electrons that are "seen" as ion impact by the detector when it happens during the detection windows. Because the background noise is random and time dependent, analysis conditions have an influence on the mean level and especially it is important to note that a higher detection rate will always reduce it.

In principle, the background noise could always be subtracted to estimate the compositions more accurately. But it is important to note that in reconstructed volume ions are identified from their mass over charge ratio and it is basically impossible to differentiate a "noise" event from a real ion.

In some analysis conditions, some correlated background noise or time-dependent background noise correlated to peaks on the mass spectrum could be observed. For example, a nonappropriate laser energy or wavelength (see Chapter 8: Laser-Assisted Field Evaporation) could create a long tail of peaks leading to strong overlaps.

Mass Resolution

The quality of the mass spectrum is assessed by the mass resolution $M/\Delta Mx\%$ where $\Delta Mx\%$ is the width of the mass peaks at x% of the maximum peak high. The mass resolution is generally measured at 50% (full width at half maximum (FWHM)), 10% (full width at tenth maximum), and 1% of the maximum of the peak. The mass resolution depends on geometrical factors (flight length and dispersion in flight length), the accuracy of measurements (time of flight and voltage), and physical factors (sample temperature and energy deficit). The optimization of the mass resolution can be carried out by hardware (like reflectrons) or postanalysis treatment (software).

Detection Limit

The detection limit is defined as the minimum concentration that can be detected in the analyzed material. It depends of course on the mass resolution and also on the background noise. Since ions with different mass over charge ratios and eventually a large variety of isotopes are detected, a detection limit of 20 at ppm does not mean that any element with such a low amount could be detected.

OPTIMIZING THE MASS SPECTRUM IN ATOM PROBE TOMOGRAPHY

Origins of the Mass Spectrum Degradation

Following Eq. (5.1) deriving the measured $M = m/n$ ratio as a function of time of flights, where the applied voltage is V, and L is the distance between the sample and the detector, the expected error in M is straightforward and follows this equation:

$$\frac{M}{\delta M} > \frac{1}{\sqrt{\left(\frac{\sum \delta V^2}{V^2}\right) + \left(\frac{4\sum \delta t_{\mathrm{f}}^2}{t^2}\right) + \left(\frac{\sum \delta L^2}{L^2}\right)}} \tag{5.2}$$

Note that we assume that the different errors are independent in this calculation. This simplification is not completely true as it will be shown for some errors, but this formula will give us a minimal value for the estimate of the mass resolution. We assume different sources of errors, linked to measurement performances or intrinsic physical limitations. Three different kinds of errors are listed, energy or voltage errors (δV), time-of-flight errors (δt_{f}), and flight length errors (δL).

Time-of-Flight Errors

Errors or dispersion in time of flights are divided into two contributions. The first is the measurement errors, related to the precision of the electronic system used. The second is the time spread induced by the nature of the emitting process on the tip, which is slightly different in voltage pulsed mode or in laser pulsed mode. At each generated pulse, a time-measurement device starts the counting for a given duration named the detection window. During the opening of the detection window, if one ion arrives on the detector, it creates a signal that enables the arrival time to be determined. A measured time of flight (t_{m}) of the ion can thus be calculated. It corresponds to the time between the opening of the detection window and the moment when the signal is measured by the detector. But, because of the existence of a time offset (t_0) between the opening of the detection window and the actual application of the pulse to the specimen, the measured time of flight (t_{m}) must be corrected to obtain the actual time of flight (t_{f}) of the ion:

$$t_{\mathrm{f}} = t_{\mathrm{m}} - t_0. \tag{5.3}$$

t_0 Calibration

The quality of elemental identification depends on the good calibration of the mass spectrum, ie, of the time offset t_0 and the flight distance L. This flight distance is considered as the distance between the specimen and the detector. It will be seen later, that the actual flight distance of the ions may slightly differ in case of wide-angle APs.

Influence of L: Changing the flight length L induces a simple translation of the mass spectrum in accordance with Eq. (5.1). Consequently, placing the mass peaks at their expected position enables the accurate estimation of the flight length.

Influence of the time offset is more complex as shown by the following equation obtained by inserting Eq. (5.3) in Eq. (5.1):

$$\frac{m}{n} = 2eV\left(\frac{t_m - t_o}{L}\right)^2 \tag{5.4}$$

Varying the value of t_0 induces a shift of the mass peaks that depends on the value of the mass to charge ratio. The distances between the different mass peaks are no more constant and equal to their expected values. Consequently, if the t_0 value is not properly estimated, it is not possible to get all the mass peaks at their expected position with a unique flight length. In the example shown in Fig. 5.3A, the mass spectrum was constructed using a wrong value of t_0. The value of the flight length has been fitted so that the Si^{2+} mass peaks are located at their expected position. But as clearly shown by the figure, in this case, the mass peaks at higher mass range (see, for example, the $56Fe^{2+}$peak) are not located at their expected position. It is not possible to find a value of L enabling the right positioning of all the mass peaks. When the correct t_0 value is used, a value of the flight length can be found to properly build the mass spectrum (Fig. 5.3).

The actual values of the calibration parameter set (t_0,L) can be estimated from Eq. (5.5) (derived from Eq. (5.4)):

$$t_m = t_0 + L\sqrt{\frac{m/n}{k\cdot V}} \tag{5.5}$$

A plot of the measured time of flight t_m against $\sqrt{\frac{m/n}{k\cdot V}}$ gives a straight line, the slope of which corresponds to the flight length and the intercept gives t_0. In practice, to extract values of t_0 and L, one has to select at least two mass peaks in a wide range of a.m.u. taking care of two important points: (1) to select a small analysis volume, ie, with a small surface and a short depth to ensure small variation of the voltage and a small spread in time of flights and (2) the mass window for each peak must be narrow and around the maximum of the peak. Fig. 5.3 presents a graph drawn considering atoms from the $28Si^{2+}$ and $56Fe^{2+}$ peaks of the mass spectrum of Fig. 5.4 and the corresponding values of the calibration parameter set. More details about the procedure are available in Ref. [19]. Conversely

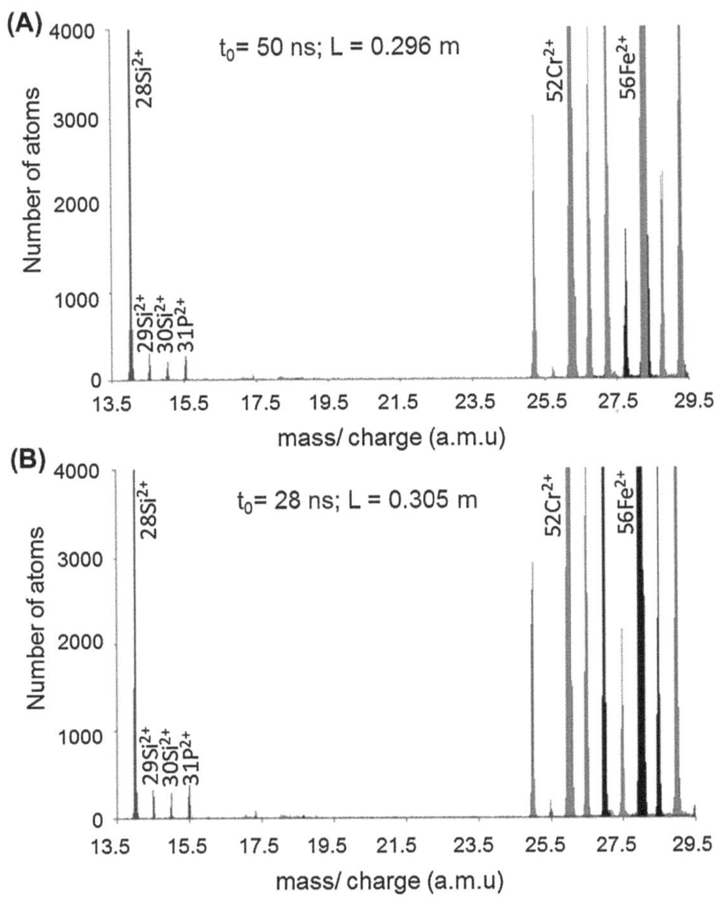

Figure 5.3 (A) Mass spectrum of the ferrite of a duplex steel obtained with an energy-compensated atom probe and constructed using a wrong value of t_0. The value of the flight length has been fitted so that the Si^{2+} mass peaks are well positioned. In this case, mass peaks at higher mass ranges are not properly located. (B) The correct value of t_0 has been used. A value of the flight length L has been found with a correct position for all peaks.

to L, the value of t_0 is the same for all experiments as long as the electronic hardware is not modified. Nevertheless, the procedure to determine the calibration parameters is generally implemented in the atom probe tomography (APT) software and a parameter set determined routinely for each data treatment.

Once the couple (t_0,L) is calibrated, which means that the mass spectrum is accurately calculated, the only remaining errors are precision errors on the

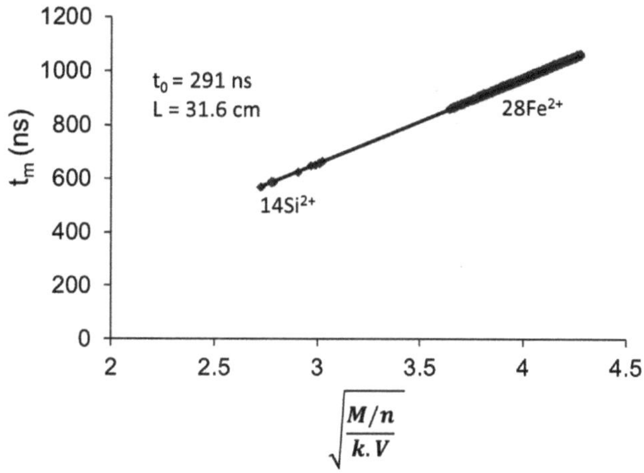

Figure 5.4 Best-fit calculation for the accurate determination of the parameter couple (t_0,L). Data points correspond to atoms from $28Si^{2+}$ and $56Fe^{2+}$ peaks of mass spectrum of Fig. 5.3.

two times t_0 and t_m. As described in the chapter related to detectors (Chapter 6: Atom Probe Tomography: Detector Issues and Technology), the precision on the time measurement system is about 25—100 ps for the measurement of a single time (FWHM). To these errors we need to add the errors induced by the physics of the evaporation process, that are different in the voltage pulsed mode or laser pulsed mode.

Dispersion in Start Time t_0

In voltage pulsed mode, the emission of ions is triggered by the application of a short voltage pulse V_P, superimposed on the DC voltage V_{DC}. The basic principle was described in "Chapter 2, Field Ion Emission Mechanisms" (section Pulsed Field Evaporation). We will assume a voltage pulse described by $V_P = \alpha_P \cdot V_{max}$, with V_{max} the peak voltage ($V_{DC} + V_P$) at the moment of evaporation, of $V_P = fV_{DC}$ ($f = \alpha_P/(1 - \alpha_P)$), and we will describe the evaporation behavior by the classical equation

$$K\left(s^{-1}\right) \propto \exp\left(- \frac{C(1 - F/F_{EV})}{k_B T} \right) \tag{5.6}$$

where F_{EV} is the evaporation field constant of the material, T the applied temperature, and C the asymptotic zero field energy of the evaporation barrier. A voltage pulse with a Gaussian temporal shape of standard deviation σ produces a shorter evaporation pulse $K_T(t)$ that can be approximated to

$$\frac{K_T(t)}{K_{T,\max}} = \exp\left[-\frac{C}{k_B T}\left(1-\exp\left(-\frac{t^2}{2\sigma^2}\right)\right)\right] \approx \exp\left[-\frac{C}{k_B T}\alpha_P\frac{t^2}{2\sigma^2}\right] \tag{5.7}$$

The width of this evaporation pulse is shorter than the voltage pulse by one range of order. The shape of this pulse is approximately Gaussian, with a standard deviation $\sigma' = \sigma\sqrt{\frac{k_B T}{\alpha_P C}}$. For instance, a voltage pulse 1 ns in width gives dispersion on the moment of evaporation of an ion of about 200 ps.

In the laser pulsed mode, the evaporation is triggered by a short temperature pulse with a temporal variation strongly depending on the illumination conditions, and on the material's physical properties. More details can be found in "Chapter 8, Laser-Assisted Field Evaporation." Assuming a temperature pulse with a Gaussian shape of standard deviation σ, we can write similarly to the voltage pulse mechanism:

$$\frac{K_T(t)}{K_{T,\max}} = \exp\left(-\frac{t^2}{2\sigma'^2}\right) \tag{5.8}$$

where σ', the standard deviation, is equal to $\sigma' = \sigma\sqrt{k_B\frac{(T_0+T_P)^2}{\alpha_P C T_P}}$.

T_P is here the temperature rise induced by the laser hit, T_0 being the base temperature. With standard values ($\alpha_P = 0.2$, $C = 1$ eV, $T_P = 200$K, and $T_0 = 80$K), the evaporation pulse width is a fraction of the temperature pulse duration. In contrast to voltage pulse, temperature pulse can be much shorter than the nanosecond in the best case, but this duration can approach several nanoseconds if bad conditions of analysis are used. Note that the real shape of the temperature pulse is not strictly Gaussian, since the shape of the temperature pulse is governed by the heating and cooling mechanisms in the tip geometry, as will be discussed in "Chapter 8, Laser-Assisted Field Evaporation," and is strongly asymmetric. Nevertheless, we use this simplification for an educational purpose.

Considering the sum of errors in time measurements, it gives us an idea of the maximal mass resolution that is achievable in APT as shown in Table 5.1.

Voltage or Kinetic Energy Errors

Assuming that the voltage applied to the sample is well calibrated, errors in voltage (or kinetic energy acquired by ions) are divided into two

Table 5.1 Mass Resolution Achievable if Only Timing Uncertainties Are Considered for Different Configurations of AP

AP Type	LAAP	LAAP	LAAP	LAAP	VPAP	VPAP
Flight length	Short ($t_f \sim 500$ ns)	Long ($t_f = 2$ us)	Short ($t_f \sim 500$ ns)	Long ($t_f \sim 2$ us)	Short ($t_f \sim 500$ ns)	Large ($t_f \sim 2$ us)
Pulse	Short	Short	Large	Large	Standard	Standard
Pulse duration FWHM (ns)	0.50	0.50	5.00	5.00	2.00	2.00
Evaporation pulse duration FWHM (ns)	0.25	0.25	2.50	2.50	0.20	0.20
Electronics errors FWHM (ns)	0.05	0.05	0.05	0.05	0.05	0.05
Total error (ns)	0.26	0.26	2.50	2.50	0.21	0.21
Maximal mass resolution (FWHM)	~1000	~4000	~100	~400	~1200	~5000

AP, atom probe; *FWHM*, full width at half maximum; *LAAP*, laser assisted atom probe; *VPAP*, voltage pulsed atom probe.
Calculations were made for a voltage of about 6 kV and a mass over charge ratio *M* of 28.

contributions, measurement errors and intrinsic dispersion errors. Electronic measurement errors are extremely small, less than 1 V in modern instruments. The main source of voltage dispersion is related to the mechanism of emission of ions from the sample apex. Eq. (5.1) where it is assumed that ions convert all the potential energy "neV" into kinetic energy. If the conversion is not fully achieved, a shift, or drop, in voltage can be observed, and subsequently a spread in kinetic energy "δE" that can be described in terms of voltage dispersion "$ne\delta V$."

Laser Pulsed Mode

In laser mode, the energy spread is generally small [2–4]. In metals, measurements vary between 0.2 and 2 V depending on the pulse fraction, and on the nature of the material, which remains not fully understood. The maximal energy resolution of laser assisted atom probe (LAAP) was measured to about $V/\Delta V \sim 4000$ at half maximum on long flight path configuration. In semiconductors or insulators, much larger spreads exist that are related to the impact of the electrical resistivity of the material.

When a material with a large resistivity (MgO, SiO_2, but also intrinsic Si at low T) is analyzed in AP, the electrical current I_{evap} induced by the ion emission (in the range of the fA) gives rise to an ohmic drop $\Delta V_{ohm} = R \cdot I_{evap}$ between the base and the apex of the sample (Fig. 5.5). R is here the global tip resistance and can be extremely large (10^{16}–10^{18} ohm). A significant drop in voltage (up to several hundreds of volt) was, for instance, measured experimentally in bulk oxides [5–7]. An additional drop, of a few volts, named band bending ΔV_{bb} is added to this ohmic drop, related to the deformation of the band energy diagram close to the tip surface submitted to a high electric field. Because of the intrinsic stochastic nature of the ion current, I_{evap} is not constant and fluctuates significantly. This results in variations of the drop ΔV_{ohm} that can be of several tens of volts in the worst case. These fluctuations can degrade the voltage resolution $V/\Delta V$ to less than 100 at half maximum for worst cases.

Induced energy dispersions may be reduced by decreasing significantly the intrinsic resistivity of the tip. Indeed, it was shown that the laser illumination reduces the value of the resistivity because it generates a high density of free carrier by the photoabsorption process and this carrier excess leads to a reduced voltage drop and ion energy deficit. In the case of silicon the simple exposition to the pulsed laser during the analysis is able to totally cancel the voltage drop. Note that the surface state of the sample plays also a role in the reduction of energy dispersion. A strong amorphization of the tip surface

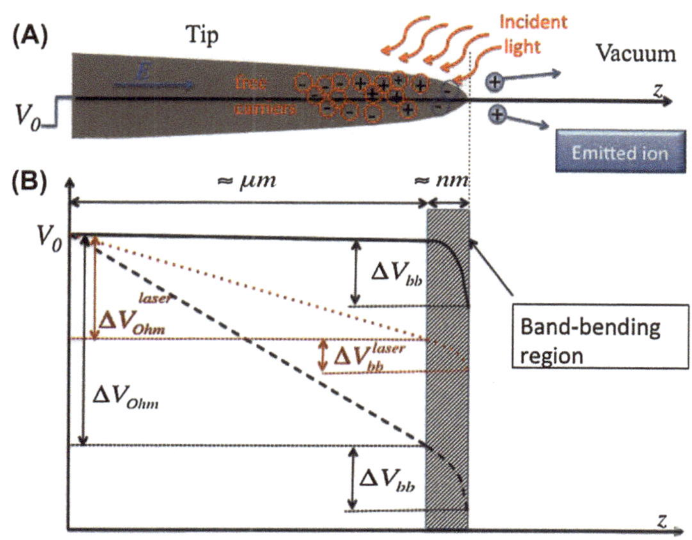

Figure 5.5 Pictorial representation of the voltage drop observed in insulating materials analyzed in atom probe. (A) Carrier generation by ion emission and laser illumination and the corresponding electronic current under the direct current field *E* inside the tip; (B) Schematic illustration of the voltage along the tip axis when the pulsed laser and the ion current are off (*solid black line*), the pulsed laser is off and the current is on (*dashed black line*), and the pulsed laser and current are on [*dotted red line*].

induced by focused ion beam milling or, worse, oxidation of the surface, increases the voltage drop significantly. Proper cleaning of the tip, or the deposition of a metal layer on the tip surface of only a few nanometers on one side of the entire shank, is enough to create a short circuit between the tip base and its apex, and cancel the voltage drop [5,8].

Voltage Pulsed Mode

In voltage pulsed mode, energy spreads are stronger. Indeed, a time-varying voltage pulse is used to trigger the emission of ions. If most of the ions are emitted at the topmost of the voltage pulse, some field evaporated ions leave the sample at slightly different times relative to the pulse maximum (start time), because of the stochastic nature of the evaporation process. This causes these ions to acquire slightly smaller kinetic energies, and therefore slightly slower velocities compared with the peak maximum, resulting in a high mass tail in the mass spectrum. This effect degrades significantly the mass resolution [9].

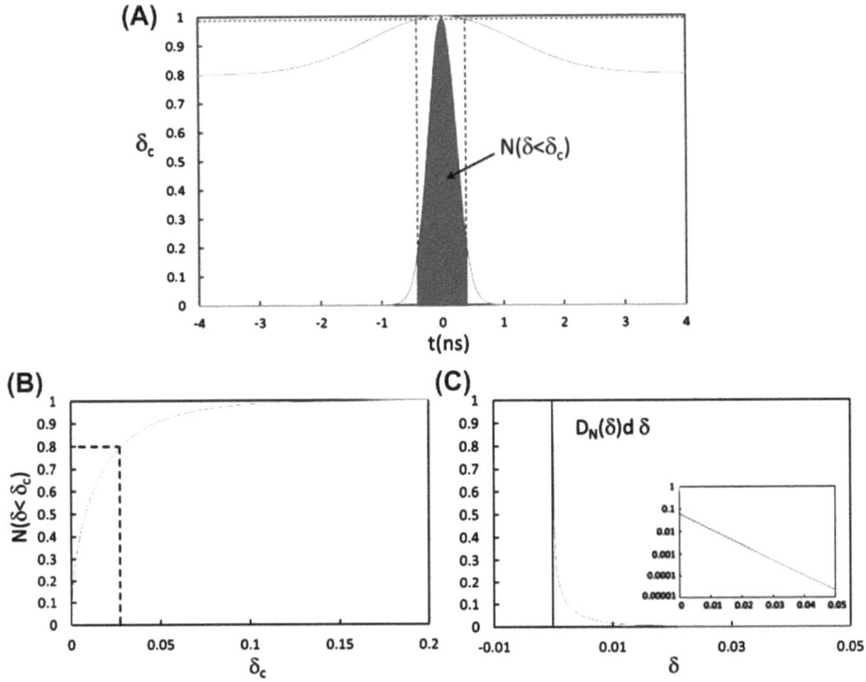

Figure 5.6 Energy dispersion in voltage pulsed atom probe. (A) Evaporation pulse when submitting the tip to $V = V_{DC} + \alpha_P V_T$ (*gray curve*). (B) Cumulated number of ions emitted with an energy deficit smaller than δ_c. (C) Calculated energy spectrum of ions, assuming a binning $d\delta = 1/5000$. Note the presence of a tail in energy, limiting the mass resolution of atom probe. This tail exponentially decreases (inset) with a slope $-C/kT$.

The dispersion in energy can be modeled considering the simple perfect Gaussian voltage pulse previously presented. Let us consider a Gaussian voltage pulse of about 2 ns in width (Fig. 5.6, $\sigma = \sqrt{2\ln(2)}$) of amplitude $V_P = \alpha_P V_{max}$ ($\alpha_P = 0.2$, superimposed to V_{DC}). We take in this example $C = 1$ eV and $T = 80$K. The evaporation rate follows Eq. (5.7). An ion evaporating with a small shift t_c compared to the voltage peak ($t = 0$) evaporates with a small energy deficit δ_c (we assume in this calculation that the ion converts all the potential energy at the time of departure into kinetic energy). If t_c small, we can write

$$\delta_c = \alpha_P \left(\frac{{t_c}^2}{2\sigma^2} \right) \tag{5.9}$$

The number of ions evaporating for $\delta < \delta_c$ is

$$N(\delta_c) \propto \int_0^{t_c} \exp\left(- \alpha_P \frac{C}{kT} \left(\frac{t^2}{2\sigma^2} \right) \right) dt \tag{5.10}$$

This can be simplified to

$$\frac{N(\delta_c)}{N_{total}} \propto \mathrm{erf}\left(\left(\frac{C}{kT} \delta_c \right)^{1/2} \right) \tag{5.11}$$

where N_{total} is the total number of evaporated ions per pulse. The voltage pulse fraction and the voltage pulse duration do not have any influence on the energy distribution (assuming that the pulse fraction is not too small, and the sample temperature not too high). This function was plotted in Fig. 5.6B. The energy deficit density distribution $D_N(\delta)$ is simply the derivative of this last equation, which is written as

$$D_N(\delta) = \exp\left(-\frac{C\delta}{kT} \right) \times \left(\sqrt{\frac{C}{kT \times \pi\delta}} \right) \tag{5.12}$$

This equation has a singular value for $\delta = 0$, which means that a perfect energy deficit spectrum cannot be properly plotted. Using $d\delta = 1/5000$ (about the intrinsic energy spread), and assuming $N(\delta \sim 0) = N(\delta < \delta c)$, it is, however, possible to plot $D_N(\delta)d\delta$ (Fig. 5.6C). At half maximum, this distribution is extremely thin ($\sim d\delta$), but the presence of a tail limits the mass resolution at lower thresholds ($\Delta\delta \sim 0.01$ at 1% of the peak maximum). To reduce this tail, it is possible to reduce the sample temperature, but this can lead to more frequent rupture of the sample, since a higher average field is required to reach the targeted evaporation rate.

This first calculation neglects other degrading effects such as dynamic effects existing in the acceleration region situated very close to the tip. Indeed, ions are accelerated through a time-varying field, which can lead to an uncertainty in their kinetic energy as well as a timing uncertainty. Many investigations of tip-plane electric field distribution in field ion microscopes and AP have been conducted to model these effects [10–15]. To make simple, when an ion leaves the surface, it is rapidly accelerated toward the detector but this acceleration is not instantaneous (Fig. 5.7). The ion is accelerated to a velocity of 10–1000 km/s in a fraction of nanosecond, over a distance of about 100 μm. The tip voltage changes slightly during this acceleration process, so that the ion does not acquire the full expected energy. The ions with the maximum kinetic energy are not the ions

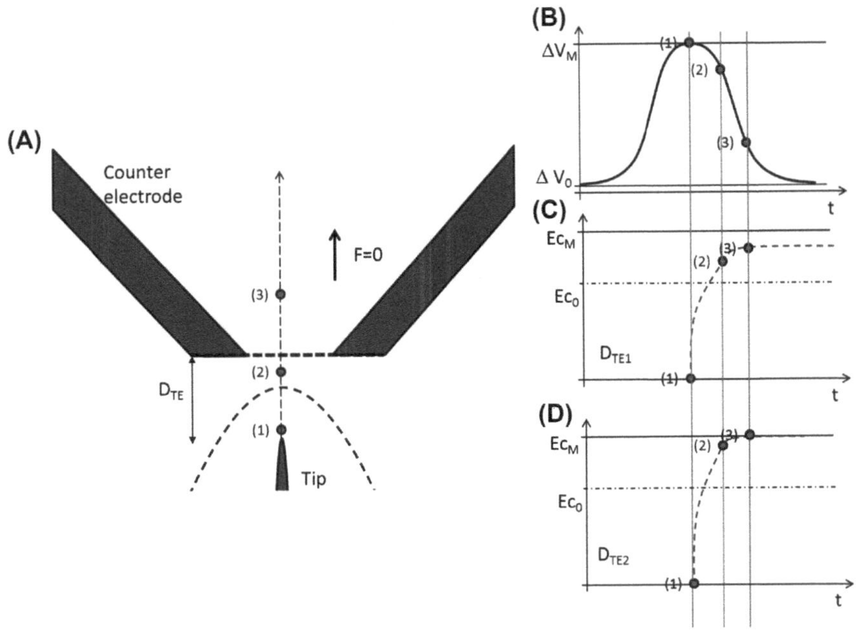

Figure 5.7 Schematic diagram explaining the dynamic effects in the energy dispersion of ions evaporated by voltage pulses. (A) Diagram with the tip *T*, situated at the distance D_{TE} from the counter electrode. The region situated inside the counter electrode is a field-free region. Three positions of the evaporated ion are considered (1) starting position; (2) in flight, in front of the tip; (3) after the entrance of the counter electrode. In (B), the tip is submitted to a varying field through the application of voltage variation $\Delta V(t)$ maximum on the top of the pulse (ΔV_M). With D_{TE} large, as seen in (C), ions are accelerated on a rather long distance, and cannot acquire the full kinetic energy of the maximum voltage since in region (2) the tip voltage significantly drops. In (D), a short D_{TE} distance induces fast acceleration of the ion, so that it acquires most of the kinetic energy very close to the tip, reducing dynamic effects.

evaporated at the topmost of the pulse but the ions evaporated a fraction of nanosecond before this maximum. This effect deforms the energy deficit density distribution and enlarges it significantly. Classical values of energy resolution obtained experimentally are about $V/\Delta V \sim 250$ at half maximum and $V/\Delta V \sim 80$ at 10% maximum.

Improving Voltage Dispersion

The negative impact of dynamic effects can be minimized using the local electrode (LE) designed with a distance between the tip and the entrance of the counter electrode of about 10–40 μm (as compared to several millimeters in classical designs). Using LE, both the accelerating distance and the acceleration duration are reduced by one order of magnitude. The maximum of the kinetic energy distribution thus coincides with the maximum of the voltage pulse, reducing the dynamic effects. Nevertheless, this setup does not reduce the energy spread tail shown in Fig. 5.6. The reduction of the voltage spread induced by dynamic effects that can be achieved with this device gives energy resolutions of about $V/\Delta V \sim 500$ at half maximum and $V/\Delta V \sim 100$ at 10% maximum.

To reduce the impact of energy spreads on mass spectra, modern instruments are generally equipped with an energy compensation device (named reflectron) [16–18]. This electrostatic device uses a static electric field in a defined region of the ion path to reverse the direction of travel of the ions entering it. In its simplest design (Fig. 5.8), a homogeneous electric field $\overrightarrow{F}$ is created between two electrodes placed on the path of ions. Note that to ensure flat potentials all along the ion trajectory, a grid at ground voltage with fine mesh and high transparency (>80%) is placed at the entrance of the device.

Let us consider an ion with an energy deficit δ. It means that the kinetic energy of the ion is $\frac{1}{2}mv^2 = neV(1-\delta)$.The amplitude of the electric field F inside the reflectron is V/P, with P the depth of the device, and V the voltage applied to the back electrode (the entrance electrode is grounded). The flight length in the regions free of field is $L = L_1 + L_2$. The time of flight t_{FF} in this region is given by

$$t_{FF} = L \times \left(\frac{1}{\sqrt{\frac{2ne}{m}V}} \times \frac{1}{\sqrt{1-\delta}} \right) \tag{5.13}$$

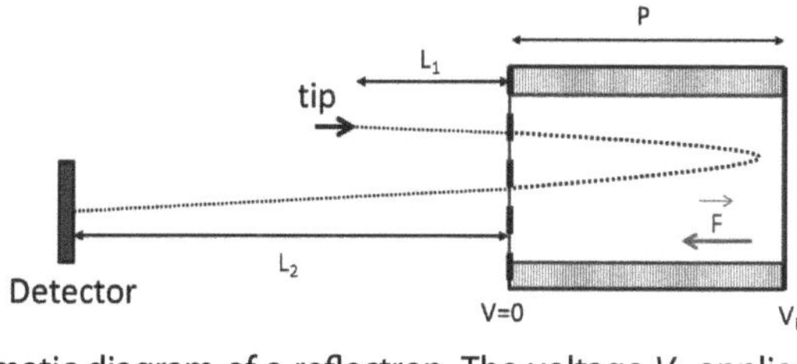

Figure 5.8 Schematic diagram of a reflectron. The voltage V_R applied to the reflectron creates a homogeneous electric field F in the device that reflects trajectories of ions. Ions with energy deficit experience a shorter flight path than the others. On the detector, energy deficits are compensated and all ions of the same m/n ratio are detected with the same times of flight.

In the reflectron, the trajectory is a parabola. Considering small angles of entrance, the time of flight inside the device is

$$t_R = P \times \left(\frac{4}{\sqrt{\frac{2ne}{m} V}} \times \sqrt{1-\delta} \right) \tag{5.14}$$

t_{FF} increases with δ, whereas t_R decreases with δ. Using $L = 4P$, a time focusing point exists where the errors in time of flight compensate themselves at the first order, which means

$$t_f = \frac{L}{\sqrt{\frac{2ne}{m} V}} \left(\sqrt{1-\delta} + \frac{1}{\sqrt{1-\delta}} \right) \approx \frac{2L}{\sqrt{\frac{2ne}{m} V}} \left(1 + \frac{\delta^2}{8} \right)$$

Using this focusing condition, a small deficit δ in energy gives rise to a small increment in time of flight of second order. For instance, an intrinsic spread of energy of 1/100, is observed in the time spectrum in a spread in time smaller than 1/10,000.

In practice the length P of the reflectron is larger than $L/4$, to avoid ion impacts on the back electrode. The voltage V_{Ref} is adjusted through a coefficient C_{Ref} so that $V_{Ref} = C_{Ref} \left(V_{DC} + V_P \right)$.

Planar reflectrons have limited angles of acceptance (about 10°), and suffer strong chromatic aberration effects [17]. But these shortcomings were solved with the invention in 2005 of the curved reflectron (or LAR for large-angle reflectron). In these designs, the device acts as a spherical mirror, producing a compression of the ion trajectories on the detector. A spherical potential distribution is produced inside the reflectron composed of two concentric spherical electrodes. The ground electrode is a spherical grid with high transparency (see Fig. 5.9).

Trajectories of ions in the reflectron are elliptical. Like in a planar reflectron, the time focusing condition can be calculated. As shown in Fig. 5.9, the geometry of the curved electrodes is calculated to minimize also chromatic aberrations onto the detector. Ions with the same entrance angle inside the reflectron but slight differences in kinetic energy are here focused spatially on the detector. Note that generally, time focusing and chromatic aberration compensation exist only on a limited zone onto the detector and variations of the mass resolution can subsist over the detector surface. It is worth noting that the presence of a grid on the path of ions may slightly degrade the spatial resolution of the APT.

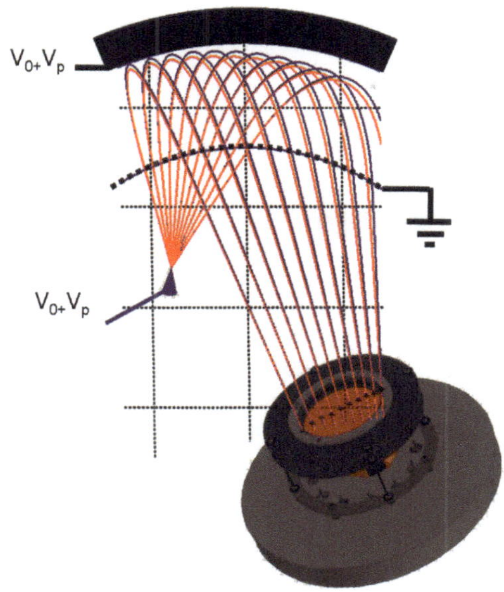

Figure 5.9 Recent generations of instruments (LEAP HR) incorporate a modified version of the reflectron, named large-angle reflectron (patented). The reflectron acts as a spherical mirror for ions entering the device, which produces a compression of the image, a time focusing on the detector, and a chromatic aberration optimization.

Each cell of the mesh of the grid can indeed be considered as a small lens that can deviate slightly the ion trajectories.

Flight Length Errors

The flight length L is defined as the distance between the sample and the impact position of ions. This distance is obviously not a constant over the detector surface. In the simplest case of a perfect planar detector situated at the distance L_0 from the sample, the actual distance that must be inserted in Eq. (5.1) is $L^2 = X^2 + Y^2$. This correction, named Pythagora correction is a first step to obtain good mass spectrometric performances (Fig. 5.10). In practice, the best calibration of the flight length by the Pythagoras theorem is insufficient to achieve an optimal correction. Indeed, the center of the detector is not perfectly aligned with the tip. A small tilt of the detector induces a significant distortion in the Pythagora correction. Moreover, detectors made of MCPs are not perfectly flat, and submillimeter deformations still exist. Furthermore, some APs use electrostatic devices to enlarge the field of view of the instrument (eg, reflectron, electrostatic lens, postacceleration

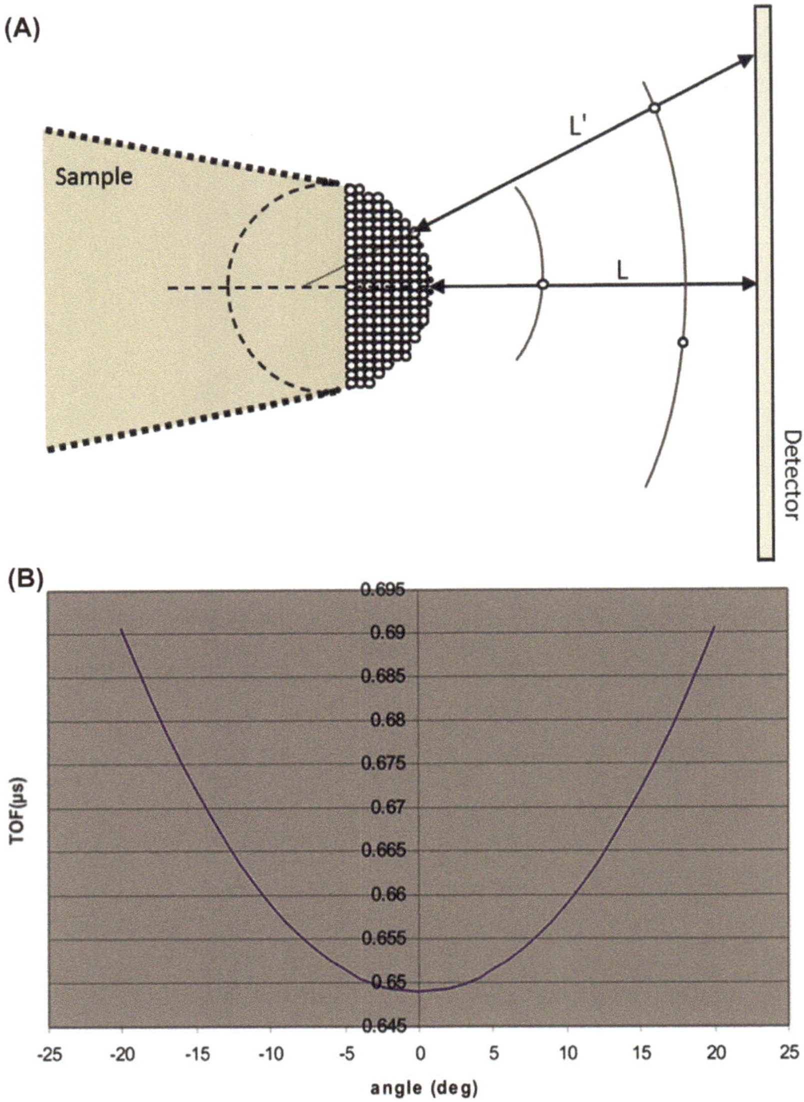

Figure 5.10 (A) The flight distance depends on the initial position at the specimen surface. Time-of-flight isosurfaces are curved. (B) Time-of-flight correction using the Pythagoras theorem.

devices). All these devices change the projection of ions between the tip and the detector.

In practice, correction of flight length variations is calculated with a pragmatic posttreatment method. The detector is divided onto a rigid (X_i, Y_j)

grid, and the mass spectrum is calculated for each cell (defined by its center). On a small cell, the variation of flight length is small, and Eq. (5.1) is valid. However, a defined flight length must be accurately estimated to adjust the mass peak (or several peaks) at the right position. A mass window ΔM is selected on a defined peak that should have an average value of mass M_s. The atoms outside this window are not considered in this first pass. A correction factor to Eq. (5.1) is found to restore the peak of average mass at its correct place M_s so that

$$M(X_i, Y_j) = Ms = C(i,j) \times |M|(i,j) \tag{5.15}$$

The complete list of atoms is then recalibrated using this correction map $C(i,j)$. A smooth variation of $C(X,Y)$ is generally used in this aim using a bilinear or a bicubic interpolation. The actual position (X,Y) of each impact is used to recalculate the mass spectrum. Note that this first correction reduces the apparent mass width of all peaks. To refine the calculation, several iterative steps of optimization, with finer correction map $C(i,j)$, can be applied up to convergence. The optimal mass spectrum is achieved when the improvement of mass resolution reaches a plateau. An example is presented in Fig. 5.11.

This correction is very efficient when the measured position (X,Y) is precise onto the detector. Any dispersion in position induces spread in mass resolution, which depends also on the slope of the correction map at the considered position (X,Y). Note that an intrinsic limitation of mass resolution exists due to MCP geometries (see Fig. 5.12). The time-of-flight measurement is obtained for the primary impact of an ion inside a channel of about 20–40 μm in diameter D_{MCP}. This channel is tilted with an angle $\alpha_{MCP} \sim 20°$ with respect to the normal to the MCP surface. It means that an ion can have its first hit on the secondary electron-emitting surface over a distance $L_{MCP} = D_{MCP}/\tan\alpha_{MCP}$. This distance is typically in a range of 100–200 μm.

Considering flight length dispersion errors, the intrinsic limitation is then about $L/\Delta L \sim 4000$ on a long flight length instrument (~40 cm) and $L/\Delta L \sim 1000$ on a short flight length instrument (~10 cm).

Summary of Performances of Different Atom Probes

The mass resolution using Eq. (5.2) that can be achieved on different instruments is summarized in Table 5.2. If good performances are obtained in LAAP, in metals, with good conditions of illumination, degraded spectra

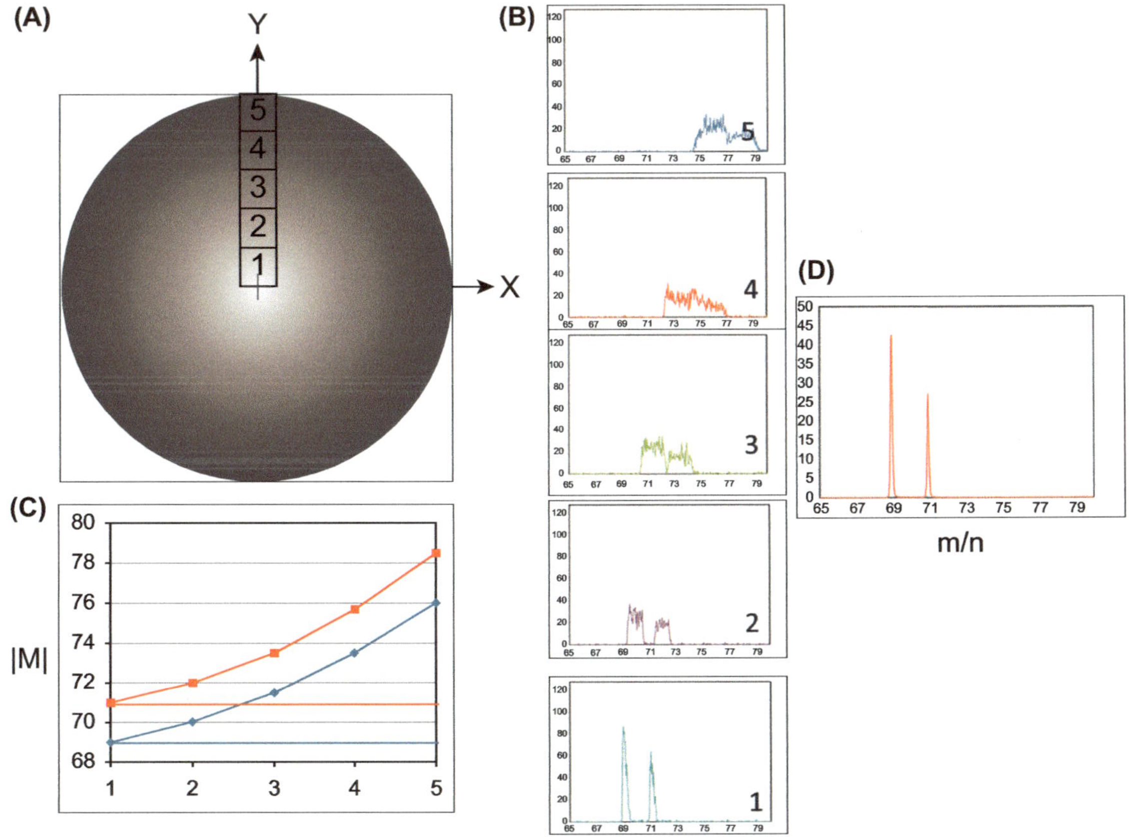

Figure 5.11 (A) Method used to calibrate the flight length as a function of the detector position. (B) The detector is divided into small squares, and the mass spectrum is calculated on each cell. The average position of mass peaks is determined on each cell (C) and gives a map of flight length (or mass correction $C(i,j)$) that must be applied to replace average mass $|M|$ at the correct places (here the first peak is $M = 69$, and the second peak is $M = 71$). A linear interpolation of the correction $C(X,Y)$ is used between the centers of cells. (D) M of all impacts are recalculated using this interpolation.

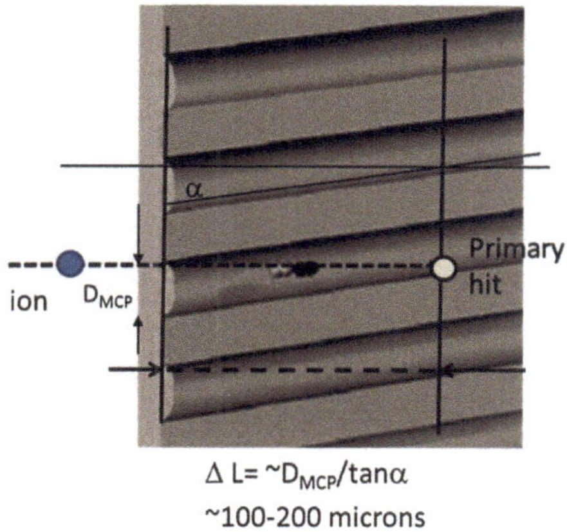

Figure 5.12 Intrinsic length spread induced by microchannel plates. Primary hit on the internal surface of a microchannel plate is dispersed over a distance $\Delta L \sim D_{MCP}/\tan\alpha$, α being the bias angle of channels (8°–20°), and D_{MCP} the diameter of channel (20–40 μm).

may be obtained if energy dispersion exists, or if evaporation pulses are too long. Optimization can be achieved using reflectron instruments that compensate energetic dispersion, and use longer flight lengths. In all cases, mass resolution M/ΔM is limited to about 2000 FWHM due to the sum of errors.

EXTRACTING INFORMATION FROM MASS SPECTRA

Definition of Mass Ranges

Not enough attention is often paid to the definition of mass ranges while this user-defined parameter has some influence on the measured concentration. Fig. 5.13 gives an illustration of a mass spectrum. In (A), an overlap occurs for two types of ions: AlH^{2+} and $^{28}Si^{2+}$. We will see later in this chapter how to deal with such a superimposition for the measurement of concentrations. In (B), dashed lines indicate the uncertainty related to the definition of the range associated to a given peak.

In the paper by Hudson et al. [20], a round-robin experience is mentioned, which was undertaken in the framework of an atom probe workshop. The same file was distributed to 20 users who were asked to define mass ranges. This file, wherein concentrations were known, was generated so as to contain a major element and three minor elements of

Table 5.2 Mass Resolution Achievable Taking Into Account All Errors for Different Analyses With Different AP Instruments

AP Type	LAAP	LAAP	LAAP	LAAP
Flight length (mm)	100	400	100	400
Configuration	Short	Long	Short	Long
Voltage (*V*)	3000	15,000	3000	15,000
Analyzed material	Metal	Metal	Metal	Metal
Timing errors (ns)	0.3	0.3	3	3
Voltage errors (*V*)	1	1	1	1
Flight length errors (mm)	0.1	0.1	0.1	0.1
m/n	30	30	30	30
Tof (ns)	720	1288	720	1288
Maximal mass resolution (FWHM)	750	1900	120	220

AP Type	LAAP	LAAP	LAAP	LAAP	LAAP With Reflectron
Flight length (mm)	100	400	100	400	400
Configuration	Short	Long	Short	Long	Long
Voltage (V)	6000	6000	6000	6000	6000
Analyzed material	Oxide	Oxide	Oxide	Oxide	Oxide
Timing errors (ns)	0.3	0.3	3	3	0.3
Voltage errors (V)	10	10	10	10	<1
Flight length errors (mm)	0.1	0.1	0.1	0.1	0.2[a]
m/n	30	30	30	30	30
Tof (ns)	509	2036	509	2036	2036
Maximal mass resolution (FWHM)	440	600	90	300	1700

AP Type	VPAP	VPAP	VPAP	VPAP With Reflectron
Flight length (mm)	100	100	400	400
Configuration	Short	Short (LE)	Long (LE)	Long
Voltage (V)	6000	6000	6000	6000
Analyzed material	Metal	Metal	Metal	Metal
Timing errors (ns)	0.3	0.3	0.3	0.3
Voltage errors (V)	18	12	12	0
Flight length errors (mm)	0.1	0.1	0.1	0.2[a]
m/n	30	30	30	30
Tof (ns)	509	509	2036	2036
Maximal mass resolution (FWHM)	300	400	500	1700

AP, atom probe; *FWHM*, full width at half maximum; *LAAP*, laser assisted atom probe; *Tof*, time of flight; *VPAP*, voltage pulsed atom probe.
[a]Dispersion in flight length is larger with reflectron due to the presence of a grid on the path of ions.

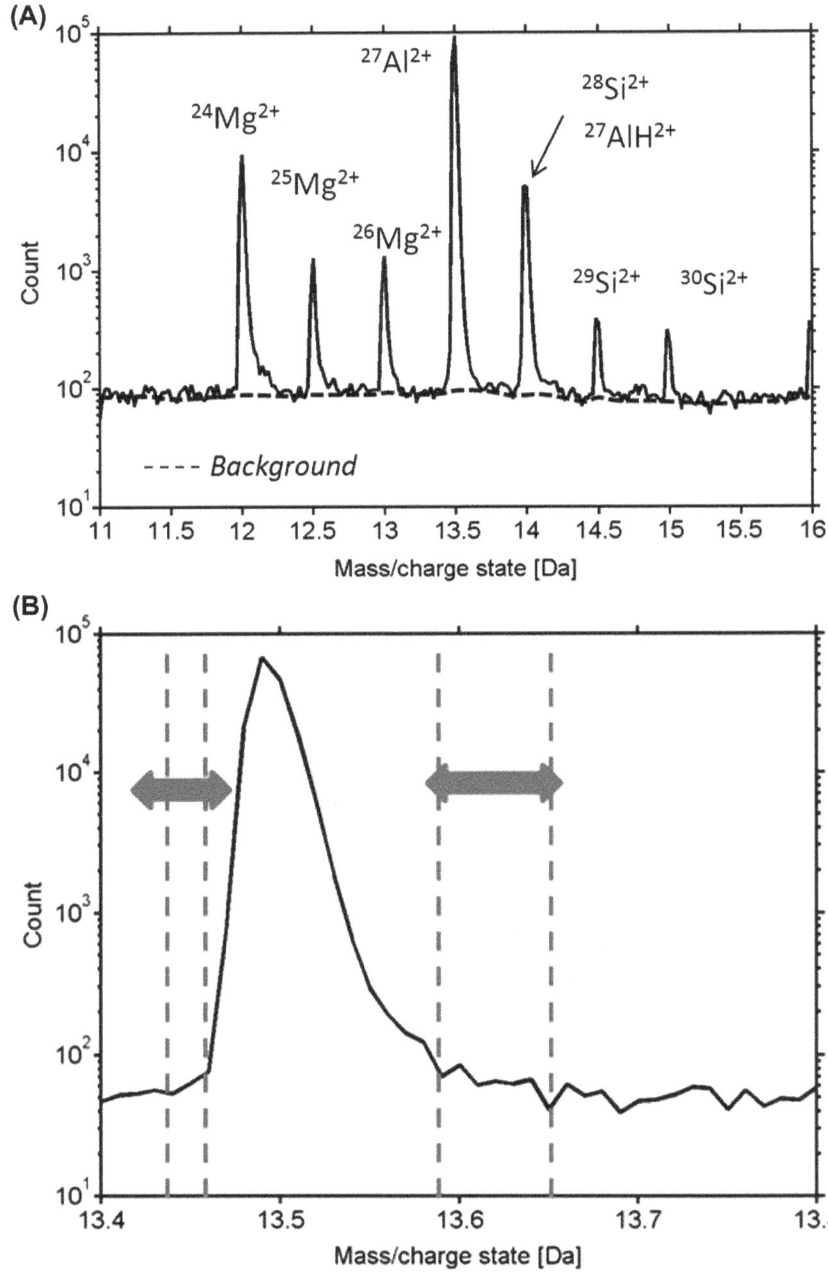

Figure 5.13 (A) Section of a mass spectrum. It was obtained from the analysis of an AlMgSi alloy. The possible superimposition of peaks at 14.0 Da is highlighted. (B) Zoom on the Al^{2+}peak at 13.5 Da. Schematic illustration of the uncertain attribution of lower and upper limits of the Al^{2+} range.

various concentrations. Statistical analysis of the results showed the following:

1. The loss of precision increased as the absolute level of the element decreased;
2. On average, major elements were overestimated;
3. The standard deviation difference between true and measured value ranged from several percentages for the major element to 20% for minor ones.

Hudson et al. [20] hence mentioned these results to highlight the need for well-defined methodologies to define mass ranges. By defining a systematic automated ranging method, they could assess a large reduction in the deviation between true and measured concentrations compared with manual ranging by multiple users on the same simulated data. Nevertheless, automated methods are not widespread in the APT community. It is consequently important to keep in mind that the manual definition of a mass range induces inevitably an uncertainty on the measured composition. In specific cases for which the composition of minor elements need to be determined with accuracy, it becomes relevant to estimate the influence of various mass ranges and to include this effect in the final uncertainty provided with the measured composition.

In another paper, published in 2015 by Haley et al. [21], an algorithm is proposed to automatically enumerate the possible identities of detected peak positions. This method addresses the problem illustrated in Fig. 5.13B. Nevertheless, this case offers the opportunity of a simple resolution, as will be shown later. But for more complex materials, especially organic materials, it is obvious that manual identification of peaks will become extremely difficult and subject to multiple errors. Indeed, the multiple combinations of carbon, hydrogen, nitrogen, and oxygen atoms and number of charges can lead to various possibilities to a given mass over charge ratio. In other fields of research, automated peak identification methods are provided [22], as discussed by Haley et al. [21]. At the time of writing, the work by Haley et al. is the first attempt in the field of APT for providing an automated peak identification and peak detection. This is justified by the fact that APT has, for a long while, been applied to materials of known composition, for which ambiguities in the identification of peaks can very often be bypassed by considering the few possible combinations allowed by the chemical species in the material. Considering the application of APT to the analysis of organic materials and soft matter, development of automated peak detection and labeling will be crucial. For this purpose, the works introduced by Haley et al. [21] will certainly be very helpful for the APT community [22–25].

Background Subtraction

In APT, compositions are directly derived from the mass ranges. All events detected in a given mass range will be identified as the element to which the peak has been attributed. As a consequence, the background level will inevitably contribute to the measured composition. This is what is illustrated in Fig. 5.13A. Although the relative contribution of the background to the peak of the major element will be weak, this will not be the case for minor elements. As long as precise measurements are sought, the contribution of the background to the peak, it is crucial to account for the contribution of noise to the measured compositions. This can be done by applying an automated background subtraction. The influence of the background subtraction procedure on the measured compositions has rarely been considered. In the paper by Hudson et al. [20], very standard methods have been tested such as:

1. Defining an average count of "noise events" per bin in the vicinity of a peak and subtracting the corresponding noise events to the contribution of the peak;
2. Fitting the noise with a decreasing curve A/SQRT(m), with A a constant and *m* the mass over charge ratio.

Such simple procedures have for long been applied by APT users and are extremely simple to reproduce. Nevertheless, they do not allow reproducing the complex variations of the background which are often encountered in LAAP and/or when small peaks are located right behind a major peak. Procedures for background subtraction are available in commercial software. Some are based on the estimated background based on a fit of a Poisson distribution [26], whereas others are based on iterative procedures [27,28]. Applying such procedures is crucial for the estimation of low concentrations. However, it is obvious that depending on the methodology used for background subtraction, small discrepancies might exist between the measured compositions. Hence, the choice of the background subtraction methodology may also contribute to the uncertainty of the measured composition.

Standard Deviation on Compositions

Based on the definition of mass ranges, let us consider that a total amount of N ions have been detected and that n of them are in the range of element A.

The measure proportion (designated as composition or concentration) of element A is hence simply derived by the ratio $C = n/N$. The question we are considering is the following: what is the standard deviation associated with this measured concentration? To answer this, we first have to consider the detection process. This detection process is binary, ie, ions hitting the detector are detected or not, with an average detection efficiency Q. Hence, when N ions are detected, the number of ions which were actually evaporated might range from N to much larger values. The probability function driving the number of evaporated ions M before N are detected is the negative binomial, with parameters (n,Q) [29]. The probability of observing $M = m$ is written as

$$P(M = m) = \frac{(m-1)!}{(m-N)!(N-1)!} Q^N (1-Q)^{m-N} \tag{5.16}$$

The mean of M is N/Q and its variance is $N(1-Q)/Q^2$. Conceptual considerations about the significance of the standard deviation on APT data have been proposed by Danoix, Grancher et al. [29,30]. These authors have considered two cases for which distinct expressions of the standard deviation have to be used. First, when the quantity measured is considered as the (whole) specimen composition, the expression of the standard deviation of the concentration is

$$\sigma = \sqrt{\frac{C(1-C)}{N}} \tag{5.17}$$

Here, the assumption is that the probed volume is representative of the specimen as a whole and that atoms are randomly distributed.

A second expression of the standard deviation will have to be used when the concentration is derived as a subpart of the probed volume. In this case, both the collecting method (eg, cluster analysis) and the detection efficiency influence the measurement process [29]. Then, the expression of the standard deviation is approximated by

$$\sigma = \sqrt{\frac{C(1-C)(1-Q)}{N}} \tag{5.18}$$

where N is the number of ions in the region of interest, inside the whole reconstruction. This latter value has to be used for plotting composition profiles, for instance, with N the number of ions in the sampling box.

Detection Limit—Detectability

The detectability of an element in APT corresponds to the lowest concentration that can be measured experimentally. This quantity is tightly related to the mass resolution, the background level of the mass spectrum, and the total ion count in the analysis. The notion of detectability is seminal and common to many fields of research. It is also inseparable from the "signal to noise ratio" concept. Considering that a peak actually corresponds to an element means that it is assumed that this peak is significantly different from the noise. Such an assumption can be turned into a trustable assertion if standard deviations are considered [31]. Let us consider the following portion of a mass spectrum, wherein a peak can be distinguished just above the background level (Fig. 5.14).

The binary process associated with the detection has been discussed previously. According to this process, the detection of N_p ions might be related to the evaporation of M atoms. The average value of M is N_P/Q and the standard deviation is

$$\sigma_P = \frac{\sqrt{N_P(1-Q)}}{Q} \tag{5.19}$$

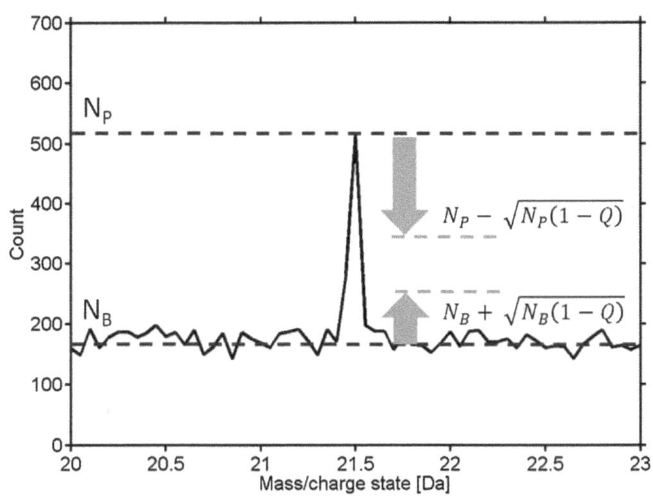

Figure 5.14 Illustration of the criterion of detectability. Portion of a mass spectrum plotted in linear scale. N_P states for the amount of counts per interval at the peak maximum. N_B states for the average counts per interval for the background.

Similarly, the standard deviation associated with the background level is

$$\sigma_B = \frac{\sqrt{N_B(1-Q)}}{Q} \tag{5.20}$$

A criterion of detectability can hence be expressed as

$$N_P - \sqrt{N_P(1-Q)} > N_B + \sqrt{N_B(1-Q)} \tag{5.21}$$

The expression here highlights the effect of the total amount of ions detected. Indeed, for twice the number of ions the respective levels of the peak and background are doubled, whereas the standard deviations grow from a factor 1.41.

Isotopic Correction

In some cases, overlap exists between peaks. For instance, the portion of mass spectrum shown in Fig. 5.13A illustrates an overlap between two peaks, namely $^{27}AlH^{2+}$ and $^{28}Si^{2+}$. Consequently, it is impossible to determine the concentration of Si without any prior correction. Accounting for the natural abundance of Si isotopes, the correction of concentration is straightforward. Assuming that the peaks at 14.5 and 15.0 Da are not subjected to overlaps, the amount of ions detected in these two peaks can be considered as a starting point. From these values, the expected amount of ^{28}Si isotopes in the peak at 14.0 Da can be calculated for the natural abundance values (which are ^{28}Si: 92.22%, ^{29}Si: 4.69%, ^{30}Si: 3.09%). The respective amount of AlH^{2+} contributing to the peak at 14.0 is simply derived from a subtraction.

This seminal example allows presenting the classical method for the correction of overlaps. Nevertheless, many cases exist when favorable conditions are not met and for which other assumptions or additional measurements with another technique are necessary. It must also be mentioned that despite the corrections based on the natural abundance, some inconsistency between results is obtained. Indeed, when a main isotope of a major element is detected, the probability of multiple hits on the detector increases with the concentration of this element. Therefore, its concentration might be underestimated.

REFERENCES

[1] Müller EW, Panitz J, Mc Lane SB. Rev Sci Instrum 1968;39:83–8.
[2] Tsong TT. Phys Rev B 1984;30:4946.
[3] Tsong TT, Kinkus TJ. Phys Rev B 1984;29:529.

[4] Sévelin-Radiguet N, Arnoldi L, Vurpillot F, Normand A, Deconihout B, Vella A. Eur Phys Let 2015;109:37009.
[5] Arnoldi L, Silaeva EP, Vurpillot F, Deconihout B, Cadel E, Blum I, et al. Ultramicroscopy 2015;159:139–46.
[6] Arnoldi L, Silaeva EP, Gaillard A, Vurpillot F, Blum I, Rigutti L. J Appl Phys 2014; 115:203705.
[7] Chen Y, Ohkubo T, Kodzuka M, Morita K, Hono K. Scr Mater 2009;61:693.
[8] Kellogg GL. J Appl Phys 1982;53:6383–6.
[9] Müller EW, Krishnaswamy SV. Rev Sci Instrum 1974;45:1053.
[10] Blavette D. [Ph.D. thesis]. France: University of Rouen; 1986.
[11] Larson DJ, Camus PP, Kelly TF. Appl Surf Sci 1993;67(1):473–80.
[12] Moy CKS, Ranzi G, Petersen TC, Ringer SP. Ultramicroscopy 2011;111(6): 397–404.
[13] Smith R, Walls JM, Phys J. D Appl Phys 1978;11(4):409.
[14] Nishikawa O, Kimoto M. Appl Surf Sci 1994;76:424–30.
[15] Zhao L, Normand A, Delaroche F, Ravelo B, Vurpillot F. Int J Mass Spectrom 2015; 386:47–53.
[16] Sijbrandij SJ, Cerezo A, Godfrey TJ, Smith GDW. Appl Sur- Sci 1996;94/95:428–33.
[17] Bémont E, Bostel A, Bouet M, Da Costa G, Chambreland S, Deconihout B, et al. Ultramicroscopy 2003;95:231–8.
[18] Peter Panayi reflectron. US Patent 8,134,119; March 13, 2012.
[19] Sebastian JT, Hellman OC, Seidman DN. Rev Sci Instrum 2001;72(7):2984–8.
[20] Hudson D, Smith GDW, Gault B. Ultramicroscopy 2011;111:480–6.
[21] Haley D, Choi P, Raabe D. Ultramicroscopy 2015;159:338–45.
[22] Gibb S, Strimmer K. Bioinformatics 2012;28:2270–1.
[23] Pluskal T, Castillo S, Villar-Briones A, Oresic M. BMC Bioinforma 2010;11:395.
[24] Sauer S, Freiwald A, Maier T, Kube M, Reinhardt R, Kostrzewa M, et al. PLoS One 2008;3:e2843.
[25] Karas M, Hillenkamp F. Anal Chem 1988;60:2299–301.
[26] Oltman E, Ulfig R, Larson D. Microsc Microanal 2009;15:256–7.
[27] Morháč M, Kliman J, Matoušek V, Veselský M, Turzo I. Nucl Instrum Methods Phys Res Sect A Accel Spectrom Detect Assoc Equip 1997;401:113–32.
[28] Bostel A. Deconvolution of mass spectra in atom probe tomography – internal report. France: University of Rouen; 2006.
[29] Danoix F, Grancher G, Bostel A, Blavette D. Ultramicroscopy 2007;107:734–8.
[30] Danoix F, Grancher G, Bostel A, Blavette D. Ultramicroscopy 2007;107:739–43.
[31] Bostel A. Detectability in atom probe tomography – internal report. University of Rouen; 2008.

CHAPTER SIX

Atom Probe Tomography: Detector Issues and Technology

G. Da Costa
Groupe de Physique des Matériaux, University and INSA of Rouen, Normandie University, France

INTRODUCTION

Atom probe tomography (APT) is a nanoscale investigation technique that gives a three-dimensional (3D) image of the position and the chemical nature of each atom inside a material [1–5]. The interest of the technique is based on the ability to provide a global and a local view of the content of the material at the atomic scale. The global view of the chemical distribution of atoms is useful to define different phases included in the material for instance, whereas the local view gives the ability to quantify the number of atoms included in each specific phase with a resolution close to angstrom.

Fig. 6.1 shows the principle of APT. A sample is prepared as a short tip with a radius of curvature of about 50 nm, placed in a vacuum chamber at cryogenic temperature. Atoms are field evaporated (in the form of ions) from the specimen surface and their trajectory is intercepted by a position-sensitive detection system placed in front of the specimen. This detector must have high timing measurement abilities. The chemical nature of

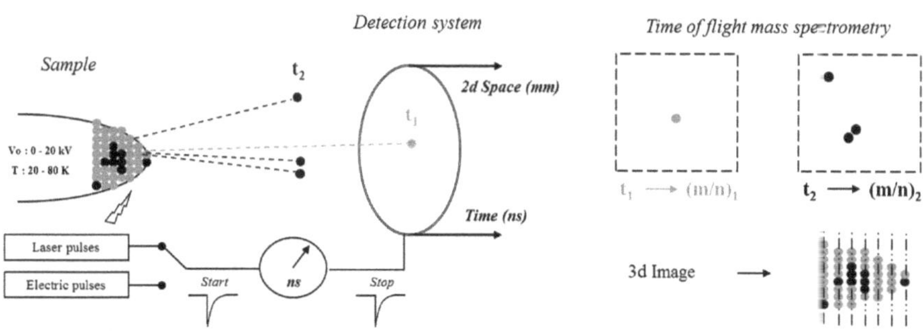

Figure 6.1 Principle of the atom probe technique. Atoms at the surface of the specimen are evaporated by field effect and fly toward a detection system, which determines XY positions and the chemical nature of each atom.

Atom Probe Tomography
ISBN: 9780128046470
http://dx.doi.org/10.1016/B978-0-12-804647-0.00006-1

each detected atom is defined by time-of-flight mass spectrometry and the original 3D position of each atom inside the material is calculated from the position information measured by the detection system and the geometrical properties of APT.

Time-of-flight mass spectrometry consists in using the calculated mass over charge ratio of each detected atom to define its chemical species. The notion of time of flight is very important in APT because it defines the resolution of the technique. During the evaporation process, a direct current (DC) voltage is applied to the specimen to reach a very high electric field that is a little bit lower than the evaporation field of the surface atoms. Nanosecond voltage pulses or subpicosecond laser pulses are superposed to the DC voltage to reach the evaporation field and control the exact moment at which atoms are extracted from the specimen. This moment, considered as the origin of the evaporation and then the origin of the timing measurement, is called the "start time." After the evaporation process, atoms fly toward a detection system placed in front of the specimen that gives the relative arrival timing information, called the "stop time," of each detected atom. The combination of "start time" and "stop time" information gives access to the time of flight and allows one to define the chemical nature of each atom by using the formula: $m/n = 2\mathrm{e} \cdot V_{\mathrm{specimen}} \times t^2/L^2$.

In this formula, the mass over charge ratio is deduced from the total voltage applied to the specimen during the evaporation of an atom, the related time of flight of the atom, and the real distance traveled by the atom between the specimen and the detection system, called the "flight length." As the detection system is position sensitive, the flight length is deduced from the combination of the physical distance specimen–detector and the relative XY position of the atom. The definition of the XY position of the detected atoms is then important for the mass spectrometry but it becomes crucial to calculate the original position of the atoms inside the material. Indeed, APT technique is a projection microscope with a projection law and a magnification value that depends on the flight length. The atomic resolution achieved on each atom position depends on the intrinsic performances of the detection system. As an illustration, Fig. 6.2 shows two images obtained from the same specimen with two different flight lengths. Image A, obtained with a conventional atom probe (flight length ~40 cm), is quite accurate for the definition of local composition but, in a global view, it is only representative of a part of the original image. This is due to the magnification value that is very large. Image B, obtained with a wide-angle atom probe (flight length ~10 cm), is more representative of the original image

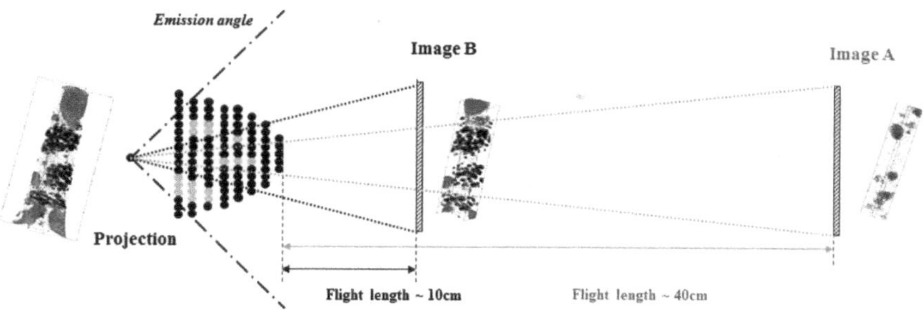

Figure 6.2 Influence of the flight length over the accuracy of atom probe technique in terms of magnification and atomic scale definition.

and still quite accurate for the local analysis of composition. To obtain this quality of image with the same accuracy at any point, detection system B must be much more powerful than detection system A.

Most of the detection systems, used in APT, are based on a coupling of one microchannel plate assembly and a specific two-dimensional (2D) anode-based system [6,7]. The performances of these systems depend on the ability to determine the mass over charge ratio (m/n) by time-of-flight mass spectrometry and the corresponding 2D XY position of each detected atom with the best accuracy. This is the definition of mass resolution and spatial resolution.

MICROCHANNEL PLATE ASSEMBLY

In APT, atoms from the surface of the specimen are evaporated combining DC voltage and electric or laser pulses. The total voltage applied to the specimen during the analysis varies from 1 to 15 kV in the best case. The total energy of the evaporated atoms is then close to some kiloelectron volts in average. The role of the detection system is to define the chemical nature by mass spectrometry and the XY position of all the detected atoms. From ionized atoms with such small energy, it is difficult to extract information without any amplification. A microchannel plate assembly (MCP) [8–11] is then used to provide this amplification (Fig. 6.3).

An MCP is an active charge to electron conversion system that allows converting any particle impact at the input into a cascade of secondary electrons at the output. It is constituted of a thin disc made from a high resistive material with a regular array of tiny tubes, called microchannels, leading across the two faces and densely distributed over the whole surface. The diameter of the microchannels is close to 10 μm and their spacing is about

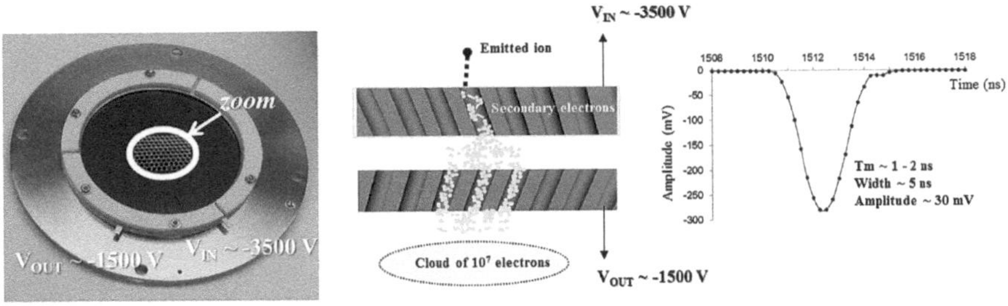

Figure 6.3 Microchannel plate assembly.

15 μm. They are placed parallel to each other and oriented from the input surface by an angle that varies between 8° and 20°. The quantity of secondary electrons is totally insensitive to the chemical nature of the detected particle if the related energy is greater than 1 keV and depends only on the polarization applied between the two faces of the MCP (ranging typically from 0 to 1200 V maximum).

An assembly of two stages of MCP placed in "chevron" is used to transform emitted atoms in electron clouds. The polarization applied between the input face of the first MCP and the output face of the second MCP is used to control the number of created electrons inside the clouds (typically 10^7 electrons in average) for any detected particles. The polarization applied on the output of the MCP assembly is used to control the size of the radial distribution of the electron cloud. When an ion enters a channel at the input of the first MCP, it creates a quantity of secondary electrons. As the channels are all oriented, the same quantity of electrons is created each time a previously created electron touches the border of the channel. A spacer of 100 μm depth is placed between the two stages of MCP to enlarge the size of the electron cloud at the input of the second MCP, and the phenomenon is the same for the second stage. At the end of the detection process, a cascade of electrons is then available at the output of the second MCP.

The advantages of the MCP assembly are the amplification of the particle in an electron cloud, which is used to define the impact position, and the related analogic signal that can be used to extract the timing information for the time-of-flight mass spectrometry. The two main disadvantages of MCP assembly are the background noise, linked to the fact that the system is polarized and then active (less than 1 ghost shot/s/cm^2), and the total efficiency linked to the physical transparency that can reach a maximum of 70–80% in the best case.

CONVENTIONAL DELAY LINE DETECTOR

Position-sensitive detection systems based on a delay line detector (DLD) have been proposed in 1987 [7,12–15]. This system is made of two independent delay line anodes or wires, disorientated 90° and placed at the output of the MCP assembly (Fig. 6.4). The first delay line is used to define the X position and the second is used to define the Y position.

The detection of an atom at the input of the MCP assembly generates an electron cloud at the output. The first part of the electron cloud is collected by the first delay line X. Two short analog signals are then created and travel the X line until each ends. The size of the signals, time width, and amplitude depends on the quantity of collected electrons. The rest of the electron cloud is collected by the second delay line Y and, as the first one, two short analog signals are created and travel the Y line until each end. As the quantity of electrons detected by the delay line Y is twice lower than those detected on the delay line X, the amplitude of the relative analog signals are then lower.

Five timing information are extracted from the analog signals collected on the delay line and the MCP. The lengths of the delay lines X and Y are fixed. So the relative propagation times (respective times required for the analog signal to go from one extremity of the line to the other) of the delay lines are known. The combination of the four timing information collected on each extremity of the delay line and the relative propagation time values gives access to the X and Y positions of the detected atom. The last timing information collected on the MCP gives access to the time of flight of the detected atom and then to its chemical nature. Note that the time of flight can also be calculated from the four timing information collected on the delay line. Even if the MCP timing information is not

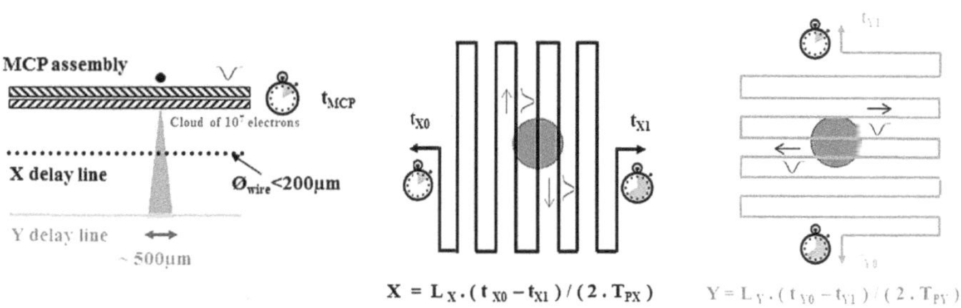

Figure 6.4 Principle of two-dimensional delay line detection system. *MCP*, microchannel plate.

crucial, it gives access to additional information that can be useful when timing information is missed on the delay line. With a detection system based on a DLD, there is a conversion of position (impact on MCP) to time (timing information extracted from analog signals) to position (calculation over the four timing information). So mass resolution and space resolution depend on the ability of the system to extract timing information from an analogic signal with the best precision. This is the definition of timing accuracy.

CONVENTIONAL MEASURE OF TIMING INFORMATION

In APT, a relative measured time linked to a detected atom is in fact a time difference between start time and stop time values. The position of the atom is the combination of four time differences measured on the delay line and the chemical nature is defined by a time difference measured on the MCP. The start time value is extracted from an analog signal linked to the evaporation pulse and the stop time values are extracted from the analog signals collected by the detection system (delay line or MCP). The accuracy of the method used to extract timing information from analog signal is then very important because it defines the accuracy of the time of flight and the position calculation and, by consequence, controls the mass resolution and the spatial resolution of the atom probe. Most of the electronic systems used to extract timing information from the collected analog signals are based on a digital conversion. Fig. 6.5 shows the three combined stages used to make this conversion.

The typical amplitude of the analog signals collected on the detection system is 10–50 mV. This amplitude is high compared to the background noise of the electronic system but it needs to be amplified before any transformation. An amplifier stage is then used to increase the signal over noise ratio.

The second stage is a discriminator or a constant fraction discriminator (CFD). A simple discriminator (Fig. 6.6A) generates a digital signal if the

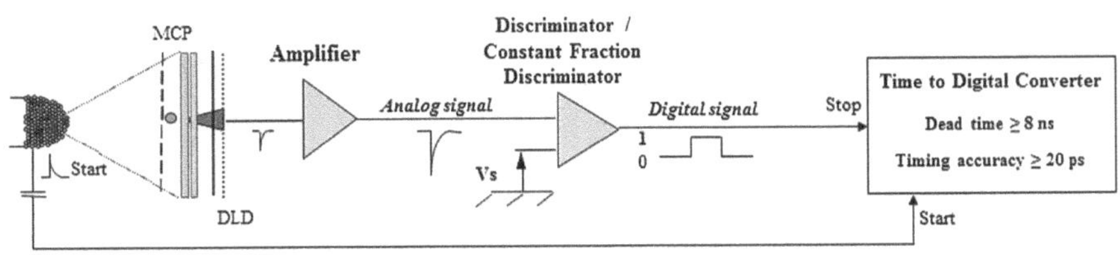

Figure 6.5 Conventional electronic system used to measure time of flight in mass spectrometry. *DLD*, delay line detector; *MCP*, microchannel plate.

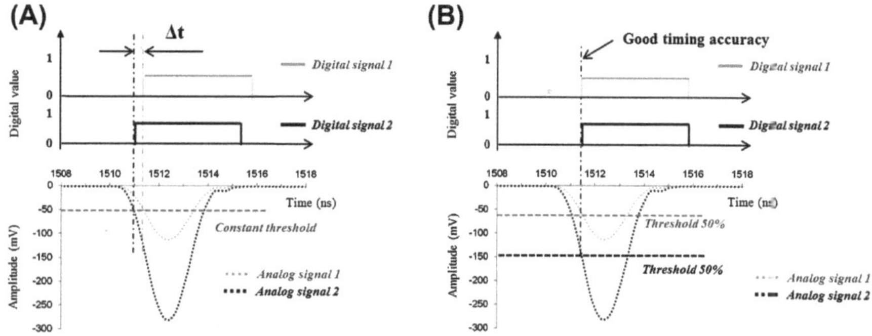

Figure 6.6 (A, B) Influence of the standard discriminator and constant fraction discriminator performances over the timing accuracy.

amplitude of the input analog signal becomes more than a fixed threshold. The accuracy of the timing information extracted from the digital signal is then controlled by the variation of the amplitude of the analog signal. In this case, the amplitude variation induces an uncertainty Δt around the timing information. With a CFD (Fig. 6.6B), the digital signal is generated when the amplitude of the analog signal is higher than a threshold corresponding to a relative fraction of the maximum amplitude of the analog signal. In this case, the uncertainty Δt around the timing information becomes insignificant and the accuracy of the timing information extracted from the digital signal is excellent.

The third stage is a time to digital converter (TDC). Each time an evaporation pulse is applied to the specimen, the relative analog signal is converted into a digital signal and sent to the TDC to start the timing measurement. When an atom is detected, the digital signal created by the CFD from the analog signal collected on the DLD is used to stop the relative input of the TDC. So the TDC gives a time value that corresponds to a timing difference measured between the start and stop digital signals with a timing accuracy that can reach ±10 ps. Note that this accuracy is only that of the TDC and not of the entire detection system.

ACCURACY OF CONVENTIONAL DELAY LINE DETECTOR FOR SINGLE EVENTS

Depending on the configuration of the detection system, four or five TDC inputs are used to collect timing information: two inputs for the delay line, two other inputs for the Y delay line, and the last one for MCP. Fig. 6.7

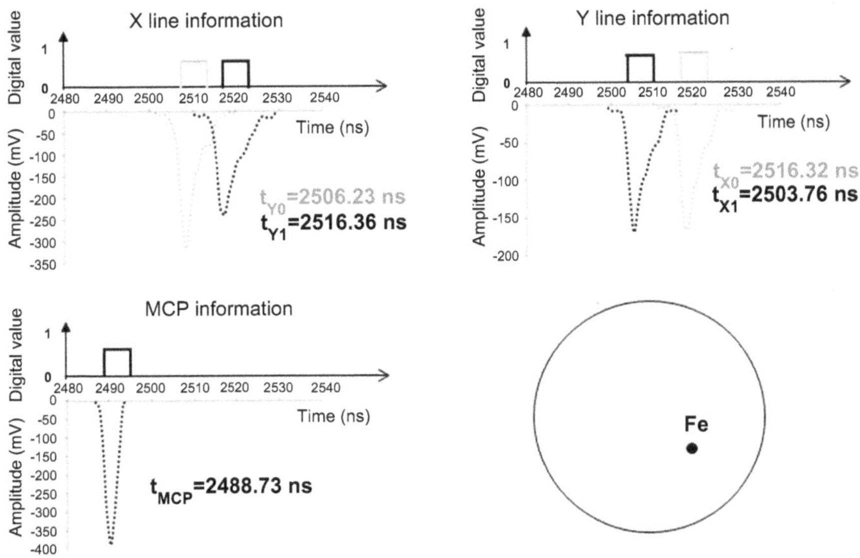

Figure 6.7 Digital information collected by conventional delay line detector. *MCP*, microchannel plate.

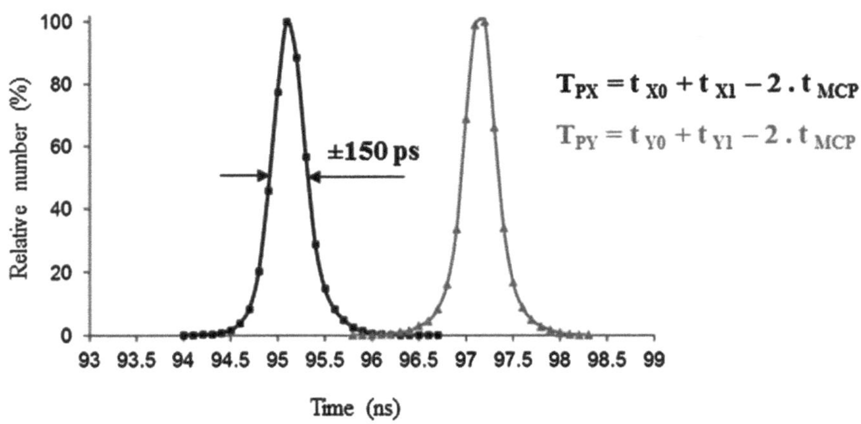

Figure 6.8 Distribution of the propagation time calculated on X and Y delay lines. The calculation is done with the time measured on the extremities of the line and the related microchannel plate time.

shows the collected information when one atom is detected on a conventional DLD system.

Fig. 6.8 shows the distribution of the propagation times calculated on the two XY lines, by taking into account only the single atom detection. For each delay line, the relative propagation time can be calculated from the timing information measured on both ends of the line and the timing information measured on the MCP. The performances of DLD detection system, deduced from the distributions, show a timing accuracy close to 50 ps and a spatial resolution close to 80 μm on the detection system. The spatial resolution of the detection system is then close to 0.02 nm at the scale of the specimen for a flight length of 10 cm that gives a magnification close to $1{,}4.10^6$.

COEVAPORATION EFFECT

Most of the materials studied using APT are composed of different phases with atoms of different chemical natures. Depending on the phase emerging at the specimen surface during the analysis, the electric field set to the specimen, and the local shape of the tip, the probability to evaporate one or more atoms "simultaneously" (ie, with the same pulse) becomes very high. As the atoms are evaporated by combining high voltage or laser pulses, the estimation of this probability is given by the distribution of the number of detected atoms per pulse.

During analysis, a detection flux can be estimated by calculating the number of detected atoms over the total number of pulses applied to the specimen. In routine experiments, the detection flux is close to 0.01 atoms/pulse. Most of the pulses are then considered as null (ie, no atom is detected), and for the pulses that generate a detection event, the distribution of the number of detected atoms per pulse is not a random distribution (Fig. 6.9). The number of multiple events, ie, events for which several atoms are detected on the same pulse, is not negligible even if the detection rate is very low. In some cases, when the material is

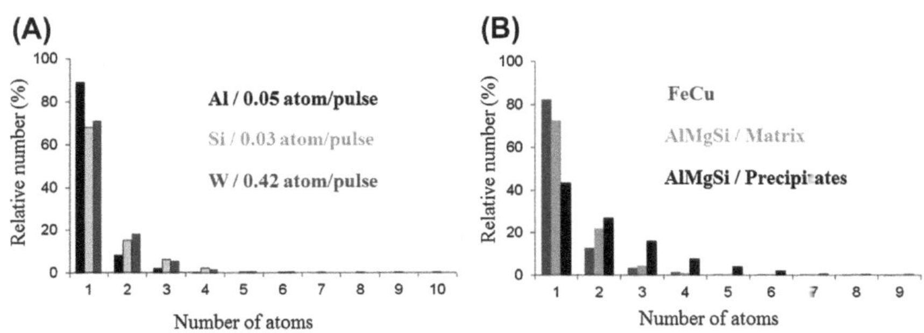

Figure 6.9 Distribution of the number of detected atoms per pulse for pure materials (A) and mixed materials, (B) both for different evaporation flux.

composed of phases with a high difference of evaporation field or phases with nonmetallic bonds, a catastrophic evaporation is observed and the number of multiple events can be significantly larger for one phase compared to others.

To avoid any fluctuation of the detection efficiency linked to this phenomenon, the detection system must be able to discriminate a maximum number of atoms received on the same pulse. And for each discriminated atom, the detection system must define the chemical nature and the position with the best accuracy.

Influence of Coevaporation Over Mass Spectrometry

The chemical nature of atoms is defined by time-of-flight mass spectrometry. When the specimen is composed of different elements with different charge states, the mass spectrum of the encountered mass over charge ratio (m/n) shows different pics (Fig. 6.10).

The detection system needs to deal with serial and simultaneous multiplicities. The serial multiplicity is defined by atoms, detected on the same pulse, with different mass over charge ratio. The simultaneous multiplicity is defined by atoms, detected on the same pulse, with the same mass over charge ratio. The definition of "the same" or "different" mass over charge

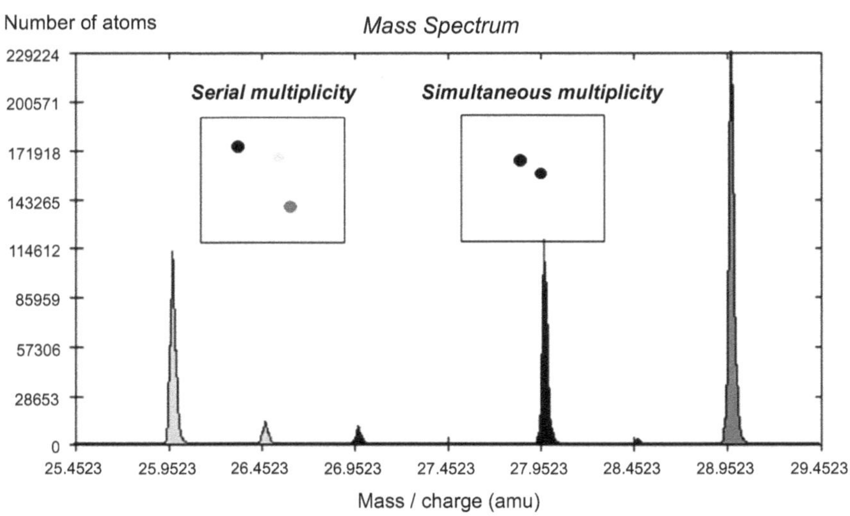

Figure 6.10 Mass spectrum of an Fe-based material. On the same pulse, serial and simultaneous multiplicities, ie, atoms with the same or a different interval of mass over charge ratio, must be detected, identified, and positioned.

ratio is very important because it depends on the mass resolution of the instrument. If the mass resolution is perfect, the shape of the pics will be close to a Dirac distribution. Unfortunately, because of the errors cumulated on the voltage applied to the specimen, the energy of the atoms, the timing and the space accuracies, the shape of the mass pics is close to a Gaussian distribution with a standard deviation equal to the cumulated errors. Consequently, for a defined mass resolution, the limit between serial and simultaneous multiplicities is defined by the ability of the detection system to discriminate atoms with very close interval of mass over charge ratios. This is the definition of mass resolving power (MRP).

In mass spectrometry, the relative *m/n* value depends on the voltage applied to the specimen, the distance between the specimen and the position of the detected atom on the detection system, and the relative measurement of the time of flight. For a fixed distance and a fixed voltage applied to the specimen, the MRP of the detection system will depend on the time resolving power (TRP) and the mass resolution will depend on the timing accuracy. The TRP of the detection system is defined by the ability of the electronic system to discriminate two analog signals with the shorter time difference. This is also the definition of dead time.

Fig. 6.11 shows two cases of a resultant analog signal measured on MCP when two atoms are detected on the same pulse with a very short time difference. In Fig. 6.11A, the time difference between the two signals is close to 5 ns and the amplitude of the second signal is two times lower than that of the first signal. The threshold of the CFD (CFD 50%) is fixed by the maximum amplitude of the analog signal that is the amplitude of the first

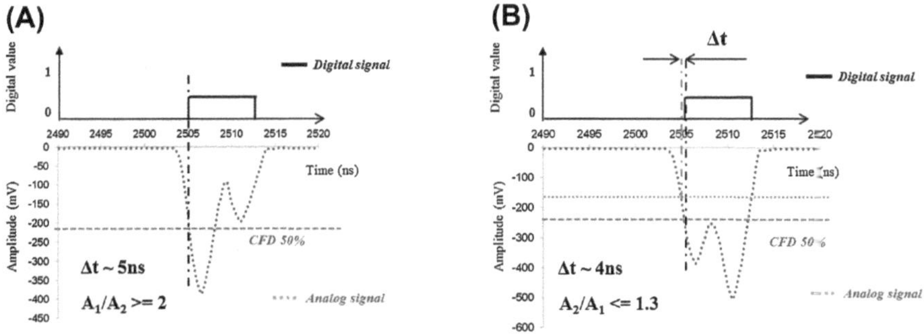

Figure 6.11 (A, B) Time resolving power of the conventional constant fraction discriminator (CFD) system in the case of detection of two atoms. Analog signal shows clearly the presence of two time information instead of digital signal induced by CFD that shows only one time information.

signal in this case. With this threshold value, the CFD creates a single digital signal and then extracts a single time information from the analog signal instead of two time information.

In Fig. 6.11B, the time difference between the two signals is close to 4 ns and the amplitude of the second signal is higher than that of the first. In this case, the threshold is fixed by the amplitude of the second signal and the CFD creates only one digital signal and then extracts only one time information instead of two. Moreover, for this case, the time information becomes worse because of the threshold value that is not representative of the amplitude of the first signal. Note that in the two cases, the second digital signal is not created because the analog signal never crosses back and again the threshold of the CFD. This effect is only due to the width of each analog signal that composes the resulting analog signal. So even if the CFD electronic is able to create very narrow digital signals (~2 ns), it will not be able to create a new digital signal in this configuration and unfortunately, the second timing information will always be lost. That means that there is a time delay during which the system cannot detect a second time information when a first one has been detected. This is the definition of system dead time.

With DLD detection systems, there is a conversion of position (impact on MCP) to time (timing information extracted from analog signals) to position (calculation over the four timing information). So the TRP and then the MRP are linked and directly depend on the dead time of the electronic system used to extract timing information from analog signals.

Fig. 6.12 shows the influence of TRP over MRP as a function of the atom probe geometry. Indeed, the easiest way to enlarge the size of the analyzed volume is to decrease the flight length by a large factor. In this configuration, the time of flight of the detected atoms becomes shorter because of the mass spectrometer definition. To maintain the same performances in terms of ability to separate serial and simultaneous multiplicities, the TRP must be improved by the same factors. In the case of a conventional atom probe (flight length ~50 cm), the detection system will be able to discriminate two atoms with a MRP of 0.2 amu if its TRP or dead time is close to 8 ns. In the case of a wide-angle atom probe (flight length ~10 cm), the TRP or dead time of the detection system must be close to 1.5 ns to keep the same efficiency over the MRP. So there is compromise to find on the atom probe geometry when the detection system is limited in terms of dead time to be sure to keep the same efficiency with the atom probe.

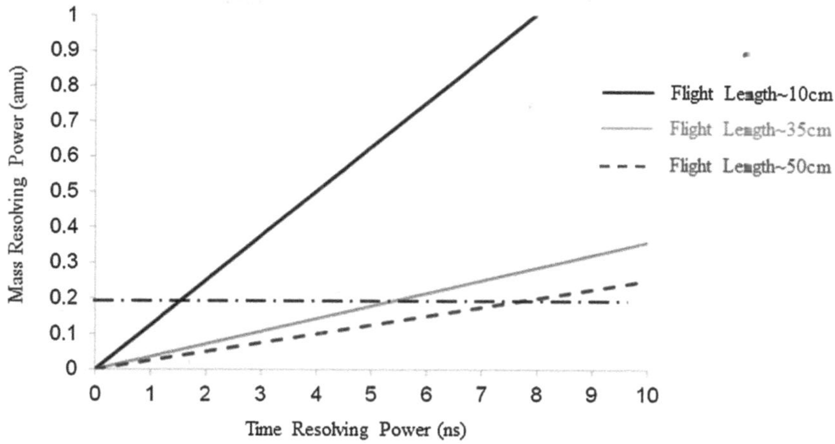

Figure 6.12 Influence of the flight length and the time resolving power of the detection system over the mass resolving power of the atom probe.

Influence of Coevaporation Over the 2D Positioning Space

Simulations have been performed to understand the mechanisms that control the evaporation of atoms and their relative trajectories during an analysis [16,17]. These simulations show that when an atom is evaporated from the surface of the tip, the probability to evaporate its first neighbors is very high. The distribution of the distance separating atoms on the tip during analysis when they are evaporated on the same pulse is shown in Fig. 6.13A.

During the analysis, the distribution of the distance separating each discriminated atom on the detection system, for multiple events, shows the same proximity that varies in function of the atom probe geometry (Fig. 6.13B). This is linked to the fact that atom probe technique is a

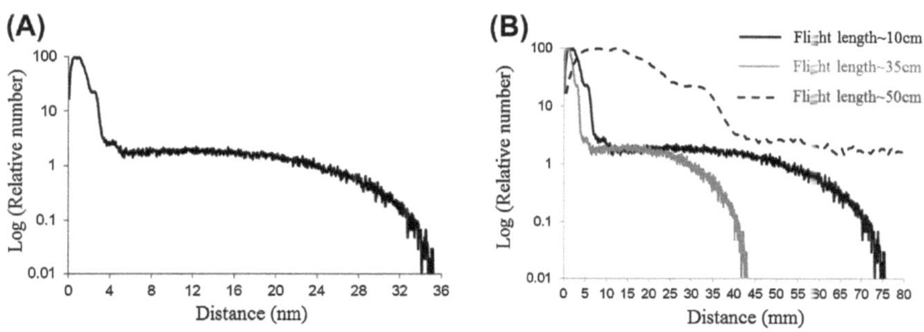

Figure 6.13 (A, B) Distribution of the distance between atoms evaporated on the same pulse on the tip scale and on the detector scale in function of the flight length.

projection microscope and the magnification value depends on the size of the detection system and its distance from the specimen. In any case, the detection system needs to discriminate all the detected atoms without any restriction and assign a 2D space XY position to each atom with the best precision. This is the definition of spatial resolving power (SRP) and spatial resolution.

In a conventional atom probe (flight length ~50 cm/angle of view ~10°), the analyzed surface is low and the magnification is high. So the distribution of the distance between atoms detected on the same pulse shows a coevaporation effect that covers the half size of the detection system and a minimum distance between two atoms close to 4 mm. In this configuration, a detection system with a poor SRP will not affect the performances of the atom probe.

In wide-angle configuration (flight length ~10 cm/angle of view ~44°), the distribution of the distance between atoms detected on the same pulse is the result of the convolution of two distributions. The first distribution represents the coevaporation effect with a higher compression factor due to the lower magnification. The second distribution is linked to the higher analyzed surface that makes possible detection of random multiple events that come from different emission sites. In this configuration, the detection system must have the best SRP to be able to discriminate each atom even if their distances are close to 0 mm.

For a conventional delay line of $80 \times 80\ mm^2$ square section, the propagation times measured on X and Y lines are close to 80 ns and the TRP of the electronic system used to collect the timing information is higher than 4 ns. As this system is based on a position to time to position conversion, the SRP or the dead space of the conventional DLD is estimated to be 4 mm, which is quite high compared to the need in wide-angle configuration.

ACCURACY OF CONVENTIONAL DELAY LINE DETECTOR FOR MULTIPLE EVENTS

In the case of multiple events detection, the limitations linked to the dead time or the TRP of conventional DLD system induce biases over the atom discrimination and some uncertainties over XY positions and times of flight calculated for each discriminated atom. Fig. 6.14A shows these biases as an equivalent dead volume in space and mass over charge ratio in function of the atom probe geometry.

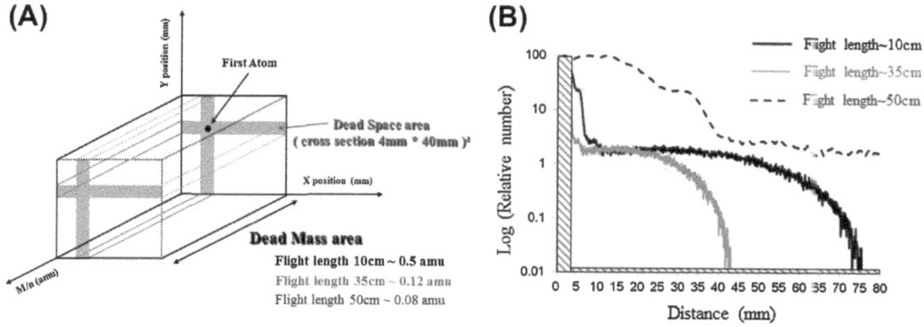

Figure 6.14 (A, B) Relative mass–space dead volume and relative dead space area versus atom probe flight length for the conventional DLD.

To explain the limitations that lead to the existence of a "dead volume," it is necessary to consider a simple case of two atoms detected on the same evaporation pulse. This implies two different XY positions and two different times of flight. If the time difference between the atoms is higher than 4 ns, the DLD system discriminates the two atoms and calculates the time of flight and the XY position of each atom. If the time difference between the atoms is lower than 4 ns, the result of the discrimination depends on the relative XY position of the atoms. If the distance between atoms is lower than 4 mm on the detection system, the DLD system discriminates only one atom and the accuracy of the relative time of flight and XY position is worse. Moreover, the second atom is not discriminated and then lost.

Now, if the position of the second atom is inside a cross-section of 10×80 mm^2 centered on the first atom, the DLD system cannot discriminate the two atoms because the calculation of the XY position of each atom cannot be solved. So in this case, the two atoms are lost, ie, they are not detected. Finally, if the position of the second atom is inside a cross-section of 10×80 mm^2 centered on the first atom, the DLD system discriminates the two atoms and calculates the time of flight and the XY position of each atom. As a function of flight length of the atom probe, this incapacity to detect one or two atoms during a given time interval induces a "dead mass" interval around each first detected atom that can reach 0.5 amu in wide-angle configuration.

Fig. 6.14B shows the result of the bias induced by the DLD system over the 2D positioning space for conventional and wide-angle configuration. The number of lost atoms is high and induces a large variation in the detection efficiency. To avoid this problem, the dead time of the DLD-based system must tend toward 0 ns to be able to extract multiple events composed of atoms with very close m/n values, so times of flight, and very close XY positions.

ADVANCED DELAY LINE DETECTOR

Similar to the conventional DLD detector, the advanced delay line detector (aDLD) [18,19] is physically based on the MCP assembly coupling with two independent delay line anodes disoriented of 90° to define X and Y positions. When an atom is detected, one analog signal is collected on MCP and four analog signals are collected at each extremity of the two XY delay lines. The difference of the aDLD system relies in the electronic system used to extract the timing information from analog signals. Based on fast digitizer modules with high sampling frequency and high bandwidth, this electronic system allows one to reduce the dead time, increase the ability to discriminate two very close signals, and improve the TRP of the detection system.

Fig. 6.15 shows a typical example of what happens during an analysis when a multiple event is measured by the aDLD system. A digitizer with one channel, 4 Gs/s sampling frequency, 1 GHz bandwidth is used to record the analog signal measured on MCP and a digitizer with four channels, 1 Gs/s sampling frequency, 1 GHz bandwidth is used to record the analog signals measured on the delay lines. In this case, the analog signals extracted from delay lines and MCP and recorded by digitizer show clearly that four atoms have been detected.

The access to the real shape of the analog signals in a computer memory gives the possibility to apply data mining on the recorded information. This allows first, increasing the signal over noise ratio of the recorded analog signal with a Fourier space filtering method. Second, it allows applying a specific deconvolution method to extract the accurate number of individual analog signals that compose the recording and filtering signals. Finally, by applying the best timing interpolation method, it enables extraction of the relative timing information of each signal. At the end, the combination of these timing information gives access to the accurate number of atoms and for each atom, the original XY position and the time of flight can be calculated.

The filtering method is based on a bandpass filter that is applied to the analog signal in the Fourier space. During the measurement and the digitalization of the analog signal, different sources of electronic background noise are cumulated and can modify the shape of the analog signal. To filter these noises, a Fourier transform is applied to the analog signal and, as a function of the frequencies linked to this noise, the bandpass filtering value is fixed and applied to the signal in the Fourier space. After the filtering, a reverse Fourier

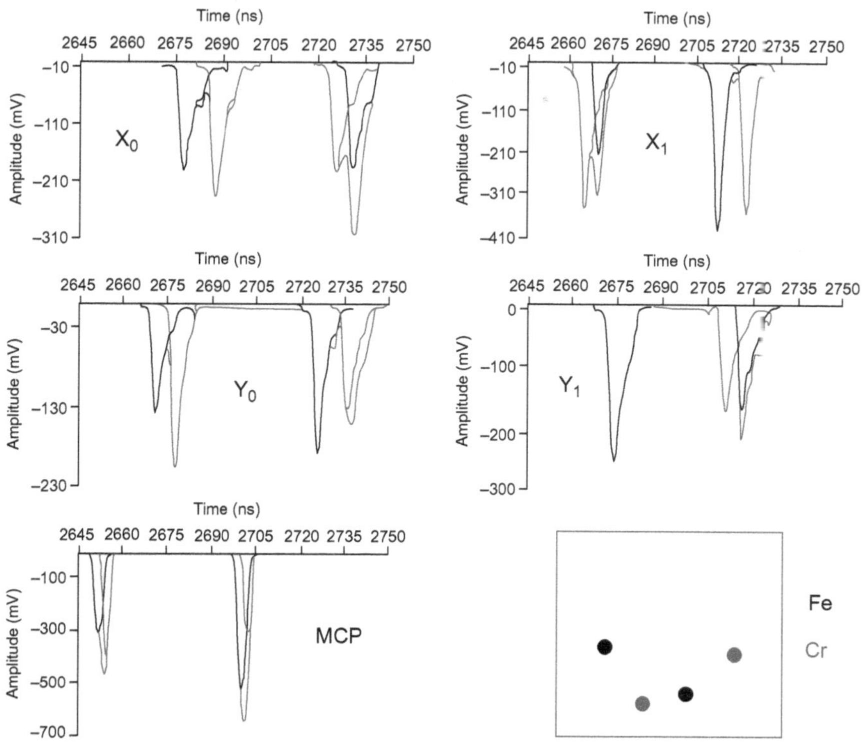

Figure 6.15 Analog signals collected on the delay line and the microchannel plate (MCP) and recorded by advanced delay line detector. The main advantage compared to a conventional delay line detector is the fast recording of the analog signal that gives access to the real shape of the signals in the computer memory.

transform is applied and the resulting filtering analog signal is available in the computer memory for the deconvolution method.

The deconvolution method is based on the recognition of the modal shape of individual analog signals that compose analog signals available in memory. The modals used to define the average shape of each signal for the four ends of delay lines and the MCP are extracted from the measured signals over all the single event detection. The performances of this method only depend on the accuracy and the relative variation of the shape of the calculated modals compared to the measured real signals. Fig. 6.11 shows clearly the efficiency of the deconvolution method applied on XY lines and on the MCP. The dead time or the TRP of the aDLD system is then close to 1.5 ns and by consequence, the SRP is close to 1.5 mm.

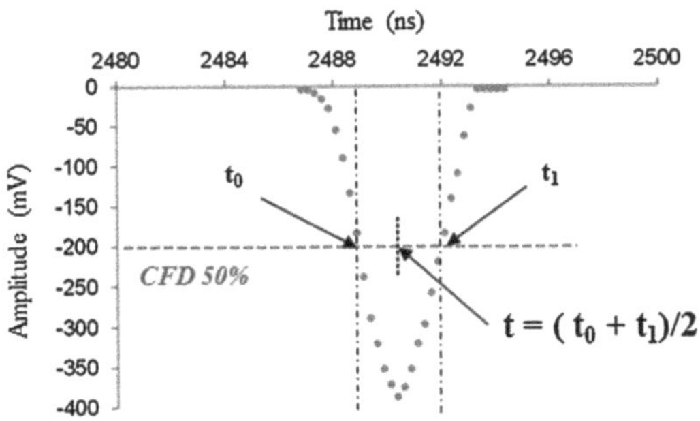

Figure 6.16 Timing interpolation method applied to analog signal recorded by advanced delay line detector.

When an individual analog signal is extracted with the deconvolution method and available in the computer memory, a lot of algorithms can be applied to extract the timing information from the signal. Some methods like constant discriminator, CFD, or new CFD 50% have been tested to see their influence over the timing accuracy. Fig. 6.16 shows the new CFD 50% method based on an average of the two timing information obtained from CFD at half of maximum method applied on the falling edge and the rising edge of the individual signals.

Accuracy of Advanced Delay Line Detector for Single Events

As for the DLD system, there is no problem to extract time information for the single atom detection and then to calculate the relative time of flight and XY position. Fig. 6.17 represents the distribution of the propagation time on

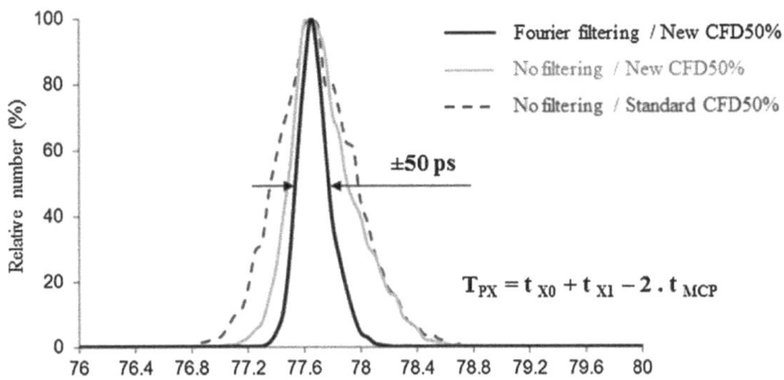

Figure 6.17 Timing accuracy deduced from the measured distribution of advanced delay line detector propagation time. The influences of the filtering method and the time interpolation method are shown.

the X delay line calculated for the single event. This calculation is obtained by the combination of the two time information measured from the analog signal collected on each extremity of the line and the relative time information measured from the analog signal collected on the MCP. The effect of the different methods used to filter and calculate the timing information from the discriminated analog signal over the timing accuracy is shown. The timing accuracy of the aDLD system is then close to 50 ps and the spatial resolution is close to 80 ps.

Accuracy of Advanced Delay Line Detector for Multiple Events

In the case of multiple events detection, the limitations linked to the dead time or the TRP of aDLD system induce biases that can be represented as a dead volume in space and mass over charge ratio in function of the atom probe geometry (Fig. 6.18A). To explain the limitations that drive to this dead volume, it is necessary to take a simple case of two atoms detected on the same evaporation pulse. That means two different XY positions and two different times of flight. If the time difference between the atoms is lower than 1.5 ns and the distance between atoms is lower than 1.5 mm on the detection system, aDLD system discriminates only one atom and the accuracy of the relative time of flight and XY position is not very precise. In all the other cases, aDLD system discriminates the two atoms and calculates the time of flight and the XY position of each atom with the same accuracy than single events.

As for the DLD system, this incapacity to detect one or two atoms during a time interval induces a "dead mass" interval around each first detected atom that varies with the flight length of the atom probe. Anyway, in the

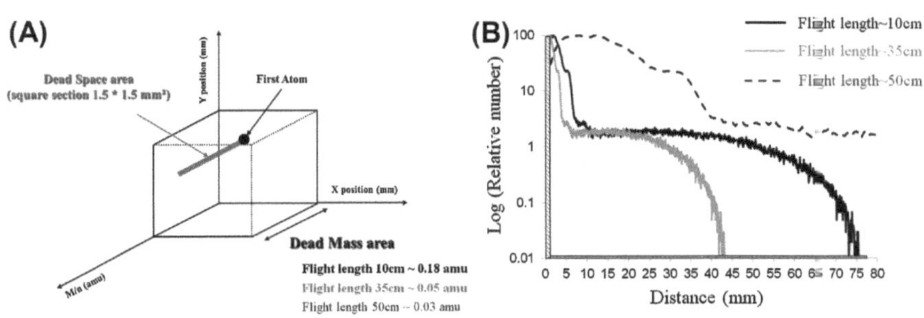

Figure 6.18 (A, B) Relative mass—space dead volume and dead space area of advanced delay line detector versus atom probe flight length.

case of an aDLD system, this interval is lower (0.18 amu in wide-angle configuration $L \sim 10$ cm). Fig. 6.18B shows the result of the bias induced by the aDLD system over the 2D positioning space for different flight length values. The number of lost atoms is low and does not affect the detection efficiency.

INFLUENCE OF DETECTION SYSTEMS PERFORMANCES ON LOW-ANGLE ATOM PROBE ACCURACY

Most of the materials studied in APT are composed of different phases containing atoms of several chemical natures. During analysis, the equivalent evaporation field at the surface of the tip varies with the composition of the encountered phase that can lead to a huge variation of the evaporation sequence in terms of multiple events (huge number of evaporated atoms on the same pulse). In any case, the stress induced by this coevaporation effect must have no impact on the performances (accuracy and efficiency) of the detection system used to collect evaporated atoms.

To study the impact of these mechanisms over the detection systems, an AlMgSi-based alloy composed of two phases, one matrix phase of Al and some precipitation phases composed of a huge quantity of Mg and Si element with some Al element, has been used to build several simulation files. By using the huge number of analyses obtained on this material, several measured distributions and associated mass spectra have been calculated by filtering data for the matrix and precipitate phases. Fig. 6.19 shows the result of the data mining applied under several acquisition data files. The distribution of the number of atoms per pulse, the mass spectrum, and the relative distribution of the distance between atoms on the detection system in the case of multiple events detection have been calculated for the two phases.

Simulations have been performed from these curves to study the influence of the performances of the detection systems over the percentage of lost atoms and the impact of this loss over the composition measurements. By an iterative process, the number of detected atoms per pulse is randomly selected from the distribution of multiple events. For each atom founded in this event, the mass over charge ratio is randomly selected in the relative mass spectrum and the position of each atom is calculated from the relative distribution of the interatomic distances. As the DC voltage is considered constant and the geometry of the atom probe is fixed, the time of flight of each atom is known.

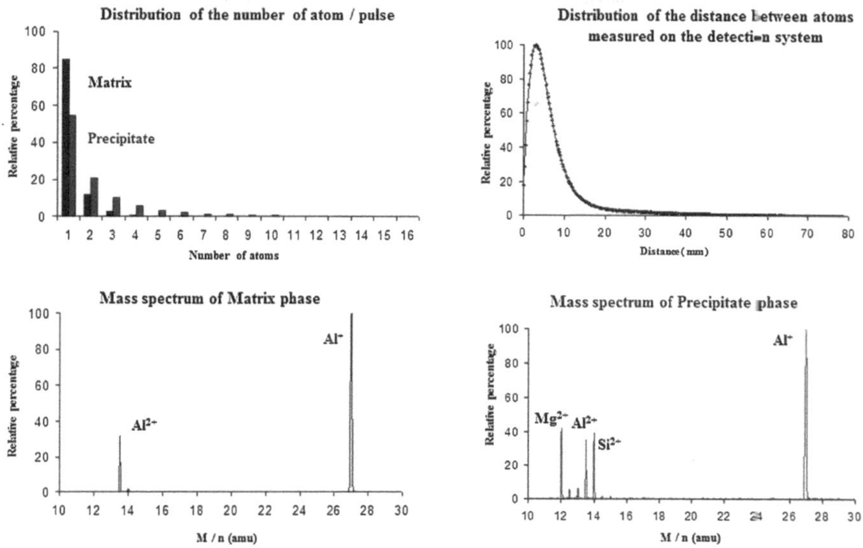

Figure 6.19 Measured distributions of AlMgSi alloy in low-angle AP configuration used to build simulated data files.

At the end of the process, a list of simulated atoms is available for each phase and saved in a simulated data file. This simulation gives access to a sort of "nominal" composition that can be calculated from the atoms for the matrix and precipitate phases. Indeed, as we consider that all the created atoms have been detected, identified, and positioned without any loss, they can be used as a reference to calculate a sort of nominal composition. In this case, the efficiency of the detection system is assumed to be 100%.

The same process has been done with several configurations of detection systems and the influence of their performances over the number of lost atoms and the induced variation of the composition have been reported in Fig. 6.20. Several generations of detection systems, tomographic atom probe, optical tomographic atom probe, conventional delay line, and advanced delay line detectors have been tested.

It is clear that the performances of the detection systems, in terms of TRP and SRP, have a huge influence on the atom probe efficiency and the accuracy of the calculated composition measurement. Moreover, even if the efficiency is poor for some detection system, the critical point is that this efficiency varies in function of the encountered phase that is a real problem for the accuracy of the atom probe technique.

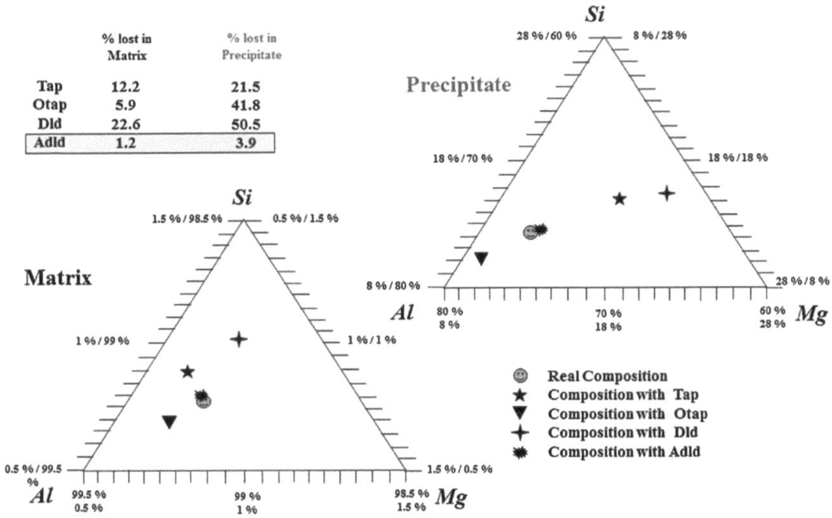

Figure 6.20 Influence of several detection systems over efficiency and the calculated composition in low-angle AP configuration. *aDLD*, advanced delay line detector; *DLD*, delay line detector; *OTAP*, optical tomographic atom probe; *TAP*, tomographic atom probe.

INFLUENCE OF DETECTION SYSTEMS ON ATOM PROBE ACCURACY IN WIDE-ANGLE AP

The same simulation has been done on wide-angle configuration AP (Figs. 6.21 and 6.22). The main difference is the higher number of multiple events that lead to different proportions inside the distribution of the number of atoms per pulse and a change of the shape of the distribution of the interatomic distances measured on the detection system (Fig. 6.21).

As for conventional atom probe, the detection system has a huge influence on the efficiency and the composition measurement accuracy. However, the difference of efficiency observed between matrix and precipitate phases is low. This is linked to the fact that the coevaporation of the atoms induces a more random distribution of interatomic distances. Indeed, in wide-angle configuration, the surface covered by the detection system is higher and the probability to detect many atoms that do not come from the same emission site on the same evaporation pulse become higher than in conventional configuration.

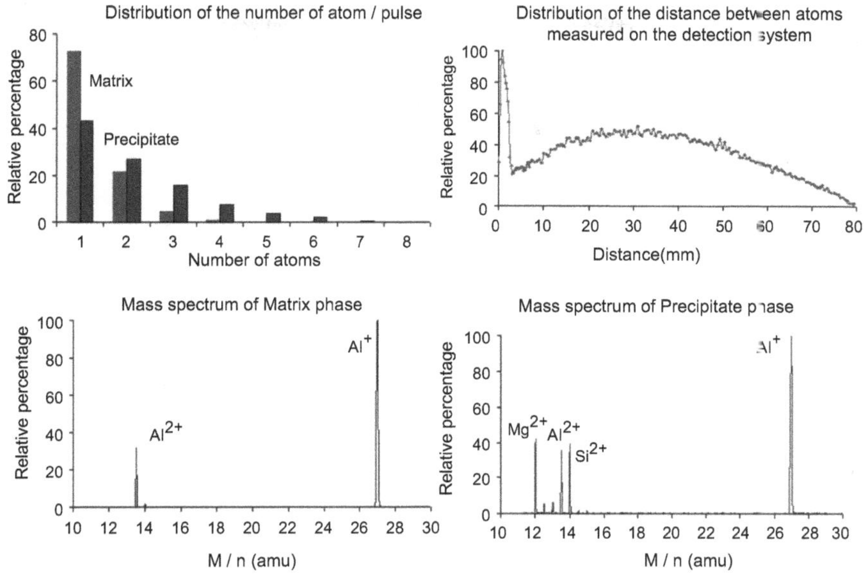

Figure 6.21 Measured distributions of AlMgSi alloy in wide-angle AP configuration used to build simulated data files.

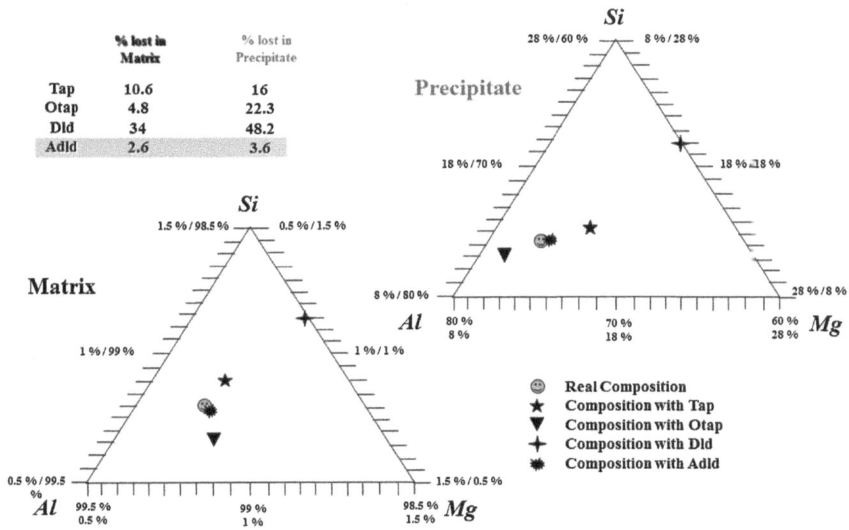

Figure 6.22 Influence of several detection systems over the efficiency and the calculated composition in wide-angle AP configuration. *aDLD*, advanced delay line detector; *DLD*, delay line detector; *OTAP*, optical tomographic atom probe; *TAP*, tomographic atom probe.

EFFECT OF MULTIPLE EVENTS DETECTION ON COMPOSITION

Biases in the composition measurement have been also observed in some materials with enriched crystal defects like highly doped implanted silicon containing enriched boron precipitates [19]. Because of the difference between the evaporation field in the precipitate and the matrix phases and the fact that the links between the different chemical natures are not the same in the two phases, a catastrophic evaporation of boron atoms has been observed when the analysis crosses the precipitate phase (Fig. 6.23A). This effect induces a huge number of multiple events that need to be discriminated, positioned, and calculated in mass by the detection system and those, with the same efficiency than single events.

In the case of this material, the distribution of interatomic distances for boron atoms is wider than those of silicon (Fig. 6.23B). This effect is favorable for the discrimination of the multiple events linked to the boron precipitates. To see the influence of the ability to treat multiple events over the detection efficiency and then the composition measurement, the same analysis has been performed on two different states of SiB material none aged and aged to 1000°C with a specific aDLD detection system that can treat delay lines and MCP information with a maximum limit of 6 times of flight and a maximum limit of 30 times of flight on the same pulse.

Fig. 6.24 shows the depth profiles obtained with the secondary ion mass spectrometry (SIMS) technique and the analysis performed with the two maximum limits and those for the two aging treatments of the material. The profiles show clearly the huge influence of the efficiency over the composition values when the ability of the detection system to detect

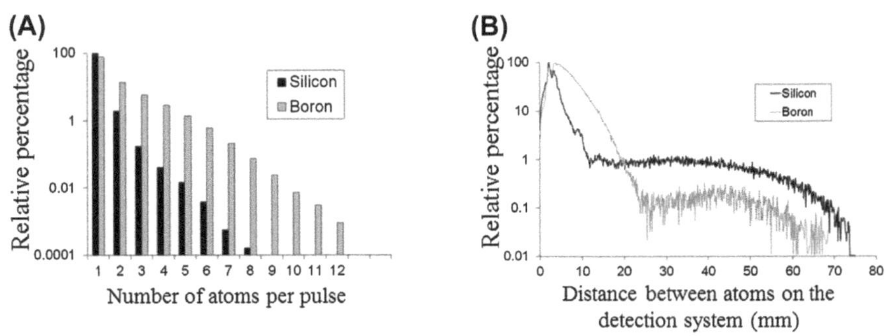

Figure 6.23 (A, B) Distribution of multiple events and distribution of interatomic distances for multiple events detection.

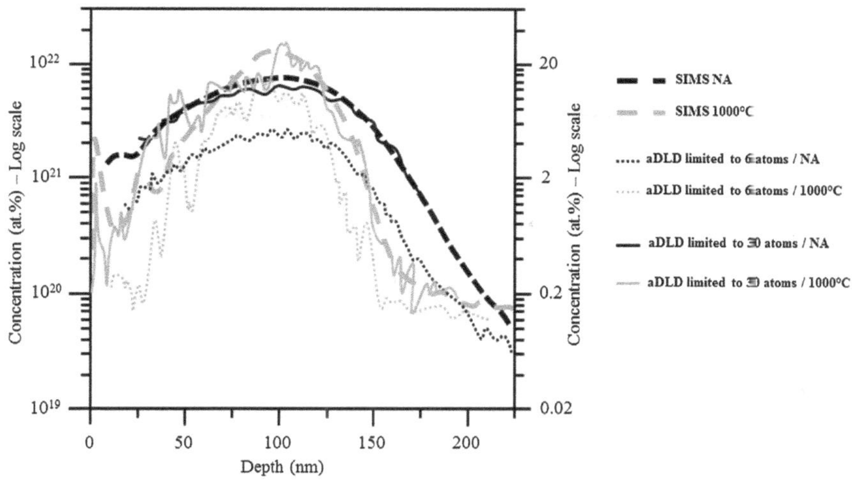

Figure 6.24 Comparison of SIMS and atom probe tomography boron concentration profiles for two aging treatments. *aDLD*, advanced delay line detector.

multiple events is limited. In the case of a maximum of six atoms on the same pulse, the depth profile is lower than it should be and it becomes close to that of SIMS when the limitation attend 30 atoms at the maximum.

CONCLUSION

In atom probe technique, detection systems used to collect the evaporated atoms are based on MCP systems. Because of the technology of MCP, between 20% and 50% of these atoms are lost. For the rest of the atoms, the efficiency depends on the detection system used to discriminate all the atoms collected on the same evaporation pulse and to assign their individual chemical nature and XY position. The MRP, mass resolution, and spatial resolution are the key parameters to obtain the best accuracy concerning the composition measurements.

Concerning the DLD-based detection system, all these parameters depend on the ability of the electronic system used to extract the timing information to deal with events (analog signals in this case) with a time difference that is very short. The impact of this dead time becomes more important when the flight length of the APT geometry is low. Because of the technology, it seems difficult to decrease this dead time to a value that should be lower than 100 ps. It is only linked to the quality of the analog signals collected on the delay line and the MCP and the ability to digitalize or discriminate these signals at the correct bandpass with enough precision. The fact that the dead time is close to 1 ns is a limitation for some APT applications when two different element species are very close in terms of mass over charge ratio. This is the case, for example, for SiN-based materials or Fe-based alloy with a lot of additional elements. Some procedures to extract composition values from the same mass pic for two different species have been developed and give very good results but the information concerning the position of the related atoms from each species in the 3D image is not available. To overcome this limitation, the development of new technology of detection system efficient to the charge state of the energy of the detected atoms would be an alternative. By this way, the multiple events could be discriminated by energy or charge state and the ability to discriminate these elements in the time space would not be the only solution.

Another limitation of the actual detection systems is the ability to detect short clusters of 10 or less atoms with good efficiency. In some materials, it is necessary to study the first stage of atom precipitation and because of the MCP efficiency, even if the density of these short clusters is high, it is quasi-impossible to have any statistical information with the atom probe. An alternative solution would be to develop a new generation of MCP with efficiency close to 100% or find an alternative technology to replace MCP. The problem, in this case, would be to keep the same performances in terms of mass spectrometry and XY space definition.

REFERENCES

[1] Cerezo A, Godfrey TJ, Smith GDW. Rev Sci Instrum 1988;59.
[2] Cerezo A, Godfrey TJ, Smith GDW. J Physic 1989;49:C6–25.
[3] Miller MK. Surf Sci 1991;246:428–33.
[4] Blavette D, Deconihout B, Bostel A, Sarrau JM, Bouet M, Menand A. Rev Sci Instrum 1993;64(10):2911–9.
[5] Kelly TF, Gribb TT, Olson JD, Martens RL, Shepard JD, Wiener SA, et al. Microsc Microanal 2004;10(3):373–83.
[6] Martin C, Jelinnsky P, Lampton M, Malina RF, Anger HO. Rev Sci Instrum 1981;52.
[7] Jagutzki O, Cerezo A, Czasch A, Dorner R, Hattas M, Huang M, et al. IEEE Transact Nucl Sci 2002;49:2477.
[8] Wiza JL. Nucl Instrum Methods 1979;162:587–601.
[9] Eschard G, Manley BW. Acta Electron 1971;14(1):19–39.
[10] Chamelton V, Chevalier P. Acta Electron 1971;14(1):99–106.
[11] Guest AJ. Acta Electron 1971;14(1):79–97.
[12] Morton JM, Parkes W. Acta Electron 1973;16(1):85–100.
[13] Kellog E, Henry P, Murray S, Van Speybroeck L, Bjorkholm P. Rev Sci Instrum 1976; 47.
[14] Keller H, Klingelhfer G, Kankeleit E. Nucl Instrum Methods Phys Res A 1987;258(2): 221.
[15] Kelly TF, Mancini DC, McCarthy JJ, Zreiba NA. Surf Sci 1991;246.

[16] De Geuser F, Gault B, Bostel A, Vurpillot F. Correlated field evaporation as seen by APT. Surf Sci 2007;601(2):536–43.
[17] Renaud L. [Ph.D. thesis]. Université de Rouen; France 2001.
[18] Da Costa G, Vurpillot F, Bostel A, Bouet M, Deconihout B. Rev Sci Inst 2005;75: 013304.
[19] Da Costa G, Wang H, Duguay S, Bostel A, Blavette D, Deconihout B. Rev Sci Inst 2012;83:123709.

CHAPTER SEVEN

Three-Dimensional Reconstruction in Atom Probe Tomography: Basics and Advanced Approaches

F. Vurpillot
Groupe de Physique des Matériaux, University and INSA of Rouen, Normandie University, France

The term tomography, which is borrowed from the ancient Greek τόμος (tómos) and from the French graphie (inherited from the Ancient Greek verb γράφω (gráphō)), means a representation by a superposition of layers or slices. Microscopy techniques that produce tomographic reconstruction are based on this simple principle. Most analytic (or nonanalytic) three-dimensional (3D) microscopes produce at the end a 3D map composed of a pile-up of two-dimensional (2D) maps of intensities that is generally used to deduce the structure, and the organization of the elemental species.

Atom probe tomography (APT) occupies a singular position. Democritus could have been turning in his grave because "tomography" is juxtaposed (representation of slices) to "atom" ("that cannot be sliced"). The information extracted from the analysis is reduced at the end to the discrete positions of atoms (and the elemental nature of these atoms) contained in a small and defined volume of matter. The dataset itself is therefore not a tomogram. However, the atom positions were obtained through a tomographic reconstruction algorithm. Indeed, the atomic positions in 3D are deduced from the in-depth pile-up of virtual layers, each layer having a subatomic thickness. It explains the apparent oxymoron of the name of the technique.

The reconstructed atomic positions are deduced from the ordered list of impact coordinates of ions resulting from the process of field evaporation of the sample. We will show that, in APT the properties of the ion projection are directly determined by the specimen itself. This is not the norm for most microscopies, and as the specimen is progressively analyzed, the field evaporating surface evolves in a complex manner. The consequence of this evolution can be global. Beyond a distance corresponding to a few radii of curvature of the specimen, the ion trajectories are influenced by the

Atom Probe Tomography
ISBN: 9780128046470
http://dx.doi.org/10.1016/B978-0-12-804647-0.00007-3

meso- and microscale parameters, such as the average radius curvature of the specimen, its shank angle, and the overall electrostatic environment of the specimen (ie, presence of a substrate, distance to an extraction electrode, etc.). Deviations of ion trajectories may also arise from atomic scale variations of the specimen surface that may result in distortions in the reconstructed image. Understanding how the image is reconstructed in 3D requires revisiting the imaging process in APT. Concepts of magnification, spatial resolution, or lens distortions must be revisited when dealing with APT datasets.

Despite impressive progress in instrumentation development, specimen preparation methods, data treatment methods, and overall analytical capabilities, the core of the methodology for reconstructing atom probe data relies on early work in field ion microscope (FIM), one-dimensional atom probe, or early designs of the microscope. The classical reconstruction algorithm, as well as modifications of the method will be presented to understand the spatial metrology limitations of the instrument, in terms of spatial accuracy and precision. Starting from the basic hypotheses of the emission process, the general protocol for atom probe reconstruction is detailed.

In the current work, we will first review efforts in modeling the imaging process in APT, discuss the main projection laws underpinning the imaging process, detail the various reconstruction protocols used (and their intrinsic limitations), and finally introduce potential pathways to move forward from the current standard methods.

CLASSICAL METHODS OF TOMOGRAPHIC RECONSTRUCTION

From Atom Probe Field Ion Microscopy to Atom Probe Tomography: A Historical Point of View

The roots of APT are in the early works of Müller on FIM (see Chapter 3: Basics of Field Ion Microscopy for more details) and the work of Müller and Panitz on the development of the atom probe in an FIM [1,2].

Indeed, in field ion microscopy, even if a contrast can exist between two different atoms observed on the imaging screen, no elemental information is directly available. Two concepts of instruments have been invented to overcome this shortcoming, the "probe-hole" atom probe and the imaging atom probe [3–6].

Imaging atom probe records the positions of a given elemental species on a spatially resolved detector, the same size of the FIM screen (Fig. 7.1A). As a

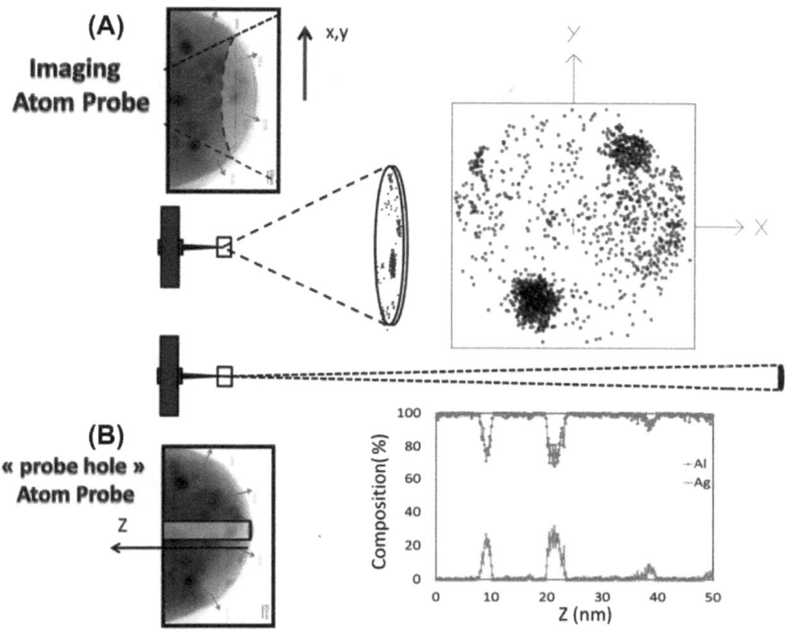

Figure 7.1 Schematic image illustrating the ancestors of atom probe tomography, the imaging atom probe (A), that record the elemental distribution of impacts relative to one elemental species (here Ag in an AlAg alloy) and the "probe-hole" atom probe (B), that probes the depth of the material, and produces a one-dimensional composition profile.

result, a 2D image of the distribution of this species is obtained. The image is the projection of the seen surface of the tip (of the considered elemental species) on the detector. The magnification $\boldsymbol{M}$ was first estimated by the ratio between the detector size and the observed size. A linear projection was used in this aim. Features of nanometer size are easily imaged and scaled with this technique.

With "probe-hole" atom probe (Fig. 7.1B), the tip is gradually evaporated and a small circular region of the tip is probed with a detector sensible to ion hits. No lateral information is obtained but the depth d_z of the core probed was roughly estimated by considering first a fixed value for the small analyzed area S_A and by calculating the volume of the probed core. Indeed, assuming a detection efficiency Q, and considering the atomic density ρ (atom m^{-3}) of the analyzed material, the volume V of the core is $V = N/(Q\rho) = S_A\, d_z$. This results from a simple rule of thumb, deduced from volume conservation laws. A linear relationship exists between the depth and the atom number in the list of detected events. In "probe-hole" atom probe, atomic scale resolution was demonstrated, especially if the analysis is conducted close to a low-index pole. The lateral spatial resolution is controlled by the size of the probe aperture, which is a function of the instrument and by the tip radius [7].

These two former modes of operation of atom probe were in fact combined in tomographic atom probe, as well as the algorithms to scale the lateral position (referred as x,y positions) onto the detector and the depth position (roughly the tip axis and z direction).

Inspection of the Dataset: Ion Impacts and Additional Information

Let us consider the information gathered by the APT user (let us call him an "atom-prober") after the process of evaporating a sample in the instrument. As described previously, the atom-prober has prepared a sample with the required pointed shape aspect, and generally knows from another microscopy technique some details of the geometry of the sample. If the atom-prober has stopped the analysis before any fracture of the sample, an image of the tip can be obtained and can offer additional information about the analyzing process. At the end of the analysis, a data file is generated which is composed of a list of information related to successful events, which means events where at least one impact was detected. This impact is defined by a position stored generally as two floating values for the (X,Y) positions (in millimeters or centimeters), a time-of-flight value (in nanoseconds or microseconds), the voltage to which the ion was submitted at the moment of evaporation (in volts or kilovolts), and the number of blank pulses applied since the preceding successful event. Some other information such as temperature applied to the sample, laser pulse energy, and pressure inside the chamber are also available.

A set of parameters is also related to the instrument itself. Each instrument is defined by a detector with intrinsic limitations in terms of size (diameter D) and detection efficiency (Q). More details about the detector can be found in Chapter 6, Atom Probe Tomography: Detector Issues and Technology. Instruments can be relatively simple, the flat detector situated at a known distance L from the tip apex, the field of view of the instrument being determined by geometrical relationships (straight flight configuration). Some instruments incorporate an ion optic device in the path of ions, to optimize the mass resolving power of the instrument (such as electrostatic lenses, postacceleration devices, or electrostatic mirrors, as described in

Chapter 5: Time of flight mass spectrometry and composition measurements). We will not discuss the consequences of these devices in the reconstruction approach, but we may note that an equivalent set of positions in a straight flight configuration can be extrapolated from the magnification properties of the used device. Fig. 7.2 gives a schematic representation of the geometric configuration of a standard atom probe that we will use in the rest of this chapter.

Zeroth Order of the Tomographic Reconstruction Process in Atom Probe Tomography

We will consider in this chapter an ideal analysis of a sample composed of several phases as shown in Fig. 7.3. The simplest approach of reconstructing the volume from the list of positions recorded in the dataset consists first in evaluating the magnification of the instrument. In the first steps of flight, the ions are accelerated in the field distribution **F**, which is in a first approach mostly radial. If the atoms being field evaporated from the specimen surface

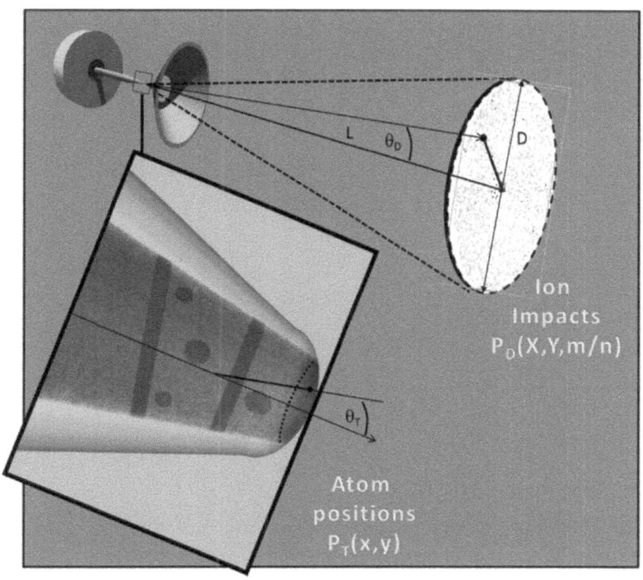

Figure 7.2 Schematic image illustrating the basic projection features in atom probe tomography. The detector situated at the distance L from the sample is described by its diameter D. Impact positions $P_D(X,Y)$ are specified in the spherical coordinate system of the tip by the position by its polar angle θ_D (or inclination angle) and its azimuth angle φ measured from the tip axis direction (z axis). This impact is related to a position $P_T(x,y,z)$ on the tip, that is often specified by its polar angle θ_T and its azimuth angle φ_T.

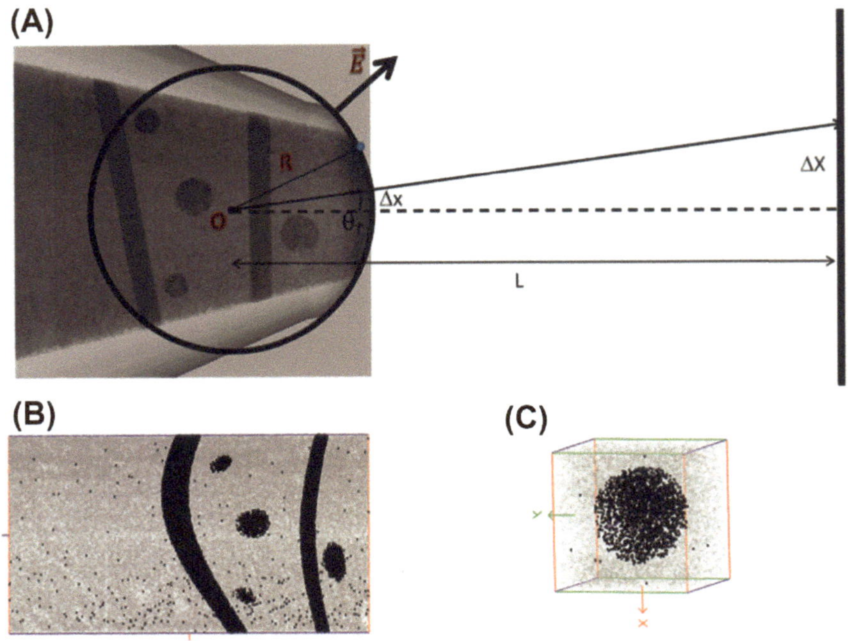

Figure 7.3 (A) Schematic image illustrating the basic projection features in atom probe tomography and a reconstruction that assumes a constant linear magnification and the conservation of the volume evaporated to reconstruct the volume in (B). A zoom on the central particle is shown in (C).

followed perfectly straight trajectories normal to the surface then the magnification $\boldsymbol{M}$ would be given by the ratio of flight path L to tip radius R

$$\boldsymbol{M} = \frac{\Delta X}{\Delta x} = \frac{L}{R} \tag{7.1}$$

However, because the trajectories curve inward slightly toward the detector, the magnification is reduced by an image compression factor (ICF) ξ:

$$\boldsymbol{M} = \frac{L}{\xi R} \tag{7.2}$$

An impact at a position P_D (X, Y) on the detector gives rise to an atom at the position P_T (x,y) in the reconstructed volume with

$$\begin{aligned} x &= \frac{X}{\boldsymbol{M}} = \frac{X}{L} \times \xi R \\ y &= \frac{Y}{\boldsymbol{M}} = \frac{Y}{L} \times \xi R \end{aligned} \tag{7.3}$$

Note that the detector area $S_D = \pi D^2/4$ corresponds to a surface $S_A = [\pi D^2/4]/\boldsymbol{M}^2$ on the tip.

A depth calibration can be found by considering that N atoms, each defined by a volume V_{at}, have been detected, which indicated that a volume $V_{probed} = NV_{at}/Q$ was probed by the detector.

The probed depth is therefore defined by

$$z_{probed} = \frac{NV_{at}}{Q \times \frac{\pi D^2}{4\boldsymbol{M}^2}} \tag{7.4}$$

The atoms can be linearly distributed in the z axis in function of the order of arrival, so that

$$z_i = i\frac{V_{at}}{Q \times \frac{\pi D^2}{4\boldsymbol{M}^2}} = i\frac{V_{at}}{Q \times \frac{\pi D^2}{4}}\left(\frac{L}{\xi R}\right)^2 \tag{7.5}$$

This extremely simple approach (for an educational purpose) will produce the reconstruction shown in Fig. 7.3B. The dimensions of the probed volume are similar to those of the initial volume even if distortions of the objects exist. Note that V_{at} is related to the analyzed material and can be found from the average atomic density of the material. L and D values are geometric features. Q depends on the detector. Which means that only R and ξ must be estimated.

A FEW WORDS ON METROLOGICAL PERFORMANCES AND TERMINOLOGY

The simple reconstruction approach proposed in the early ages of the APT is to be able to produce a 3D image of probed material [8]. Locally the performances of this crude method are good as proven by the reconstruction of the spherical precipitates of Fig. 7.3C. The intrinsic "instrumental" lateral resolution of this method is simply $\delta X/\boldsymbol{M}$, with δX being the detector spatial resolution (Chapter 6: Atom Probe Tomography: Detector Issues and Technology). M is about 1 million, so that a detector with 0.1-mm resolution gives a subnanometer resolution. The intrinsic in-depth resolution is given by the depth increment between two individual atoms. It gives a value smaller than a fraction of a picometer. Nevertheless, it is clear that the reconstructed volume is far from being perfect. We will detail in the rest of the chapter some of the improvements of the reconstruction algorithm that are currently used in practice. To understand the limitations of any method and therefore of the different generation of algorithms that have been developed the last 25 years, it is, however, mandatory to evaluate the metrological performances of the technique.

APT is a 3D microscope. A microscope is defined by metrological parameters. We will focus on spatial parameters (metrological parameters exist also for the composition measurements). These parameters are listed hereafter:

Field of view: "the maximum size of the object than can be observed." Note that for a tomographic instrument there are three different values corresponding to *x*,*y* (lateral directions), and *z* (depth direction).

Spatial resolution: "the minimum feature that can be observed." Classical microscope refers the spatial resolution to the size of the pixel onto the camera. In APT, atom positions are represented by three floating values (*x*,*y*,*z*) for each atom, there is no pixel size. Each position is the result of a single experiment. As a result, the use of spatial resolution for the set of points is awkward; we will refer to spatial precision instead. Note that spatial resolution can, however, be used in composition maps or density maps.

Spatial precision: "this is the spatial dispersion Δp corresponding to the measurement of a position in a single experiment" (see Fig. 7.4). Spatial precision can be estimated by repeating a large number of experiments. The dispersion of the values around the average value (half width for instance) gives an estimate of the precision.

Spatial accuracy: "the spatial accuracy Δ_A corresponding to systematic distortion of the position measurement." Spatial accuracy may affect strongly the precise absolute measurement of large-scale distances of features.

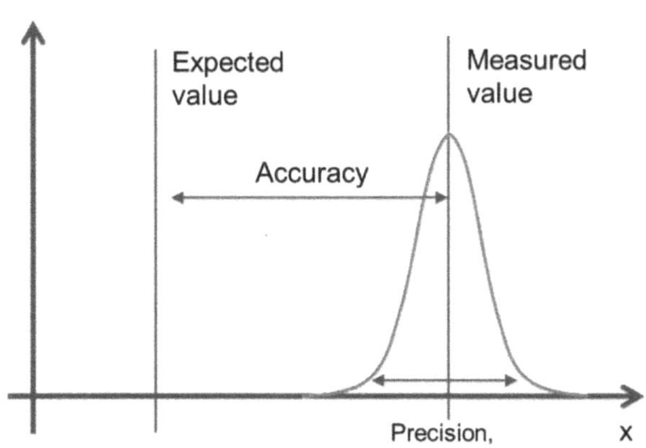

Figure 7.4 Definitions of the metrological parameters used (Accuracy and Precision).

When producing an image composed of a set of points, optimization of the reconstruction is evaluated using accuracy and precision. Accuracy degradation produces image distortion. Optimizing accuracy consists in knowing precisely the transfer function between the actual position (x_{expected}) and the detected position (x_{measured}). This can be done if the transfer function $x_{\text{measured}} = \mathbf{f}(x_{\text{expected}})$ remains bijective (see Fig. 7.5) Several solutions of reconstructed positions are therefore possible and only maximum likelihood methods can help to correct the final image. Spatial precision degradation induces image blurring and can be seen as an enlargement of the function **f**. Note that image restoration methods can be used to improve the image when sufficient information exists in the image (section Spatial Distribution Map, Local Vicinity Histogram or Autocorrelation).

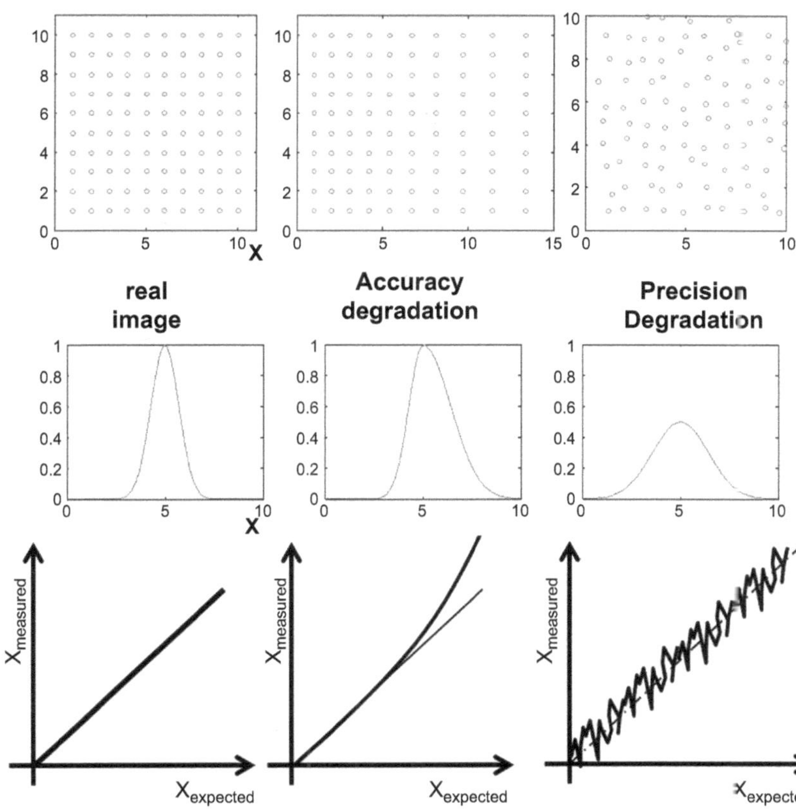

Figure 7.5 Impact of accuracy degradation and precision degradation in a perfect lattice case. Note that image accuracy distortion can be corrected if the image transfer function is bijective and perfectly known.

SPATIAL PRECISION

Experimental Determination of the Spatial Precision in APT Images

Image reconstruction in APT is derived from two interlinked physical processes, ie, field erosion which governs the in-depth inspection, and ion projection giving rise to lateral dimensions. It is therefore natural to have an image sharpness that is different laterally (in x,y) and in depth (in z). It is worth noting that the image is produced by the sample itself, so that each experiment theoretically produces its own image quality, and should be estimated. In practice, estimation can be difficult, and an estimation is only given for a given material (aluminum-based materials, steel, silicon, etc.).

Experimentally, the image precision can be measured statistically using an infinitely sharp feature, referred as a standard, to estimate its average enlargement. It can be a Dirac function of position, the metrological method being to measure the width of the reconstructed feature, or a step-like function, the steepness being used to extrapolate the precision. At the atomic scale, the regular spacing of atomic planes in crystals is often used in this aim. Numerous examples of atomically resolved images, showing well-defined atomic planes along various crystallographic directions may be found in the literature (see Fig. 7.6). Measured enlargement of these atomic planes gives access to a spatial precision estimation in this direction.

Unfortunately, only crystallographic planes approximately perpendicular to the tip surface can be resolved. It indicates that the depth erosion gives rise to a much precise estimation of the z direction than the lateral imaging process. Atomic planes not perpendicular to the normal to the tip surface appear blurred due to the superposition of the depth and lateral precision degradations [9–14].

Experimentally lattice plane enlargement can be measured by two methods that average the signal of presence of the atomic planes.

Fourier Transform Calculation of the Atomic Distribution

The Fourier transform is defined in three dimensions by the well-known expression:

$$F(\mathbf{R}) = \iiint f(\mathbf{r}) \cdot e^{i2\pi\mathbf{R}\cdot\mathbf{r}} dxdydz \tag{7.6}$$

where $\mathbf{r}(x,y,z)$ is the position in the real space, $\mathbf{R}(Rx,Ry,Rz)$ is the position in the Fourier or reciprocal space, and f is the structure function related to the atomic positions in the considered volume. This volume consists of a list of N atoms defined by their elemental identities and their positions $r_i(x_i,y_i,z_i)$. The structure function can be expressed as the sum of Dirac peaks centered on the measured atom positions

$$f(\mathbf{r}) = \sum_{i=1}^{N} \delta(\mathbf{r} - \mathbf{r}_i) \tag{7.7}$$

As a result, the Fourier transform writes simply as

$$F(\mathbf{R}) = \sum_{i=1}^{N} e^{2\pi \mathbf{r}_i \mathbf{R}} \tag{7.8}$$

And the intensity of the Fourier transform reads as

$$I(\mathbf{R}) = F(\mathbf{R}) \times F(\mathbf{R})^*, \tag{7.9}$$

where the sum is applied over every atomic position within the reconstructed volume. This straightforward method enables us to compute the complex value of the Fourier transform in any position of reciprocal space, and can be used regardless of the 3D spatial resolution in real space, the size of the real volume, or the nature of elemental species. Any zone of the reciprocal space can be explored. The main drawback of this method is the computation time.

The result of the Fourier transform calculation is a 3D field of complex numbers in the reciprocal space. The power spectrum (or intensity) was used to image the 3D diffraction pattern. The positions of diffraction spots may be observed using this method. When atomic planes are reconstructed, the Fourier diagram show the presence of a spot placed at the fundamental positions $R = 1/d_{hkl}$ where d_{hkl} is the interplanar distance. A perfect arrangement of atoms (infinitely thin sheets of atoms) would lead to all the harmonic spots regularly spaced in the z axis. Fig. 7.6C shows the Fourier amplitude along the plane direction in a silicon sample analyzed along the <111> direction. Diffractions peaks are clearly observed but the intensity of these peaks decreases with $\mathbf{R}$. Let us assume that the reconstructed positions of atoms are perturbed randomly and that errors follow a Gaussian distribution $G(z)$. The regular sequence of plane is therefore convoluted by

$$G(z) = e^{-\frac{z^2}{2\sigma^2}}, \tag{7.10}$$

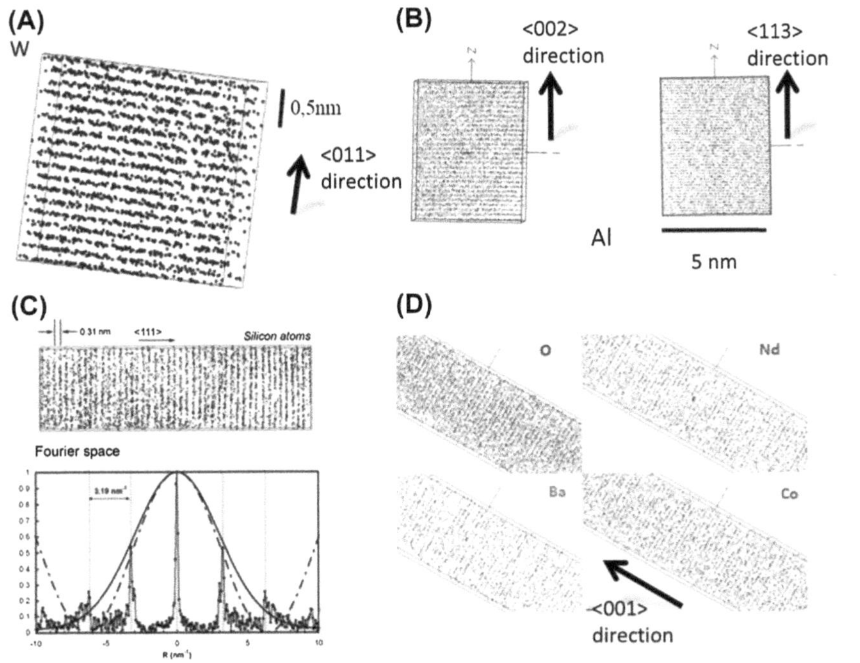

Figure 7.6 Different materials in which atomic planes along different crystallographic directions are revealed: (A) in tungsten, (B) in aluminum (two different directions), (C) in silicon along the <111> direction of the diamond structure, and the calculated Fourier transform in the z axis of the signal (D) in an ordered double perovskite NbBaCoO oxide structure.

where σ is the standard deviation of G, and the width at half maximum of the atomic distribution around planes is given by $\Delta = 2\sqrt{2\ln(2)}\sigma$.

The structure function may be expressed as

$$f(\mathbf{r}) = \sum_{i=1}^{N} \delta(\mathbf{r} - \mathbf{r}_i) \otimes G(z) \tag{7.11}$$

Using Plancherel theorem, the amplitude of $F(\mathbf{R})$ is given by

$$\begin{aligned} |F(\mathbf{R})| &= \left| TF\left(\sum_{i=1}^{N} e^{2\pi \mathbf{r}_i \mathbf{R}} \right) \right| \times |TF(G(z))| \\ &= \left| TF\left(\sum_{i=1}^{N} e^{2\pi \mathbf{r}_i \mathbf{R}} \right) \right| \times e^{-\pi^2 R z^2 \times 2\sigma^2} \end{aligned} \tag{7.12}$$

The amplitude of diffraction peaks is shown to decrease with Rz, and it is therefore possible to extract an estimated value of σ or Δ. In Fig. 7.6C, the spatial resolution along atomic planes direction is better than 1 Å.

The same method was used in an aluminum sample analyzed at 40K along different orientations tilted with respect to the (002) plane by an angle θ varying between 0° and 20° (Fig. 7.7). The width of the plane perpendicular to (022) was observed to increase from ~0.5 Å to 1.8 Å. Above this value, the planes are not resolved anymore, and it is impossible to measure the resolution. Indeed, the width at half maximum approaches the 2.04-Å interplanar distance.

This enlargement is the result of the superposition of two different contributions, the lateral precision ΔL ($\Delta L^2 = \Delta x^2 + \Delta y^2$ at the tip surface), and the depth precision Δp. The depth precision is obtained from the resolution measured at $\theta = 0°$, giving a value of $\Delta p = 0.55$ Å. The resolution of planes having an angle θ with the tip surface can be approximated to Δ with Δ

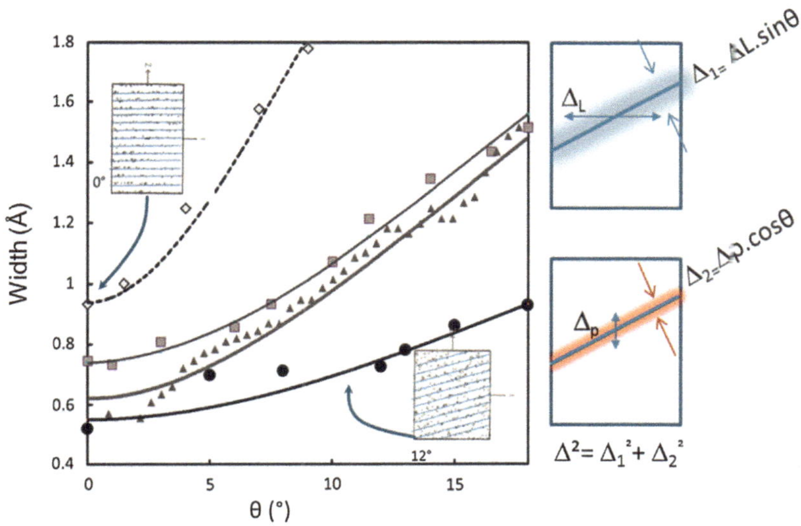

Figure 7.7 Variation of the width of reconstructed atomic planes as a function of the angle between the depth axis and the plane direction <*hkl*> for Al analyzed at 40K (*triangle*, (002) planes); Al, 80K (*square*, (002) planes); Al, 140K (*diamond*, (002) planes); and for W (20K, (011) planes). The enlargement is induced by a quadratic sum of the depth precision and the lateral precision projected on the plane direction. Curves represent the plane widths expected for Δp and ΔL. Note the impact of temperature discussed in section Origins of the Depth Precision in the Reconstructed APT Images (Al 40K, $\Delta p = 0.6$ Å and $\Delta L = 4.5$ Å; Al 80K, $\Delta p = 0.75$ Å and $\Delta L = 5$ Å; Al 140K, $\Delta p = 0.95$ Å and $\Delta L = 10$ Å; W 20K, $\Delta p = 0.5$ Å and $\Delta L = 2.5$ Å).

being the quadratic sum of $\Delta L \sin\theta$ and $\Delta p \cos\theta$, the projections of the lateral precision and depth precision on the direction perpendicular to considered planes.

$$\Delta^2 = (\Delta L \sin\theta)^2 + (\Delta p \cos\theta)^2 \tag{7.13}$$

Using this approach the lateral precision in this experiment (Al sample) is estimated between 4 and 6 Å.

Spatial Distribution Map, Local Vicinity Histogram, or Autocorrelation

The same kind of measurement can be obtained in the real space using methods such as spatial distribution map, local vicinity histogram, or autocorrelation calculation. In these three methods developed concurrently, the average environment of atoms is calculated to observe the presence of systematic arrangement of atoms [7–9,11].

Let us return to the definition of the autocorrelation function $g(\mathbf{r})$.

$$g(\mathbf{r}') = \iiint f(\mathbf{r}) \times f(\mathbf{r} - \mathbf{r}')dxdydz \tag{7.14}$$

Using the structure function (Eq. (7.11)), g can be expressed simply as

$$g(\mathbf{r}') = \sum_{i=1}^{N}\sum_{j=1}^{N} \delta(\mathbf{r}' - (\mathbf{r}_j - \mathbf{r}_i)) \tag{7.15}$$

This function is different from zero for discrete value of $\mathbf{r}'$. Thus, the value of this function at a position $\mathbf{r}'$ is obtained by determining the number of atoms situated at the end of the vector $\mathbf{r}'$ centered on the position of each atom. Numerically, $g(\mathbf{r}')$ is determined by counting the number of atoms contained within a small volume $dxdydz$ situated at the end of the vector $\mathbf{r}'$ for each position of each atom. A rigid and regular grid is then superimposed to the real space. Each cell of this grid corresponds to a vector $\mathbf{r}'$. The result is a 3D histogram composed of discrete numbers for each value of $\mathbf{r}'$. It can also be calculated by computing the inverse Fourier transform of $|F(\mathbf{R})|$ (Eq. (7.11)).

An example is presented in Fig. 7.8, calculated on the volume of W atoms of Fig. 7.6A. This volume presents a different kind of atomic planes with different noncoplanar directions, indicating a very good lateral precision. This excellent precision is confirmed by the presence of spots in the autocorrelation function, proving sufficient precision to image, in average,

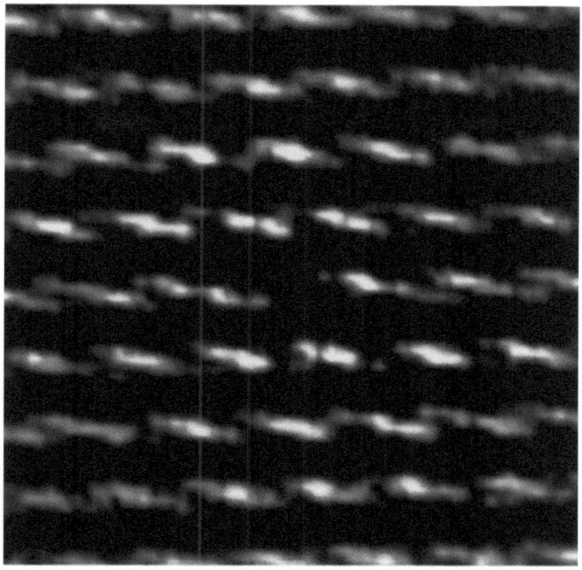

Figure 7.8 Three-dimensional autocorrelation calculation (or spatial distribution map) calculated for the tungsten volume of Fig. 7.6A. Note the presence of defined crystallographic directions and positions blurred by a three-dimensional convolution function, about 0.6 Å in depth, and 2 Å laterally.

the crystal lattice of *W*. The width of these spots in depth is about 0.6 Å. The width of these spots laterally is ~2 Å. The lateral precision is there high enough to resolve the lattice in all the directions.

Pure metals are observed to produce best depth precisions (values as small as 30 pm obtained experimentally). High-index order plane structures can be imaged in aluminum. The huge number of the types of planes that can be observed can be highlighted by calculating the Fourier transform perpendicularly to the tip surface (a discrete map of directions). When regular evaporation of atomic planes exists, a spot of high intensity is present in the Fourier space (Fig. 7.9). On the analyzed surface of the aluminum sample, about 250 types of atomic planes give a signal in the Fourier space in this case.

It is interesting to note that this kind of material enables the use of image rectification methods [12,15,16]. The average position of lattice sites can indeed be estimated in the 3D image. Atoms may be moved iteratively to the closest lattice position to obtain a better image. This method was applied to the tungsten case. Fig. 7.10 shows that the BCC (body centered cubic) lattice structure in the <011> zone axis and in the <111> zone axis is

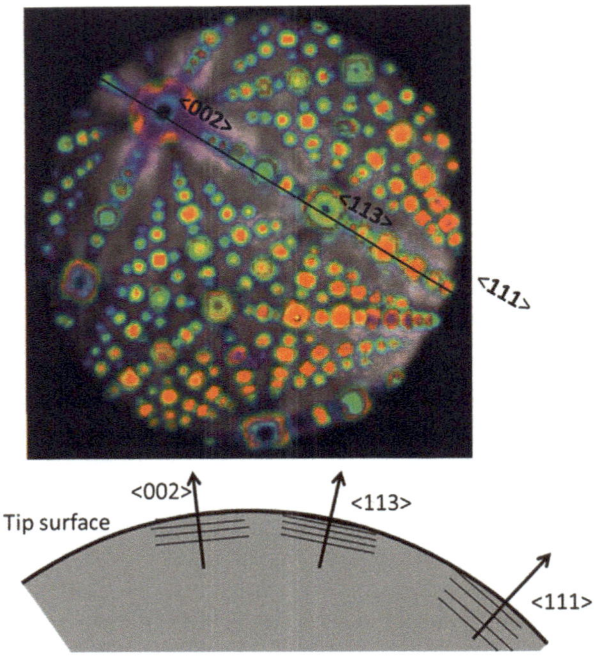

Figure 7.9 On the detector, the presence of regular sequence of planes can be detected using a Fourier algorithm. In Al a huge number of visible atomic planes can be revealed by this method onto the surface.

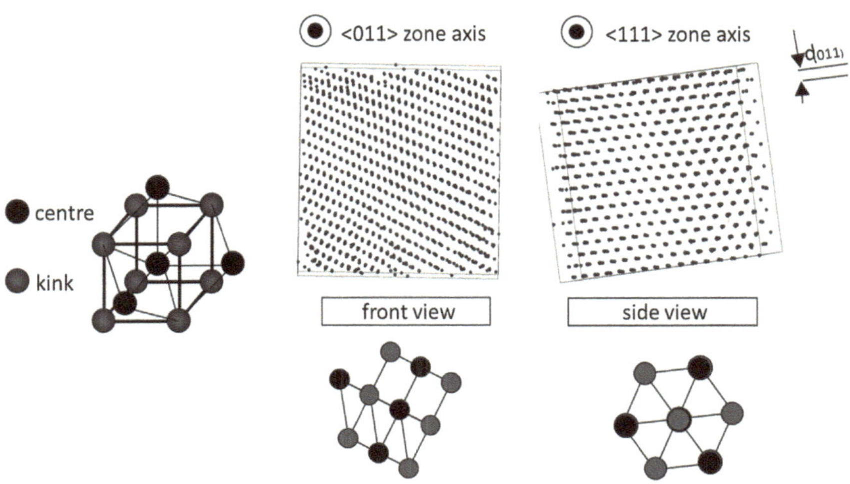

Figure 7.10 Restored image of the atomic positions of W shown in Fig. 7.6A. Image restoration was obtained by finding the average lattice positions obtained from the dataset itself.

then resolved. Less than 5% of the lattice sites are occupied by more than one atom. It is worth noting that the lattice is slightly deformed. These deformations are essentially due to the fairly strong hypotheses and assumptions used in the reconstruction of the original image. A more refined reconstruction should take into account better imaging function, but it does not affect the precision but the accuracy of the image. This kind of image improvement method was used recently and enables to couple APT reconstruction with atomic scale modeling approach of materials [15,16].

Origins of the Depth Precision in the Reconstructed APT Images

The measured dispersion in depth in the reconstruction of atom position is extremely good, but significantly larger than the intrinsic depth resolution (following Eq. (7.4), a fraction of picometer). The origin of this dispersion comes from any mechanisms that can perturb the well-ordered sequence of evaporation of atoms from the surface. Atoms are evaporated one by one following Eq. 2.2. ie, the rate of evaporation follows $K(s^{-1}) = A\exp\left(-\frac{Q(F)}{k_B T}\right)$ and $Q_{EV}(E) = C\left(1 - \frac{F}{F_{EV}}\right)$, C being in the range 1−3 eV close to the evaporation field threshold F_{EV} (see Chapter 2: Field Ion Emission Mechanisms). As a result the activation coefficient $C/k_B T$ is about 100−300. Because of the strong variation of this rate with surface field, any protruding atom may have a high probability to be expelled from the tip surface. As a result, only atoms situated at the extreme surface, which means atoms touching a virtual spherical cap of radius R will have a significant chance to be evaporated. This is the process that gives rise to the regular sequence of evaporation at the surface and to the small dispersion in depth.

Let us now assume a protruding atom adsorbed on the surface of the tip of radius R, the surface being submitted to a field F_R, close to the evaporation field. This atom will be submitted to an enhanced field F, that can in a first approach be estimated using classical electrostatics. A spherical protrusion of radius R_{at} induces an enhancement γ of about 3 ($F = \gamma F_R$, see Chapter 2: Field Ion Emission Mechanisms). This protrusion was created by eroding a complete spherical layer of thickness R_{at} around the remaining atom. So that it was necessary to remove the volume $V \sim 2\pi R^2 \times R_{at}$ of N atoms considering the surface of hemispherical end apex. This corresponds to about 10^5 atoms for a 50-nm tip. During this erosion process, we can assume that the surface field above the considered protruding atom will increase linearly from the initial field F_R to the enhanced field F_γ in function of the number of atoms being evaporated around it. Note that this relationship between the field and the degree of protrusion was used in some evaporation models [7]. As a function of n the number of atoms, the field $F(n)$ may be written as

$$F(n) = (1 + 2n/N)F_R. \tag{7.16}$$

The rate of evaporation of a delayed evaporated atom is then

$$\begin{aligned} K(n) &= A\exp\left(-\frac{C}{k_B T}\left(1 - \frac{F_R}{F_{EV}} - \frac{2n}{N}\frac{F_R}{F_{EV}}\right)\right) \\ &= K\cdot\exp\left(\frac{C}{k_B T}\left(\frac{2n}{N}\frac{F_R}{F_{EV}}\right)\right) \end{aligned} \tag{7.17}$$

If F_R is close to F_{EV}, the relative rate of evaporation of a delayed atom (shift of n atoms) exponentially depends on n following

$$K(n)/K = \exp\left(\frac{C}{k_B T}\left(\frac{2n}{N}\right)\right) \tag{7.18}$$

The width at half maximum of reconstructed atomic planes corresponds to the number of atoms n necessary to divide this rate by a factor 2, so that

$$n \approx \frac{k_B T}{2C}\ln(2)N \tag{7.19}$$

Because of the stochastic nature of the evaporation process, the uncertainty in the order of evaporation can be of several hundreds of atoms considered only the nature of the field evaporation process. It may enlarge the reconstructed width of atomic planes. This rough estimation shows the link between the in-depth precision and the degree of protrusion of a surface atom.

A better understanding of the in-depth precision may be found by calculating properly the relationship between the degree of protrusion and the local field above atoms on a realistic tip. Gruber et al. showed, for instance, than the influence of the temperature on the degradation of the regular sequence of evaporation for crystalline tips [17]. In this model the electric field is calculated above an atomically defined sample, and the evaporation is calculated using the standard equation of Chapter 2, Field Ion Emission Mechanisms. To model the stochastic nature of the field evaporation, a Monte Carlo algorithm was developed. This model shows that at low temperature (20K) the number of atoms at high risk to be field evaporated is extremely low (a few units), so that the field evaporation sequence is very deterministic. Conversely, at high temperature, the number of atoms that can be removed increases significantly which broadens the reconstructed atomic planes (as presented in Fig. 7.11). This degradation explains why tomographic images obtained in pulsed laser mode, where the evaporation is triggered by fast temperature pulses, are slightly less resolved (see Chapter 8: Laser-Assisted Field Evaporation for more details). The temperature makes the evaporation process more random at the moment of evaporation. A significant probability of evaporating atoms not in position to be evaporated (which means on a larger depth at the tip surface) exists and degrades the order of evaporation. This calculation gives a trend in good agreement with experimental observations [17–20].

Other factors such as the local variation in binding energy may also influence the evaporation of surface atoms during the imaging process. This phenomenon was pointed out by Gault and coworkers who measured a dependency of the depth resolution on the crystallographic direction of the considered atomic planes. Surprisingly, spatial resolution was found to be better for high-index crystallographic planes compared to low-index atomic directions. To explain this effect, Gault and coworkers introduced the average variation in binding energy as a function of the location onto the

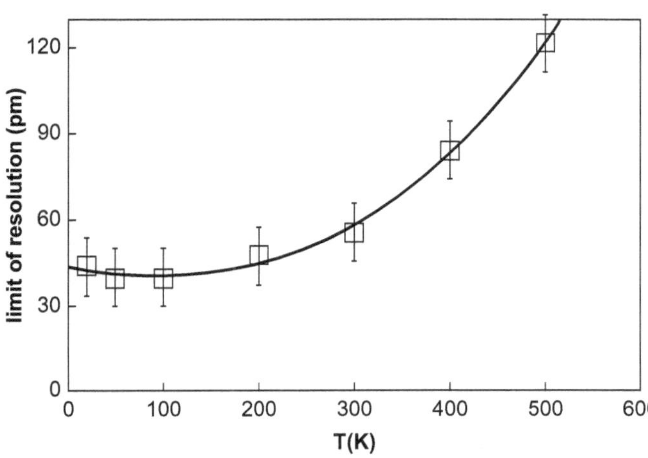

Figure 7.11 Depth resolution as a function of temperature calculated in the Fourier space of the modeled data (see Ref. [17] for more details).

tip so that the dispersion may be described by $\delta = \delta_0 + d_{hkl}\exp(-\Delta E/k_B T)$, where ΔE is the average binding energy at which atoms are evaporated and δ_0 a constant term. In this approach, atoms at the border of terrace of low-index poles have larger variation in binding energy and therefore larger dispersion in evaporation fields, giving rise to randomness in the evaporation sequence.

In alloys, the presence of elements with different evaporation fields also perturbs the perfect sequence of evaporation and the depth precision of reconstruction. Experimentally, the observation of atomic planes in crystalline alloys, and especially disordered alloy, is more difficult. This means that the depth precision is degraded to several angstroms.

Several studies on the impact of evaporation field difference between species have been performed. Fig. 7.12 shows the impact on resolution of the presence of two elements in a crystalline alloy (compound $A_{0.8}B_{0.2}$ composed of 20% of a high evaporation field atom B with the ratio of evaporation field between the elements ε_B randomly distributed within the host lattice sites A). Including this difference in the evaporation field, the sequence of field evaporation can be calculated [21,22]. The electric field is indeed estimated numerically above the atomically defined structure confined in a tip geometry. In a specimen composed of two species (A and B), each chemical species has a specific evaporation field (E_A and E_B). To take these different field strengths into account, a rescaled map of electric fields at the tip surface was built and considered for the evaporation step. The

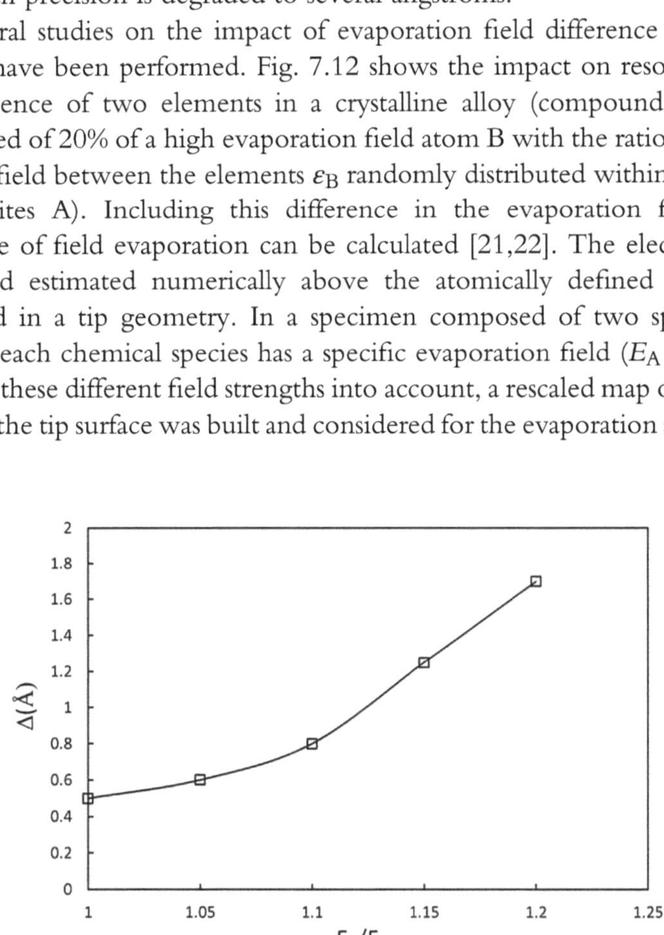

Figure 7.12 Simulated depth precision in function of the evaporation field ratio E_A/E_B in an alloy $A_{0.8}B_{0.2}$ measured with the Fourier method.

field above B atoms was artificially multiplied by the ratio E_A/E_B. The atom submitted to the highest field was chosen as the first evaporated species. This method replaces the steep evaporation rate equation by a step-like function, and therefore is only valid for low temperature field evaporation. By considering two chemically different atoms submitted to the same rescaled field, it is easy to demonstrate that the ratio of the electric fields at which they are evaporated is equal to the ratio of the evaporation fields.

This degradation of the reconstruction in depth is observed even for small differences in evaporation fields as seen in Fig. 7.12. Atomic planes separated by 2 Å are completely blurred for evaporation differences stronger than 25%. For lower values a gradual degradation of the depth precision is demonstrated in function of this parameter.

Origins of the Lateral Dispersion in the Reconstructed APT Images

The best lateral precision achieved experimentally approaches 2 Å a value close to the ultimate interatomic distance. However, this is not the general case. Since precipitates as small as 1–2 nm are commonly observed experimentally, we generally think that the lateral precision is below or close to 1 nm.

We may first find the instrumental limitation of the technique. The instrumental intrinsic precision of APT is found using the lateral magnification and the limits of the detector. Modern delay line detectors have spatial precision of the order of 50–100 μm for an open area of 4–12 cm depending on the instrument (see Chapter 6: Atom Probe Tomography: Detector Issues and Technology). In the worst case, the detector gives an image of the surface of the tip of about 200 nm in width (materials with low evaporation field, at high voltage of operation). In this case the instrumental limit is about 5 Å. It is close to our experimental value, but we may note that it is an upper bound. In the standard case, the lateral precision is limited by the physical process involved during the field evaporation.

Considering the reconstruction process, several physical mechanisms can be raised in the interpretation of the degradation of the resolution.

These mechanisms are represented in Fig. 7.13 and are listed below:

- Initial transverse velocity of the evaporated atoms:
  The early work of Chen and Seidman proposed this mechanism as one of the source of blurring in field ion microscopy. Any considered atom is not still at the atomic surface, but has an intrinsic velocity that can be derived from the thermal agitation energy. Assuming a temperature T

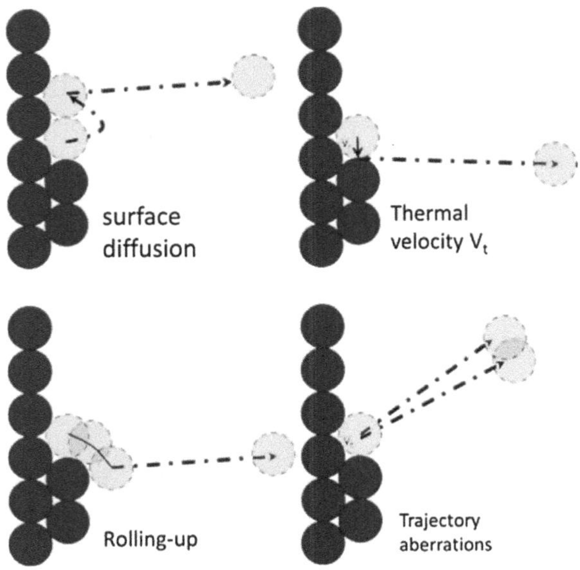

Figure 7.13 Possible physical mechanisms responsible for degradation of the reconstructed lateral position; lateral position is perturbed by atomic hopping at the surface induced by surface diffusion, by the influence of the initial lateral kinetic energy, reconstruction process at the moment of the evaporation (rolling-up), or trajectory deviations induced by variation of the electrostatic environment.

on the sample, and a distribution of velocity that follows a Maxwell distribution, the average kinetic energy of the atom is $3k_BT/2$, and the initial average transverse velocity is $\sqrt{\frac{k_BT}{m/n}}$. For a standard material ($m/n \sim 10-200$), the transverse velocity V_t is in the range 40−200 m/s at cryogenic temperatures (20−100K). This velocity gives rise to a lateral blurring of the impact position onto the detector. Because V_t carries the ion in a region where the field diverges, the lateral displacement writes $\Delta L_D \sim 2V_t\ T_{\text{fligth}}$, with T_{flight} being the time to travel from the tip to the detector. If T_{flight} is written as $T_{\text{flight}} = L\sqrt{\frac{m}{2neV}}$ and $F = V/k_FR$, assuming a magnification given by Eq. (7.3), the lateral dispersion induced by the thermal velocity is

$$\Delta L = \xi\sqrt{\frac{2k_BTR}{2nekF_{EV}}} \tag{7.20}$$

This calculation gives for aluminum a lateral dispersion of about 1 Å.

- Surface diffusion prior to field evaporation

Even if sample temperature is low, a probability to diffuse at the tip surface exists (see Chapter 2: Field Ion Emission Mechanisms). This effect depends on the energy barrier for surface diffusion compared to the barrier for field evaporation and the residence time of a surface atom prior to field evaporation. Surface diffusion barrier Q_{jump} is about 0.15–1.1 eV for most materials [23]. In voltage mode the field evaporation barrier is small (<0.1 eV), and the residence time depends on the evaporation flux. Evaporating 100 nm of tip per hour (~500 atomic planes/hour), gives the mean residence time of an atom at the surface of 7 s. Considering a classical Arrhenius expression for the diffusion, the residence time of an atom on its site is $t \approx t_c \times \exp\left(\frac{Q_{jump}}{kT}\right)$, where t_c is the inverse of the vibration frequency of atoms ($1/10^{12}$ Hz).

In voltage mode, in most cases, the residence time of an atom before evaporation is much shorter than the residence time on its atomic site before diffusion (taking $Q \sim 0.3$ eV), (about $1000\ \text{s} >> 7\ \text{s}$ at $T = 100\text{K}$). However, if the evaporation rate is very small, it can become comparable.

To conclude, surface diffusion is relevant only in particular cases, when a very small diffusion barrier exists at the tip surface, or if high temperatures of analysis are used (in the example, if $T > 200\text{K}$). In laser mode, the temperature is higher but only during a very short pulse. Only very high laser energies induce significant surface diffusion in the general case [16,24].

- Rolling-up motion

The classical field evaporation theory assumes the atom to be field evaporated to follow a trajectory normal to the surface. Some distortions in desorption maps of pure metals led some authors to assume that the paths of evaporated atoms may be significantly altered by the interaction with neighbor atoms in the first step of flight. Analyzing the experimental dependence of the evaporation field on temperature for (110) surfaces of a number of metals (including Al), Wada [25] concluded that neither the image-hump nor the charge-exchange model are able to produce satisfactory agreement. The experiment energy barrier for field evaporation was found too low for surface fields lower than the evaporation field. Wada suggested an alternative route for surface atoms to evaporate: instead of escaping directly from their original position, the atom undertakes a short trip (by means of diffusion) to locate itself where evaporation is easier. In particular, a jump of the kink atom atop the step edge was considered in Ref. [22]. The above mechanism implies only a short-range diffusion such as a jump to a metastable site (a surface atom diffuses for example, to the edge of a tip plane). The field there might be sufficient for the atom to leave the tip straight away. In such a case the diffusion becomes a rate-controlling process able to suppress the direct evaporation.

This mechanism was named rolling-up motion by Waugh because rather than leaving the surface directly, the atom experiences a path by rolling around one of its neighbor atoms to reach a position where it can be lifted out from the surface [26]. Sánchez and Lozovoi calculated theoretically the energy barriers existing at the tip surface of a (111) atom plane for an atom in position to be field evaporated to diffuse from its site to a bridge or a top site of the atomic surface [27]. A jump from one hollow site to another via the bridge site becomes almost barrierless at large fields. Even at small fields the diffusion barrier is much smaller than the evaporation barrier. Hence the diffusion mechanism of evaporation can indeed compete with the direct evaporation mechanism. The same kind of mechanism was observed by modeling the field evaporation via molecular dynamic approaches [28,29]. However, in both cases, no general rule for the local diffusion of atoms was given. In particular, if the effect is systematic, then the lateral precision will not be affected; however, if the height of the energy barrier to diffuse is similar to the evaporation field barrier, then the competition between both mechanisms will blur the position to a maximal interatomic distance. In dilute alloys, directional surface diffusion can be observed even at a low temperature. Species like P, N, or C are observed to move preferentially toward defined crystallographic locations, although the matrix atoms remain still. This behavior is not general and is not well understood at this moment. This kind of movement induces artificial chemical gradients in 3D volume [30].

- Trajectory aberrations in the first steps of flight

  As we will discuss in more detail in the next paragraph, the strong magnification of APT is related to the diverging trajectories close to the curved surface of the tip. In the first steps of flight, ions are accelerated in the electric field distribution of the apex. Any deviation of trajectories very close to the surface will be strongly magnified by the same mechanism. If we consider an atom on a given lattice position at the surface of the tip, the trajectory of this atom is deterministic and follows a path induced by the distribution of surrounding atoms that build the local distribution of field. A self-similar trajectory will exist if the local atomic distribution is always the same for successive evaporated atoms. At low temperatures, these criteria exist, since the order of evaporation is governed by the local roughness of the surface (as it was discussed for the depth precision). Two atoms evaporated from two successive layers, exactly at the same (x,y) position (but shifted in z), will have exactly the same environment in a pure crystalline metal, and trajectories will be similar. Any mechanisms that perturb the order of evaporation will produce slight variations of the electrostatic environment. The consequence will be small deviations in ion trajectories and spatial precision degradation [19,31].

As discussed for the depth precision, the temperature modifies the order of evaporation at the tip surface. The theoretical impact of the temperature on the impact position on the detector was modeled theoretically with the model presented previously [15,18]. At low temperatures only small changes are observed. Atoms evaporated from different layers hit the detector with very close positions and form an image that reveals crystallographic features as pole and zone lines. At higher temperatures (room temperature), and as observed experimentally by Gault et al. [22], the degradation blurs strongly the desorption map. When the order of evaporation is determined with a statistical component, the local geometry of the tip varies from layer to layer. This affects the local electric field distribution, resulting in a lateral displacement of the impact position.

Presence of species with different evaporation fields will also perturb the lateral positioning. It was found by simulation that a significant lateral spread exists. For instance, simulating the ion trajectories of a disordered alloy AB, evaporated considering a moderated difference in evaporation field between A and B atoms ($E_B = 1.2E_A$), the induced lateral spread was calculated to reach one interatomic spacing, in contrast with pure metals where atomic resolution was theoretically predicted. The sequence of evaporation, in a random solid solution is not periodic. Layer after layer, the tip shape changes and does not remain self-similar. Each atom has a random local environment. Thus, layer after layer, the order of evaporation of atoms in equivalent crystallographic positions is not identical. This means that ion impacts of evaporated alloys will be subjected to a shift with respect to the ideal position observed for pure metals.

It is worth noting that spatial precision degradations are observed in nonmetallic samples. In silicon, the depth precision of 1 Å is close to that of metals, but it was noted that correlated evaporation of atoms and local reconstruction of the atomic planes of atoms could prevent the correct reconstruction of the diamond crystal structure. As suggested by Kreuzer and coauthors, the crucial difference in field evaporation of metals and insulators is the basic fact that metals expel electrostatic fields and insulators allow them to penetrate [26]. This intense field first shrinks the intrinsic band-gap of the material (the surface becomes semimetallic) and reorganizes significantly the structure of the atomic surface on the first atomic layers (the crystal structure will change much akin to melting so that site specificity could be lost). This local transformation disturbs the well-organized atomic structure of semiconductors and oxides and gives rise to additional deflections of ion trajectories. For example, the depth resolution measured in an oxide with perovskite-type structure (chemical superordered along the cell axis in Fig. 7.6D) is estimated to ~2 Å, whereas the lateral resolution is measured to 20 Å using the method described previously.

IMPROVING SPATIAL ACCURACY OF APT IMAGES

Spatial precision is mostly the result of local and nonsystematic deviations of the transfer function between the initial positions to the impact positions and the initial positions onto the tip. The atom-prober can only deal with physical or experimental parameters of analysis to optimize images, such as the base temperature, the evaporation rate, or a better detector. Spatial accuracy improvement is obtained by determining the best transfer function between the initial and final positions of all the atoms collected of the entire volume. At the time of writing of this book, the ultimate transfer function, being the base of the reconstruction protocol, was not achieved. Nevertheless, significant improvements of the standard protocol were developed in the last 20 years, and are described in this chapter. Note that the term reconstruction protocol (or algorithm) refers to the methodology enabling the transformation from detector ion-hit positions (X,Y,N) (where N is the ion sequence number) into specimen spatial coordinates (x,y,z).

Starting from the zeroth order reconstruction protocol proposed previously, we can incrementally improve the complexity of the algorithm by the input of several hypotheses. The first model described previously is clearly limited to small-field-of-view experiments, and considering a small amount of atoms eroded from the sample. It is also restricted to materials with homogeneous properties of field evaporation; several adjustments will be necessary when dealing with materials of complex microstructure.

From Small Field of View to Wide Field of View

Ion Trajectories in the Atom Probe

Let us start from the simplified model of the tip that is presented in Fig. 7.14. Most reconstruction algorithms have been designed based upon a set of assumptions similar to those included here [32–47]:

- The field evaporated specimen is approximated by a spherical surface of radius R positioned on a truncated shaft that is often described as a cone with a half-shank angle (α); this shape is generally of revolution symmetry along the tip axis. Note that the sample can be tilted with respect to the tip to detector direction.

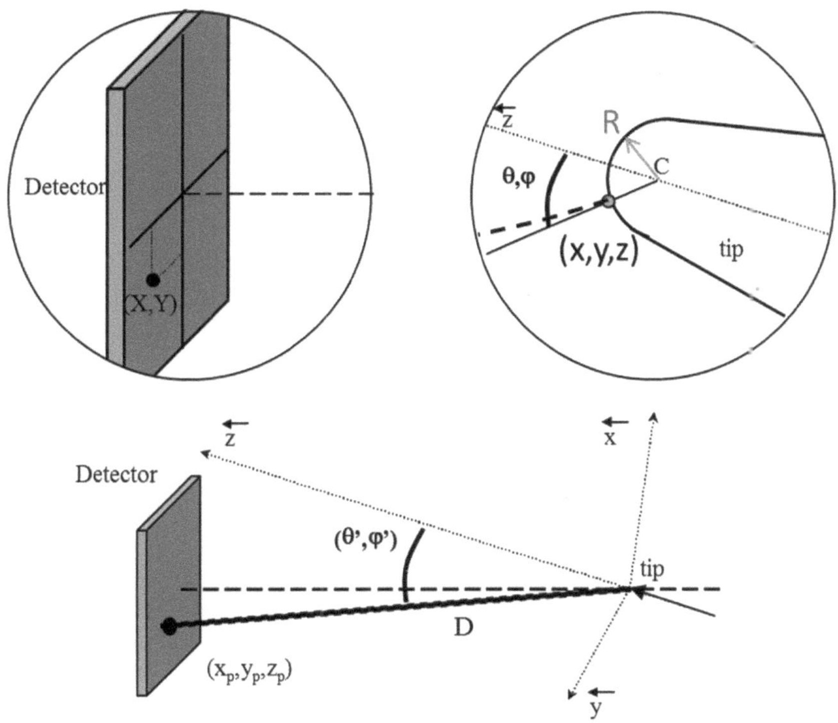

Figure 7.14 Schematic drawing of the simplified model used for designing the reconstruction algorithm.

- The highly magnified image of the specimen surface formed by the successive impact of the ions on the detector is described by the projection of ions in the electric field distribution produced by the tip in its electrostatic environment.
- The original position of the atom (x,y,z) at the tip surface may be computed, using spherical coordinates (r,θ,φ), by postulating the existence of the tip center C, located along the specimen main axis at a distance R, the curvature radius, from the apex.
- The impact position on the detector may be computed, using spherical coordinates (r',θ',φ'), by including the tip to detector distance L, and using the coordinate of the impact (X,Y) on the detector.

In the Cartesian coordinate system of the tip with **z** align with the tip, and taking the extremity of the sample as the origin of the z axis, the position (x,y,z) onto the tip is deduced by the standard relationship (r,θ,φ)

$$\begin{aligned} x &= R\sin(\theta)\sin(\varphi) \\ y &= R\sin(\theta)\cos(\varphi) \\ z &= R(1-\cos(\theta)) \end{aligned} \tag{7.21}$$

In the Cartesian coordinate system of the tip with **z** align with the tip, the position (x_p,y_p,z_p) onto the tip is deduced by the standard relationship

$$\begin{aligned} x_p &= D\sin(\theta')\sin(\varphi') \\ y_p &= D\sin(\theta')\cos(\varphi') \\ z_p &= D\cos(\theta') \end{aligned}$$

with $D=\sqrt{X^2+Y^2+L^2}$ the distance from the tip to the impact position.

The reciprocal transformation is written as

$$\begin{aligned} &\theta = \arctan\left(\frac{\sqrt{x^2+y^2}}{z}\right) \quad \text{et} \quad \theta' = \arctan\left(\frac{\sqrt{x_p^2+y_p^2}}{z_p}\right) \\ &\phi = \arctan\left(\frac{y}{x}\right) \qquad\qquad \varphi' = \arctan\left(\frac{y_p}{x_p}\right) \end{aligned} \tag{7.22}$$

To get (x_p,y_p,z_p) from (X,Y), the back transformation of the impact position in the detector coordinate system to the tip coordinate system should take into account the rotation of the tip with respect to the detector. When no rotation exists, coordinates (X,Y) are identical to (x_p,y_p) and $z_p=L$. N

being a point with coordinates (x,y,z) in the coordinate system of the tip at rest. The coordinates of this point can easily be calculated in the reference frame that has undergone successive basic rotations. These coordinates are calculated by using rotation matrices around each axis of rotation. For example, if the reference discloses a rotation angle α_1 about the z axis, the coordinates (x',y',z') of N are deduced by the relation

$$\begin{bmatrix} x' \\ y' \\ z' \end{bmatrix} = \begin{bmatrix} \cos(\alpha_1) & \sin(\alpha_1) & 0 \\ -\sin(\alpha_1) & \cos(\alpha_1) & 0 \\ 0 & 0 & 1 \end{bmatrix} \times \begin{bmatrix} x \\ y \\ z \end{bmatrix} \tag{7.23}$$

When the system makes three successive rotations, α_1, α_2, and α_3, on three independent axes $A\alpha_1$, $A\alpha_2$, and $A\alpha_3$, described by three matrices M_1, M_2, and M_3, the coordinates of the point M are derived by the relationship

$$\begin{bmatrix} x' \\ y' \\ z' \end{bmatrix} = M_{123} \times \begin{bmatrix} x \\ y \\ z \end{bmatrix}, \quad M_{123} = M_1 \times M_2 \times M_3 \tag{7.24}$$

It is then possible to deduce the position of a point $M'(x',y',z')$ defined in the reference frame that has undergone these rotations in the origin marked by the inverse transformation

$$\begin{bmatrix} x \\ y \\ z \end{bmatrix} = M_{123}^{-1} \times \begin{bmatrix} x' \\ y' \\ z' \end{bmatrix} \tag{7.25}$$

The coordinates of the point (X,Y,L) on the detector is used to calculate the coordinates of the point (x_p,y_p,z_p).

$$\begin{bmatrix} x_p \\ y_p \\ z_p \end{bmatrix} = [M_{123}]^{-1} \times \begin{bmatrix} X \\ Y \\ L \end{bmatrix} \tag{7.26}$$

The matrix M_{123} depends on how the tip is rotated in the space. The easiest way to describe this rotation is to use the goniomic transformation that tilts the tip without rotating the tip around its z axis. Rotations of angles Θ and Ψ correspond to a single Θ rotation around the direction Oy perpendicular to Oz and tilted of Ψ with respect to the Oy axis. Finally, this transformation is written as

$$M_{123} = \begin{bmatrix} \cos(\Psi) & \sin(\Psi) & 0 \\ -\sin(\Psi) & \cos(\Psi) & 0 \\ 0 & 0 & 1 \end{bmatrix} \times \begin{bmatrix} \cos(\Theta) & 0 & \sin(\Theta) \\ 0 & 1 & 0 \\ -\sin(\Theta) & 0 & \cos(\Theta) \end{bmatrix} \times \begin{bmatrix} \cos(\Psi) & -\sin(\Psi) & 0 \\ \sin(\Psi) & \cos(\Psi) & 0 \\ 0 & 0 & 1 \end{bmatrix} \tag{7.27}$$

Writing the projection relationship with angular coordinates is much more natural than using the classical magnification expression (Eq. (7.1)). Indeed, if the tip moves several microns, this displacement δ is not amplified by the projection. The initial angles on the tip remain identical, and the projected angles $\arctan(\delta/L)$ are close to zero for standard L (0.1−0.4 m). The quest to the best accuracy of APT reconstruction requires precise evaluation of the relationship between (R,θ,φ) and (θ',φ'). This relationship may be evaluated by considering how ions fly in the atom probe chamber from the tip surface to the detector. Two approaches may be used for this estimation. The first one is empirical. APT or FIM projects the surface of the tip onto the detector. This surface contains some crystallographic information that can be used to estimate the projection relationship. The second one is theoretical. The trajectories of ions can indeed be calculated assuming the geometry of the tip and of the surrounding environment.

Empirical Investigation of the Ion Projection

Because image projection in APT is similar to the projection observed in FIM, the description of the relationship between (θ,φ) and (θ',φ') has been intensively studied in the 1960s and 1970s [34,35,37−39]. When a crystalline tip is field evaporated, the end apex has an average hemispherical shape. This means that the regular crystal structure is intercepted by this shape and it creates regular structures of concentric atomic rings known as poles. The center of these poles is simply the exact position where a defined type of atomic planes (*hkl*) is parallel to the surface. The direction <*hkl*> is normal to these planes. Alignment of these poles exists along zone axes of the crystal structure (Fig. 7.15). Between two poles (*hkl*) and (*h′k′l′*), the

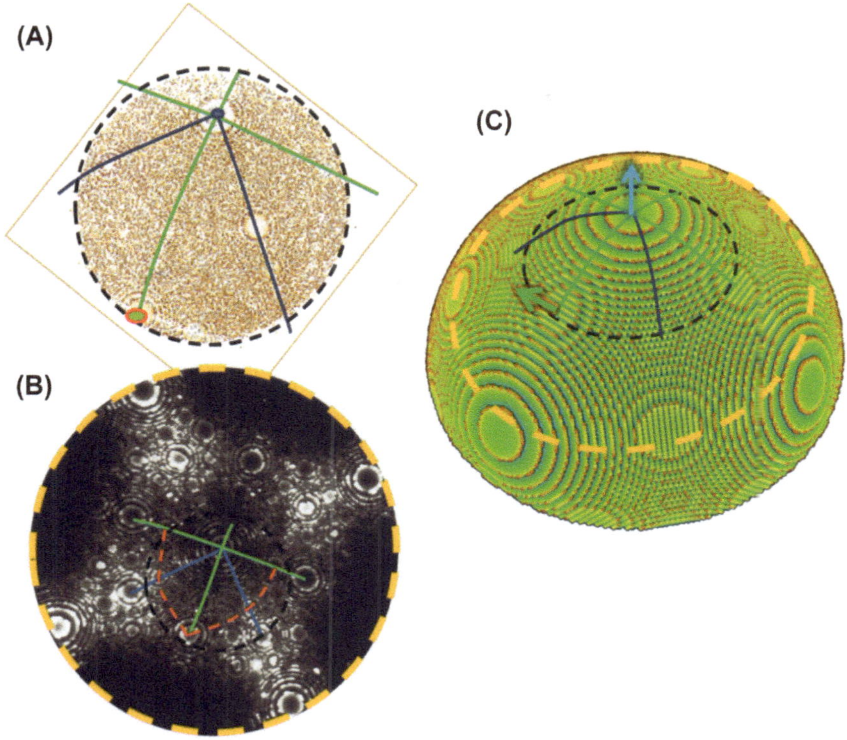

Figure 7.15 Image projections of an FCC (face centered cubic) Al surface in the desorption map of a fraction of monolayer in atom probe tomography (A), in field on microscopy (B), and representation of the surface of the tip (C). Crystallographic directions <011> and <002> are represented by green and blue arrows, respectively, on the surface. These directions appear as pole centers in field ion microscopy and atom probe tomography desorption maps. Zones axes of [001] types are drawn as green lines. Zone axes of [011] types are drawn as blue lines.

angle θ_P can be easily calculated, and writes in cubic structures (more complex equations exist for other structures):

$$\theta_P = a\cos\left(\frac{hh' + kk' + ll'}{\sqrt{h^2 + k^2 + l^2} \times \sqrt{h'^2 + k'^2 + l'^2}}\right) \tag{7.28}$$

Images obtained by the projection of the surface of the tip onto the detector reveal the same structure in 2D, which can be interpreted and indexed with an a priori knowledge of the studied structure. Poles appear in these images as concentric rings; the position of the centers can be measured as (X,Y) coordinates. Following Eq. (7.21), the projected angles and the related

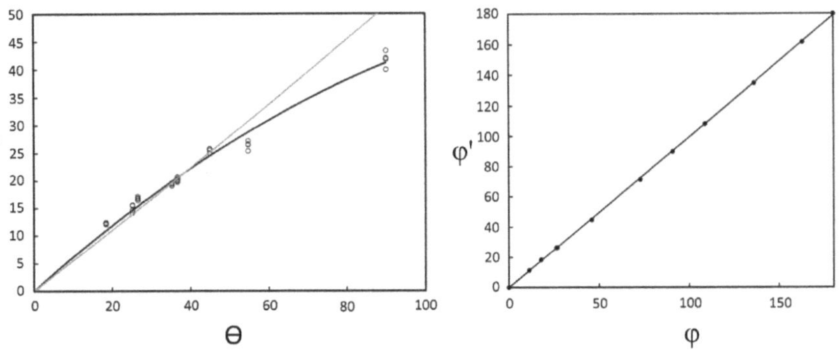

Figure 7.16 Projection relationship $(\Theta',\varphi') = f(\Theta,\varphi)$ deduced from FIM image of aluminum (Fig. 7.15). $\varphi = \varphi'$ is obvious, whereas the relationship between θ and θ' is less straightforward.

angles onto the tip can be determined experimentally. Fig. 7.16 presents an example of the empirical projection relationship between (θ,φ) and (θ',φ') for a few poles of Fig. 7.15B, taking the (002) central pole as the center of the detector.

Empirically, the azimuthal angle shows no variation. This is due to the rotational symmetry of the system around the tip axis. The projected polar angle is compressed compared to the theoretical polar angle θ on the sample. This compression is in the range 1.3−2, but varies from tip to tip and is a function of the electrostatic environment of the sample. Nevertheless, experimentally $\theta' = f(\theta)$ diverges significantly from linearity for angle larger than 45°.

Different empirical models were expressed to model this behavior, the simpler being $\theta' = \theta/\xi$, with ξ the angular compression, or image compression factor of Eq. (7.3). Indeed, in the small angle limit,

$$\sqrt{x^2 + y^2} = R\sin(\theta) \approx R\theta$$

and

$$\sqrt{X^2 + Y^2} = L\tan(\theta') \approx L\theta' \tag{7.29}$$

Some authors found a better agreement at large angle by writing directly

$$\sqrt{X^2 + Y^2} = k_\theta \theta \tag{7.30}$$

For θ small, it means $k_\theta \sim L/\xi$.

Another model treats the projection as a simple point projection. Indeed, FIM images or desorption images are similar to classical stereographic projections used in crystallography. However, to model a correct compression, it is necessary to assume that the point projection is not situated at a distance of one radius of curvature from the tip center C, but at an intermediate distance PC, which is a fraction of the tip radius (Fig. 7.17). This distance is often written as mR, m being the parameter describing the degree of compression. Using this model, we write

$$\begin{aligned} \theta' &= \arctan\left(\frac{\sin(\theta)}{m+\cos(\theta)}\right) \quad \text{or} \\ \theta &= \theta' + \arcsin(m\sin(\theta')) \\ \varphi &= \varphi' \end{aligned} \tag{7.31}$$

For small angle $\xi = m + 1$.

In FIM experiments with no counter electrode, the image compression factor is generally around 1.8; however, in APT with a local electrode in front of the tip, this factor tends to reduce to 1.4–1.6. Indeed, the compression is related to field distribution in front of the tip, and any device that modifies this distribution influences the projection of ions. Some differences

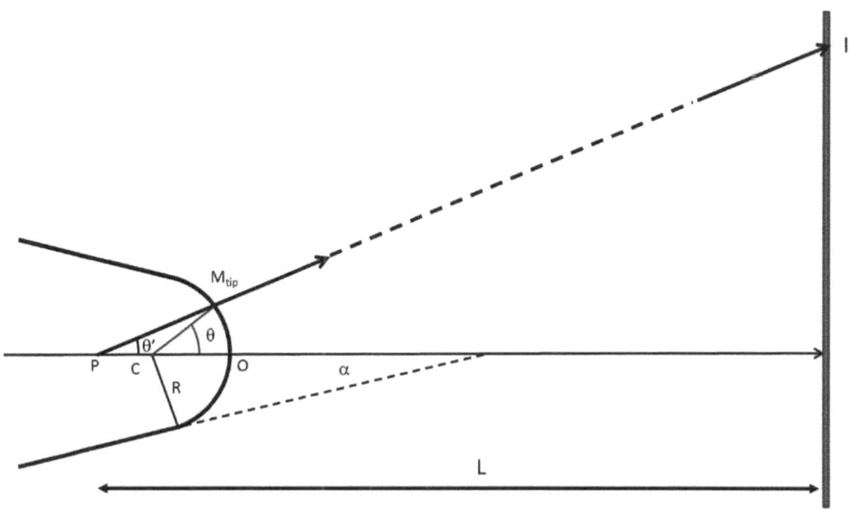

Figure 7.17 Definitions of angles in the quasi-stereographic projection model. A point M_{tip} on tip gives an impact I. Note that L is the macroscopic distance between the tip and the screen much larger than OP. The point projection P is situated at a distance $PC = mR$ from the tip apex.

are also found depending on the material studied showing the dependency of the tip shape on the global features of the ion projection.

Numerical Methods of Investigation of the Ion Projection

The understanding and prediction of the projection features is mandatory to improve significantly the reconstruction of APT images. Ion trajectories modeling was introduced in the early ages of the technique with this aim [48–55]. It has long been known that the single point-projection provides a relatively poor description of field ion micrographs and atom probe data and the influence of the shape of the sample and of the tip to counter distance must be evaluated.

The problem of image projection seems relatively simple. Ions fly from the tip to the detector following the field distribution created by the application of the voltage to the tip. Prediction of the trajectories is done by calculating classically the trajectory of charged particles in this field distribution. Note that in the electrostatic approximation, when considering ions with no initial kinetic energy, ion trajectories are identical whatever the mass, the charge, or the voltage of the tip.

Several investigators performed analytical or numerical calculations to extract the image compression factor in atom probe. For instance, assuming hyperboloidal or paraboloidal approximations of the tip shape, the field distribution is easily expressed analytically. These kinds of models were given in Chapter 2, Field Ion Emission Mechanisms.

The field factor clearly depends on the tip shape as well as the tip to screen distance. Unfortunately, no analytical expression exists for the compression factor, so that a numerical calculation of the ion trajectories is required to evaluate ξ. Using this approach in the paraboloidal geometry, a linear compression is found ($\xi \sim 1.5$) for standard parameters of tip to counter electrode distance (100 μm), and tip radius (100 nm) (Fig. 7.19). However, such a model is unable to give the relationship between the compression and the shank angle, for instance. In addition, compression can only be calculated close to the tip apex. A hemispherical apex can be defined only for a small surface of the tip, close to the axis of the tip in this conical approximation.

To take into account the tip shape (tip radius and shank angle) as well as the exact geometry of the atom probe chamber, several investigators have modeled the field distribution around the specimen with finite element methods. This kind of calculation is not straightforward since the tip must be precisely defined close to the tip surface (refined mesh) without neglecting the large scale of the atom probe chamber. A multiscale model or dedicated finite element method models are necessary in this aim [43,52,56].

Using numerical calculation of the electric field in a multiscale model, the surface field and the ion trajectories from a tip defined in cylindrical coordinates were calculated. The basic process of numerical field evaporation to obtain the stationary shape of the tip apex is described in different papers for more information [52,57,58]. The model assumes that the tip is a compact stack of atoms confined in a sharply pointed needle shape. A 3D space with revolution symmetry (r,z) is used. In this 2D cylindrical model, atoms are considered to occupy identical squares, so that the interatomic distance is defined by 1 unit cell D_{cell} ($D_{cell} \sim 0.3$ nm). The radius of curvature of the end apex is defined by R_0. The mesh is regular close to the tip apex. Beyond a few hundred nanometers, and to model the impact of the electrostatic environment, an irregular orthogonal mesh was employed. The limits of the simulation are about 10×20 cm^2. A counter electrode at ground voltage at a distance h along the z axis was placed in front of the tip. A probe-hole equivalent to a maximum angle of detection of $\pm 45°$ is modeled. The radius of this hole is adjusted to keep constant this maximum angle. The tip is submitted to a constant voltage V. The distribution of the electrostatic potential within all mesh cells is obtained by numerical solving of the Laplace equation $\Delta V(r,z) = 0$. A finite difference algorithm with iterative convergence is used in this calculation.

The amplitude of the electric field is measured above all surface atoms. To simulate field evaporation, the atom submitted to the highest field was chosen as the next atom to be removed from the surface. Due to the linear dependence between the voltage and the field, the calculated local electric field enables the evaluation of the voltage required to maintain constant the surface field $\mathbf{F_E}$. The selected unit cell is removed and the electrostatic potential is recalculated over the whole volume. The ion trajectory is obtained from a numerical calculation of the classical Newton equation using the potential corresponding to the evaporated atom as an empty cell.

With this model, the gradual evolution of the field (or of the voltage required to keep a constant field), the ion impacts on the detector, and the gradual evolution of the tip shape are all simulated. Taking into account the observed tip radius R predicted by the model, the impact angles θ' are measured as a function of the crystallographic angle θ at the tip surface. Fig. 7.18 shows an example with three tips with the same initial radius but with different half shank angles. A monotonic relationship can be drawn between the impact angle θ' and the initial angle θ of the atom at the tip

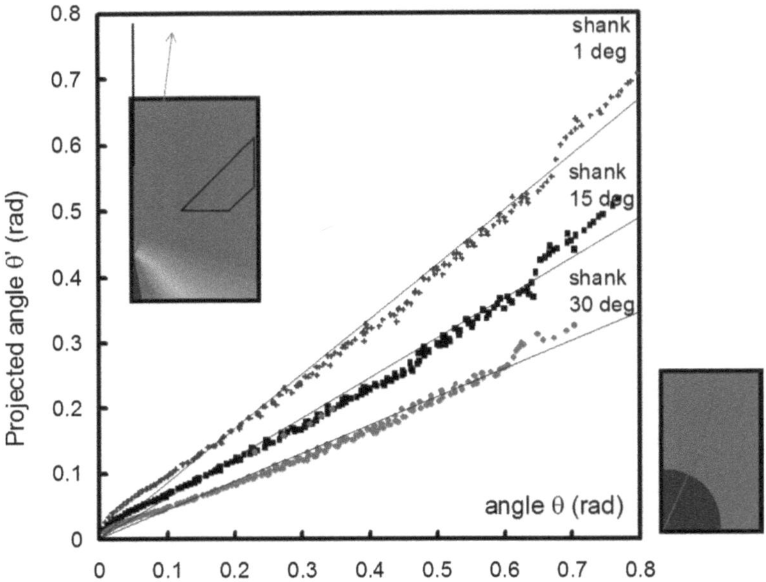

Figure 7.18 $(\Theta',\varphi') = f(\Theta,\varphi)$ deduced from numerical calculation of the field evaporation of surface atoms. Shank angle increases the compression of ion trajectories, changing ξ from 1.1 for a nanowire to 2.2 for a tip of 30° of half shank angle.

surface. Using a linear fit, a straightforward measurement of ξ is obtained. This modeling approach has shown a strong dependence in the image compression factor as a function of the geometry of the sample and also as a function of the presence of the counter electrode. This result was confirmed by finite-element simulations of idealized smooth specimens [51,54]. The numerical model was used to compute the field factor and the image compression factor with α, the half shank angle, varying between 1° and 30° and for tip to counter electrode distance from 30 μm to 1.2 mm. The variation of ξ is given in Fig. 7.19. Variations in the range 1.1–2.4 are calculated. Note that ξ evolves similarly to k_F, the field factor as described in Chapter 2, Field Ion Emission Mechanisms. It indicates that both parameters are linked together. Fig. 7.20 shows the clear relationship that exists between these parameters. The log–log plot of k_F as a function of ξ shows a straight slope of 3 ± 0.1. As a result the field factor k_F can be simply and surprisingly written as ξ^3. No clear explanation of this dependence is found, since no analytical expression of the compression factor can be found even in a simple geometry.

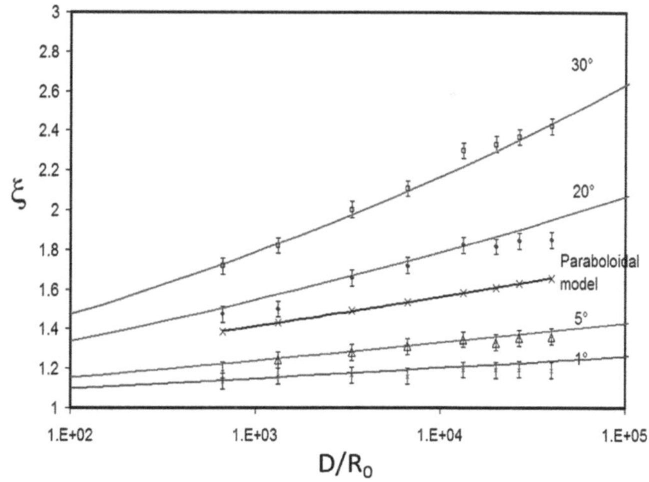

Figure 7.19 Image compression factor as a function of the specimen geometry, for different ratios between the tip to counter electrode distances D, and the tip radius R_0. Note that the paraboloidal model also gives the same trend.

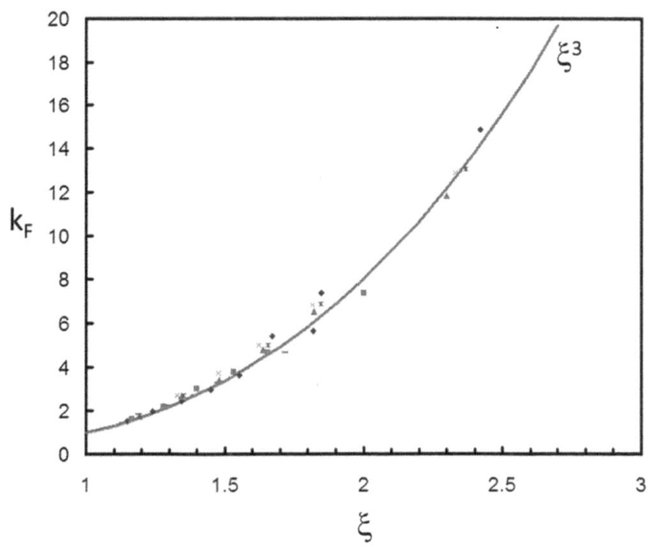

Figure 7.20 Field factor (k_F) as a function of image compression factor. k_F can be fitted by ξ^3.

Reconstructing Positions on the Tip Surface

The Need for an Estimate of the Radius of Curvature

Let us assume that a good estimate of $\theta' = f(\theta)$ was found, and that the rotational symmetry exists. As a result, finding positions on the tip is straightforward using Eqs. (7.21) and (7.22), and assuming a correct radius R for the tip apex. Indeed, we recall that (x,y,z) is written as

$$x = R\sin(\theta)\sin(\varphi)$$
$$y = R\sin(\theta)\cos(\varphi)$$
$$z = R(\cos(\theta))$$

Both experimentally and numerically, the end apex is well adjusted by a single curvature radius R up to an angle of about $\pm 35°-40°$ (Fig. 7.21). This angle is close to the actual maximal angle of collection used in most of modern APT. However, the general aspect of the tip is rather ellipsoidal for larger angles, due to the influence of the field evaporation that flattens the apex of the sample. The equilibrium radius of curvature R_{eq} deduced from the positions of surface atoms is larger than the initial radius $R_{0.}$ The presence of the shank modifies the field distribution in the close vicinity of the tip axis giving rise to a reduced electric field for high angles (agreeing with the previous literature results [47]). As a result the curvature difference is stronger

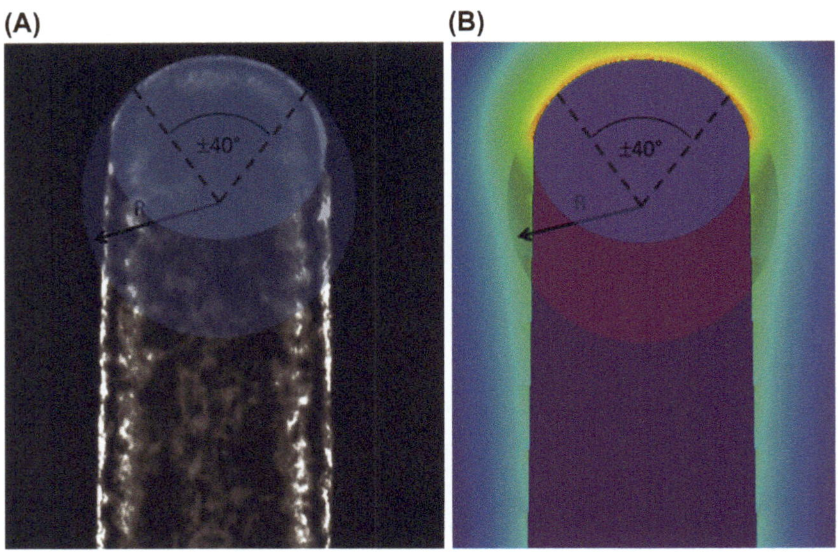

Figure 7.21 End apex morphology of a field evaporated steel tip. (A) The experimental end apex can be modeled by a spherical shape of radius R over $\pm 40°$. (B) The same shape is reproduced by simulation of the field evaporation.

between the tip axis and the outside borders of the end apex for tips with large shank angles.

Estimation of the radius can be done by several methods. We can first use a picture taken in transmission electron microscopy (TEM) or scanning electron microscopy (SEM) to determine R (Fig. 7.21A). The precision of this measurement can be as good as a few nanometers, so that the reconstructed analyzed surface accuracy may be of this range of order. A better estimate is obtained when concentric rings related to a pole is observed in the dataset, or if an FIM experiment was done on the same sample, provided a pole contrast. The most accurate and simple method of determining the APT tip radius is to count the number of net rings between a pole center and a known angle in the FIM. This method first suggested by Müller and Bahadur was formulated in FIM by Drechsler [59–62]. Considering geometric relationship (see Fig. 7.22), if i is the index of the considered ring, and θ_i the angle, the radius R is given by

$$R = \frac{i d_{hkl}}{1 - \cos(\theta_i)} \tag{7.32}$$

A linear regression of $\cos(\theta)$, or θ^2 as a function of i enables an accurate calibration of R by using

$$\begin{aligned} \cos(\theta_i) &\approx 1 - \frac{\theta_i^2}{2} \approx 1 - i \times \frac{d_{hkl}}{R} \\ \frac{\theta_i^2}{2} &\approx i\frac{d_{hkl}}{R} \end{aligned} \tag{7.33}$$

The Evolving Radius of Curvature

Obviously these two methods are only valid at a given moment of the analysis (when the TEM image was taken, for instance). If one of these two methods was used at the beginning of the analysis, a good instantaneous value of R is found, providing a good reconstruction of the position of atoms at this tip surface. It does not provide any information on the evolution of this radius during the eroding process. Indeed, due to the presence of the shank angle, as the tip is eroded, the specimen progressively blunts, and the tip radius progressively evolves (generally increases) throughout the experiment. As discussed in Chapter 2, Field Ion Emission Mechanisms, this increase in radius R will normally induce a decrease in field following equation $F = \frac{V}{k_F R}$. However, experimentally the evaporation rate is kept constant by the control of the voltage V applied to the sample. As a result,

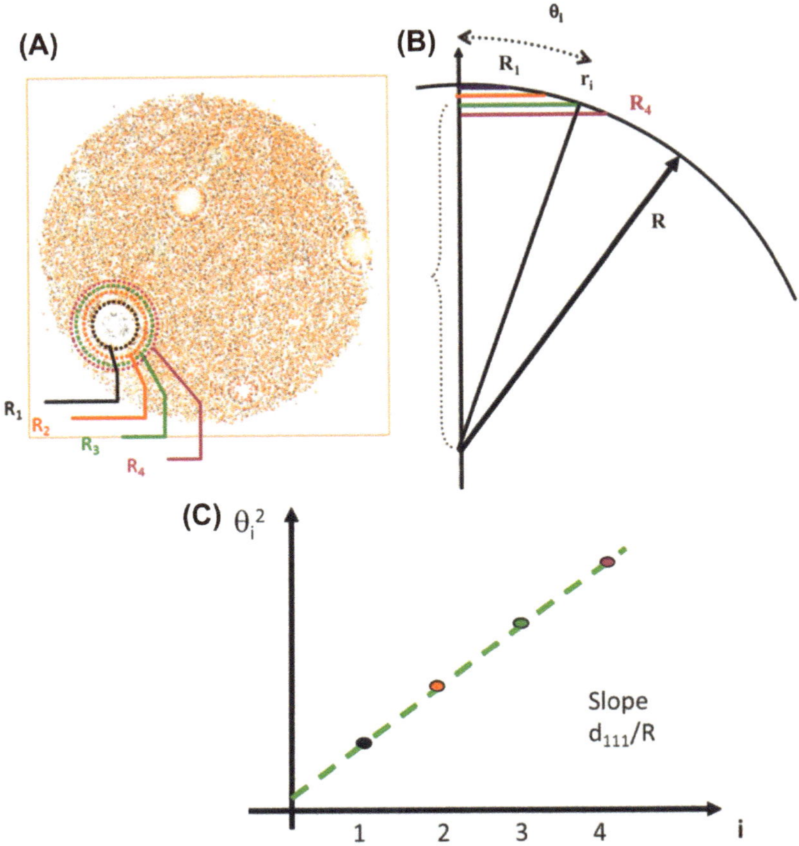

Figure 7.22 Drechsler method of measurement of the curvature radius, by counting the successive concentric rings (index *i*), and measuring the apparent angle θ_i (in (B)) on a desorption map of a fraction of a monolayer, or an FIM image (in (A)). (C) The measured angle squared is proportional to the index of the plane.

the surface field F_A during the analysis induced by the quasi-constant evaporation rate imposed by the control software is approximately constant and is proportional to the average evaporation field F_{EV} of the studied material. Assuming k_F is constant (we will discuss the validity of this assumption in the following chapter), the radius R is estimated by

$$R = \frac{V}{k_F F_A} \tag{7.34}$$

This method is useful to estimate the radius of curvature for a reasonable range of voltage around the effective value of kF_A deduced from the radius measurement.

Another way to determine experimentally the evolution of the radius of curvature is to assume a given geometry for the specimen. A geometrical model can be used to determine the relationship between the diameter of the sample and the depth z relative to the end apex position. The simpler model assumes the tip having the profile given in Fig. 7.2. The apex is described as a portion of sphere, smoothly fitted on a cone, which means a tangential continuity exists between the connected shapes. The cone is defined by its half shank angle α and the initial radius is R_0. In this case,

$$R = R_0 + z\frac{\sin\alpha}{1-\sin\alpha} \tag{7.35}$$

The term $\sin\alpha/(1-\sin\alpha)$ is also equal to dR/dz considering a small increment in z.

Note also that the radius R_{cone} of the cone at the connection with the sphere is $R_{\text{cone}} = R\cos\alpha$, so that the diameter of the end apex varies following

$$D = D_0 + z\frac{\sin\alpha\cos\alpha}{1-\sin\alpha} \tag{7.36}$$

More generally, the variation of D with z (depth axis) can be easily measured from a SEM or a TEM image of the sample prior to the field evaporation (Fig. 7.23).

Note that an additional refinement was proposed by Geiser et al. to take into account that the tip is not strictly spherical on its apex [63].

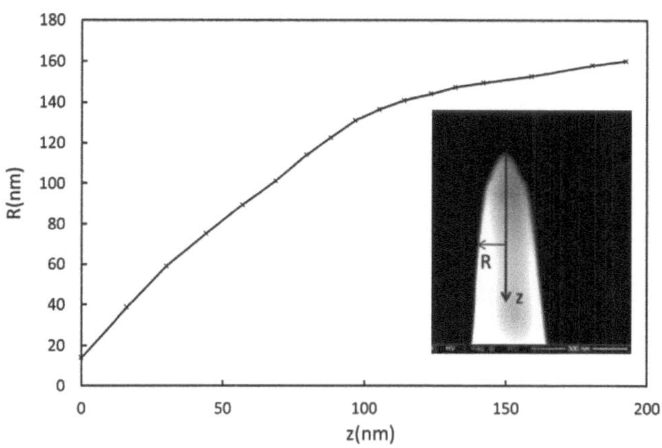

Figure 7.23 Measurement from a scanning electron micrograph of the evolution of R with the probed depth (z). Between two points, the tip has a quasi-constant shank angle and variation of the relationship in Eq. (7.34).

An additional parameter named tangential discontinuity was added to allow the hemispherical approximation of the tip to not have a tangential continuity with the sustaining cone. This implementation allows taking into account the flattening of the apex induced by field evaporation mechanisms. In this model, a fixed angle between the normal to the sphere and the normal to the cone surface is added as an input parameter. This angle relaxes the condition $R_{cone} = R\cos\alpha$ to the more general expression $R_{cone} = R/f_{s/c}$, $f_{s/c}$ being a constant.

In this model the evolution of R with z was found similarly by taking a virtual shank angle α' that satisfies to the tangential continuity. This angle is slightly larger than real shank angle and follows

$$\sin\alpha' = \frac{\sin\alpha}{\cos(\alpha)\Big/f_{s/c} + \sin(\alpha)\sqrt{1-\frac{1}{f_{s/c}^2}}} \tag{7.37}$$

Using this angle, the relationship in Eq. (7.35) is fulfilled.

Accurate Depth Calibration of the Image

An accurate estimation of R may be obtained if z is correctly estimated. This value is not straightforward but can be deduced from the application of the conservation of the mass or volume. Let us start from the initial shape of the tip. The end apex is defined principally by its curvature radius R_0. If a homogeneous evaporation is considered at the tip surface, N atoms evaporated correspond to a volume V removed from the sample. The simplest approach (following Eq. (7.3)) enables scaling the eroded depth $z_{probed} = NV_{at}/(QS_D)$. Optimization of this estimation can be obtained by considering a better approximation of the evaporated volume V. Assuming the tip as a truncated cone of half shank angle α with an end apex of radius R_0, the volume evaporated v_{probed} during the analysis was calculated by Walck et al. for imaging atom probe (and by Geiser et al. for the wide-angle 3D atom probe [41,45]) and reformulated in Ref. [55]. V_{probed} is geometrically given by

$$v_{probed} = \pi R_0^2 \sin^2(\theta_{max}) \times z_{probed}\left[1 - \left(\frac{\sin\alpha}{1-\sin\alpha}\right)\left(\frac{1-\cos\theta_{max}}{\sin\theta_{max}}\right)^2\right] \times \left[1 + \frac{z_{probed}}{R_0}\frac{\sin\alpha}{1-\sin\alpha} + \left(\frac{z_{probed}}{R_0}\frac{\sin\alpha}{1-\sin\alpha}\right)^2\right] \tag{7.38}$$

For a reasonable half shank angle and a small depth variation δz, which corresponds to an amount of atoms Δn, the volume is well approximated by

$$\delta v \approx \pi R_0^2 \sin^2(\theta_{\max}) \times \delta z \left[1 - \frac{\sin\alpha}{1-\sin\alpha}\left(\tan\frac{\theta_{\max}}{2}\right)^2\right] = \Delta n \times \frac{V_{at}}{Q} \tag{7.39}$$

Note Q is here the detection efficiency, which depends on the specific detection system, but which usually varies between 35% and 80%. For tips with low shank angle α, or in case of small-field-of-view instruments, the relationship is identical to Eq. (7.3). On the z axis, starting from initial assumptions z_0, R_0, α, and ξ, the z position can be found iteratively. Considering a gradual variation of R as a function of n, the evaporation of one atom increases the z by

$$dz = \frac{V_{at}}{Q\pi R_0^2 \sin^2(\theta_{\max})\left[1 - \frac{\sin\alpha}{1-\sin\alpha}\left(\tan\frac{\theta_{\max}}{2}\right)^2\right]} \tag{7.40}$$

The radius $R(z_0 + dz)$ is found from Eq. (7.34), or from the curve using the method in Fig. 7.21, or from the voltage required to evaporate the next atom using Eq. (7.33). The new value of α or dR/dz is used in the further step to obtain an accurate calibration of $R(n)$.

In an off-axis position, this depth calibration for any atoms in the sequence of evaporation must be added to the z position calculated from Eq. (7.20).

If the tip has a constant shank angle all along its profile, and using Eq. (7.35), the radius R depends on the number of atoms n

$$R(n) = \sqrt[3]{\frac{3\frac{\sin\alpha}{1-\sin\alpha}\frac{V_{at}}{Q}}{\pi \sin^2(\theta_{\max})\left[1 - \frac{\sin\alpha}{1-\sin\alpha}\left(\tan\frac{\theta_{\max}}{2}\right)^2\right]}n + R_0^3} \tag{7.41}$$

The voltage variation is then given by

$$V(n) = \sqrt[3]{\frac{3\frac{\sin\alpha}{1-\sin\alpha}\frac{V_{at}}{Q}(Fk_F)^3}{\pi \sin^2(\theta_{\max})\left[1 - \frac{\sin\alpha}{1-\sin\alpha}\left(\tan\left(\frac{\xi\theta'_{\max}}{2}\right)\right)^2\right]}n + V_0^3} \tag{7.42}$$

The interest in this method is that the comparison of the real voltage with the virtual voltage can help to optimize the free parameters introduced in the model.

Influence of Physical Parameters in the Accuracy of the Reconstruction

Most of the reconstruction parameters are generally straightforward to evaluate or measure. The detector efficiency and the detector field of view are linked to instrumental features. Atomic volumes are deduced from material crystallographic data. Evaporation fields are obtained from tabulated values or deduced from measurements of charge states. It is worth noting that evaporation fields depend on physical constants of the studied materials and not on any geometrical features. Assuming V_{at}, Q, and F as defined constants, the depth calibration is thus dominated by the accuracy of the two parameters ξ and k_F, or the variation of R with z. Remember that the ICF, in the general case, is in the range $1 < \xi < 2$; the field factor k_F is in the range $2 < k < 8$.

Any errors in the estimation of these parameters influence the accuracy of the reconstruction. The magnitudes of these errors are evaluated considering the evolution of the lateral position $L_p = \sqrt{x^2 + y^2}$, and Δz for a small number of atoms evaporated (less than one radius of curvature). Considering Eq. (7.20), (7.34), and (7.37), the equations are simplified to

$$
\begin{aligned}
L_p &\approx \frac{V}{kF} \times \xi \times \theta' \\
z &\approx \frac{V}{k_F F} \times \xi^2 \frac{(\theta')^2}{2} + \frac{\Delta n \times V_{at}}{Q\pi \left[\frac{V}{kF}\right]^2 \sin^2\theta_{max}}
\end{aligned}
\tag{7.43}
$$

Using the voltage curve for reconstructing the data or

$$
\begin{aligned}
L_p &\approx R \times \xi \times \theta' \\
z &\approx R \times \xi^2 \frac{(\theta')^2}{2} + \frac{\Delta n \times V_{at}}{Q\pi R^2 \sin^2\theta_{max}}
\end{aligned}
\tag{7.44}
$$

using the knowledge of $R(z)$.

Fig. 7.24 illustrates the different reconstructed volumes that could be obtained using these standard algorithms on a virtual sample which contains well-separated layers. Errors in detection efficiency perturb the second

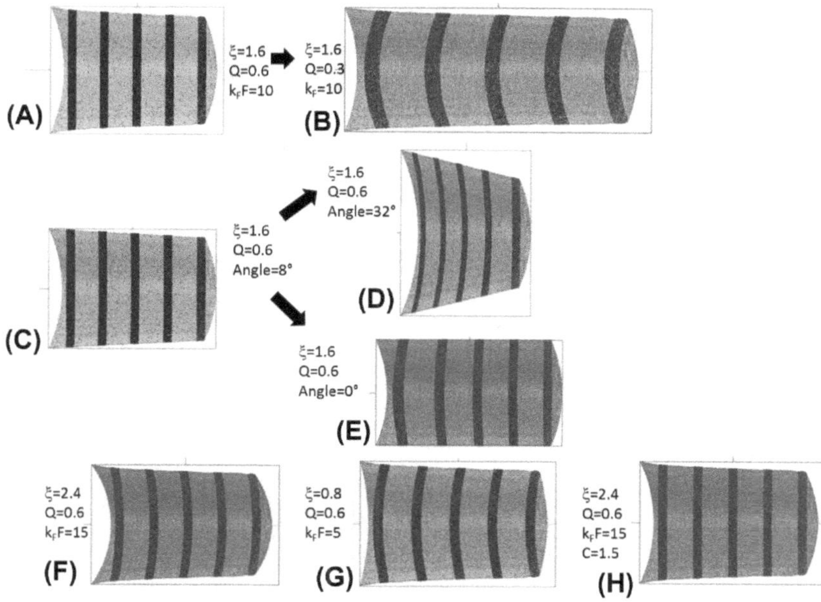

Figure 7.24 Reconstruction of a multilayer sample with different reconstruction parameters. (A) Perfect reconstruction using voltage mode ($\xi = 1.6$, $Q = 0.6$, $k_F F = 10$ *V/A*); (B) incorrect detector efficiency ($Q = 0.3$); (C) cone angle reconstruction with $\alpha = 8°$; (D) incorrect angle $\alpha = 32°$ or (E) 0°; (F) incorrect value of $k_F F$ and ξ, but correct value of $k_F F/\xi$ ($\xi = 2.4$, $k_F F = 15$ *V/A*); (G) $\xi = 2.4$, $k_F F = 15$ *V/A*; (H) incorrect value of $k_F F$ and ξ, but correct value of $k_F F/\xi$, and curvature correction factor $c = 1.5$.

term in the depth axis reconstruction equation. Distortions appear giving incorrect distances between layers and curvatures problems. Note that two atoms situated at the same depth, but with one atom at the center ($\theta = 0$) and one atom at the detection angle $\theta_{\max}$ are shifted in the evaporation order by $\Delta n \approx \frac{Q\xi^2 \pi R^3 \sin^2\theta_{\max}}{2 \times V_{at}} \theta_1^2$. The curvature correction corrects this parabolic distortion.

When using shank angle reconstruction, incorrect shank angles give rise to nonstationary distortions in depth. For instance, using $\alpha = 32°$ rather than 8°, the distance between layers reduces for the last evaporated layers.

In voltage mode reconstruction the most important parameter is the ratio $\xi/k_F F$, controlling the magnification. A good estimate of this value ensures a correct average reconstructed volume (if Q is also correct). However, if $\xi/k_F F$ is correct but ξ (and therefore $k_F F$) is too large, some deformations of the layers occur. This effect can be empirically corrected dividing the curvature correction term by a curvature correction factor c, which will not change the global volume of the reconstruction. An example with incorrect values of ξ and $k_F F$ corrected by the factor c is presented in Fig. 7.24H.

In general, the presence of features in the image that can be correlated to other techniques such as TEM is a good way to reconstruct fair APT images. Numerous examples exist in the literature [64–67]. Ideally, a TEM tomography image of the same object before analysis can serve as a model to adjust step by step the global parameters (Fig. 7.25). Q is generally constant, so that only $\xi/k_F F$ must be tuned significantly. In pure metals or homogeneous materials, the ICF can be closely determined using field ion microscopy or field desorption images, and the average field factor is deduced using a measurement of the tip radius R. The presence of visible atomic planes is often used to optimize the parameters with reasonable accuracy. However, the visibility of crystallographic features such as depleted poles or zone lines or atomic planes in the reconstruction is not systematic. In the general case, there is no straightforward calibration of the parameters.

It should also be pointed out that the algorithm assumes that the geometric parameters k_F and ξ used to build the 3D atom map are constant over the course of an analysis. This assumption is obviously an approximation, as shown previously. The ICF and the field factor vary with the global morphology of the electrodes. The tip blunts under field evaporation, and features such as the final shank angle, or the tip radius of tip to counter electrode ratio evolve significantly during the analysis. A strategy was proposed

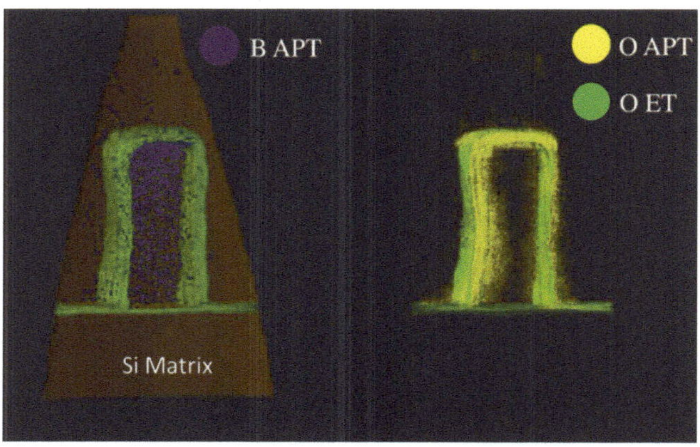

Figure 7.25 Three-dimensional reconstruction of a FinFET transistor after segmentation in electron tomography combined with APT reconstruction of the same device [62]. An accurate mapping of the boron distribution inside the gate [in blue] is obtained by adjusting the reconstruction parameters to fit the oxide layers in both images.

by Gault et al. to alleviate this problem through the implementation of an algorithm that dynamically accommodates variations in the tomographic reconstruction parameters in pure metals or diluted alloys. This requires the automatic recognition and indexing of crystallographic poles and associated atomic planes. This enables the extraction of the variation of the ICF all along the analysis. The reconstructed interplanar spacing, through Eq. (7.37), enables one to find also the correct evolution of the curvature radius as a function of the number of evaporated atoms [68].

Ideally, the dynamic reconstruction parameters could also be predicted using simulations, so as to enable the application of this protocol to data in which less crystallographic information is retained (such as in amorphous or nanostructured specimens). However, the trends extracted from electrostatic simulations do not always exactly reflect experimental measurements. Exact representations of the tip, the specimen holder, the counter electrodes, and the electrical voltage applied on each electrode are required to accurately predict the evolution of the reconstruction parameters, which is extremely cumbersome.

Artifacts Degrading Spatial Accuracy: Beyond Classical Standard Algorithms

The hypotheses underpinning the standard reconstruction algorithm are extremely simple. Assuming a homogeneous evaporation field of the sample, the specimen is well approximated by a spherical surface of radius positioned on a truncated cone with a shank angle, giving rise to a deterministic ion projection. Even if the algorithm was designed for limited field-of-view reconstructions, and for homogeneous materials, reconstructions based upon this algorithm yields "good looking" 3D images in a variety of systems. However, a deeper inspection of the data quality in numerous examples of materials from pure metals to complex nanostructured materials shows the obvious presence of image distortions that cannot be corrected by adjusting the global parameters.

Beyond the Spherical Hypothesis of Homogeneous Materials (Constant Evaporation Field)

A simple, first-order representation of the specimen surface as a spherical cap is limited for capturing the surface details that affect reconstruction accuracy. It is particularly true in atom probes that have a wide field of view. Because the surface morphology of the tip is linked to the evaporation field through the equation $F = V/k_F R$, with R the local tip radius, any anisotropy or inhomogeneous evaporation field of the materials surface will generate deviations to the spherical surface. Due to the correlation of the field evaporation threshold and the bounding interaction between surface atoms, the field evaporated end form of pure metals is not a perfectly smooth surface. Instead, it has many edges of the lattice steps and bump and flat facets. This prediction was observed experimentally in pure metals or semiconductor samples [69–74]. Local radii of curvature may change by more than 20% between two crystallographic directions that may induce accuracy degradation of several nanometers. In addition, these variations are temperature dependent, which prevents the use of simple methods for correcting the images.

To improve the dataset, the input of additional morphological information is required. For instance, the presence of well-identified planar layers in the volume that were characterized outside the atom probe can be used as landmarks to improve the reconstruction [55,75]. For depth profiling of flat layers, the reconstruction method could be improved by including this information.

Let us consider a sample with an unknown end shape (Fig. 7.26A). In this case, the standard model applies a curvature correction to shift the atom positions along the z axis following Eq. (7.44). If a planar feature exists in the volume, ie (dashed line in Fig. 7.26A), the last atom A of this feature that is analyzed is the atom situated on the tip axis. The z position (z_A) of this atom defines the actual z position of the entire feature. A correct reconstruction method would adjust the z to replace all the atoms of the feature (atom B in the example) at this defined position (z_A). The curvature correction normally compensates for the difference in evaporation order between nonaxial interface atoms and the central atom (A). In the standard reconstruction procedure, the curvature correction is limited to a hemispherical correction. If the tip shape is complex, with variations of local curvature, distortions will be observed. In the case of Fig. 7.26, Eq. (7.44) can be rewritten separating both terms in the z calculation.

$$\begin{aligned} \Delta z_B &\approx f_S(L_B) \\ \Delta z_A &\approx \frac{\Delta n \times V_{at}}{QS_D} \\ \Delta z_A &= \Delta z_B \end{aligned} \tag{7.45}$$

f_s is here an unknown function that describes the shape of the tip along z as a function of the position of the impacts onto the detector. Note that we will

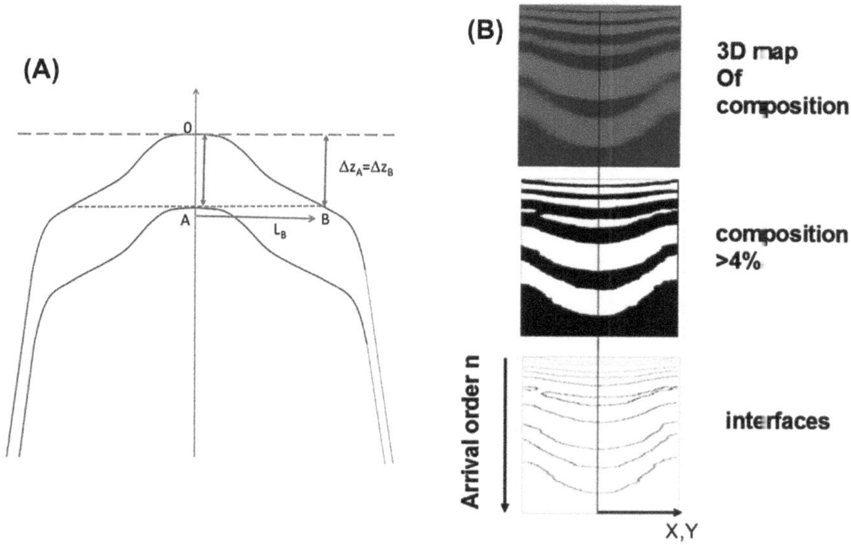

Figure 7.26 (A) Schematic of the equivalence evaporation order/curvature correction during field evaporation of planar features. The difference in evaporation order between an atom A of the center of the tip and an atom of the border B is equivalent to the required curvature correction in reconstruction. (B) Experimental evidence of the variation of the evaporation order on a nonspherical tip shape. The positions of straight interfaces, obtained by applying threshold criteria in the composition map are wavy. The interface shapes are used to deduce the correct correction to be applied.

consider simply that the position onto the detector linearly depends on the position of the atom laterally onto the surface. This equation shows that, the curvature information is contained inside the order of evaporation of atoms Δn.

The presence of interfaces in the dataset represented with coordinates (X, Y, n) with X, Y the positions onto the detector and n the order of evaporation is highlighted by plotting a 3D composition map in an experimental case in Fig. 7.26B. Interfaces are defined as boundary regions between phases of interest. A shift along the n axis between the tip center and the 3D image border is related to the actual difference in depth of the tip surface between the tip axis and the tip border.

By detecting the locations of planar interfaces, it is possible to deduce the function f_s for different depths inside the volume. A linear interpolation of this function is used to extract this information for any impact of the dataset. Finally, a tomographic reconstruction of the new dataset is performed.

An example of the reconstruction that can be achieved using this method is shown in Fig. 7.27. A multilayer test structure composed of

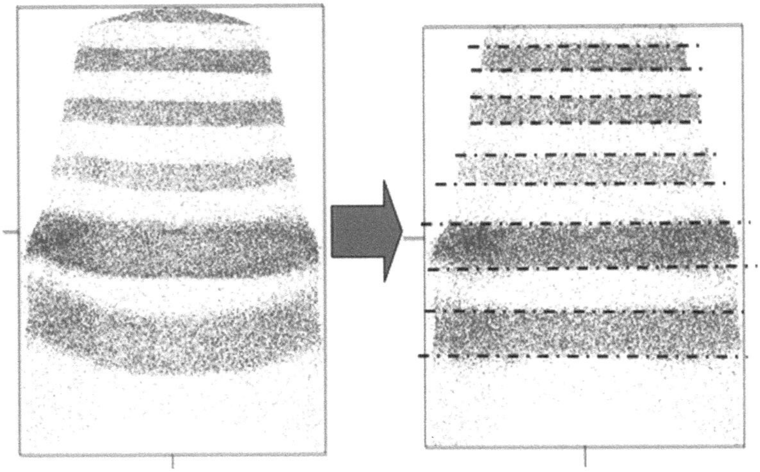

Figure 7.27 Comparison between the standard reconstruction method on an GaN/InGaN/AlGaN multilayer example and the interface-based reconstruction method. Note the variation of the curvature of layers with the standard method.

successive ~10-nm layers of InGaN/GaN followed by AlGaN/GaN layers was analyzed in APT. The standard reconstruction fails to reconstruct straight interfaces over the total analyzed surface. The interfaces show strong variations of curvature for wide detection angles, which illustrates the flaws in the reconstruction. Using the alternative method, a new reconstructed volume was obtained. Only eight continuous interfaces were used to correct the curvature aberrations. As expected, planar interfaces are reconstructed by the method. Losses of resolution are only observed close to the volume border due to the influence of lateral resolution for these large angles. This reconstruction enables composition profiles across the complete sequence of layers with a higher degree of accuracy [57].

Lateral Distortions in Homogeneous Materials (Constant Evaporation Field)

In one-dimensional atom probe, and also in imaging atom probe, two ancestors of APT, it was noted that the ion trajectories of field evaporated surface atoms of pure metals could differ significantly from the ion trajectories experienced by ionized gas atoms in field ion microscopy [76—79]. The image of ion impacts from pure crystalline materials shows unexpected density variations that are highly correlated with the crystallographic nature of the studied material. Low-index poles are often highly depleted as well as zone axes (Fig. 7.26). Sometimes the reverse situation is observed with a high density of impacts close to these regions. Local focusing and/or defocusing of these trajectories can explain these observations in correlation with field evaporation mechanisms. Since field evaporated end form of pure metals is not a perfectly smooth surface but has many edges of the lattice steps and bump and flat facets, it can perturb the ideal radial projection. Close to high curvature regions, divergence of the ion trajectories could be observed. To explain dense regions, it was proposed by Wada and Waugh that during field evaporation, the ion path can experience subtle deviations induced by the interaction with the closest neighbors [23,24,74]. This idea was confirmed by simple geometric simulations provided by Moore et al. [80], but the physical nature of the proposed local interactions and the impact of the temperature on the sharpness of the observed features were rather unclear at this time. Already shortly in the history of atom probe, people started to develop simulation tools to compare with experimental micrographs. It was proved by calculating the deterministic trajectories of ions during field evaporation that most of the aberrations are induced by subtle deviations from the ideal radial projection of ion trajectories close to the surface atoms. Low-index poles appear as locally very rough, and the presence of periodic terraces perturbs significantly the ion trajectories. Note that these zone lines tend to disappear at high temperature when the evaporation order becomes less deterministic (or more random) [11,15,20,78]. The local morphology of the surface becomes less self-similar, which prevents the observations of systematic ion trajectory deviations.

In materials composed of grains or nanograins, the deviations can be significantly different at both sides of a grain boundary. Density artifacts are therefore observed at the location of the grain boundary, which can be of either low density or high density depending on the direction of the global deviations from both grains at the tip surface (Fig. 7.29).

A possible way to correct these distortions is to insert the projection relationship calculated by simulation to extract the exact conversion between the lateral position on the tip and the detected impact positions. It was done, for instance, to correct the density in a simple case around a (001) pole of an FeAl alloy, using cylindrical symmetry assumption [81]. This method was able to correct for the huge depletion of atoms at the pole center. Obviously, the cylindrical symmetry used was unable to correct weaker zone lines effects. In this aim a real 3D model is mandatory. Such kind of input was proposed by Larson et al. by computing an image compression factor function that changes with the angle (Θ,φ) in a BCC structure [78,79]. However, the application of these methods to real data is extremely

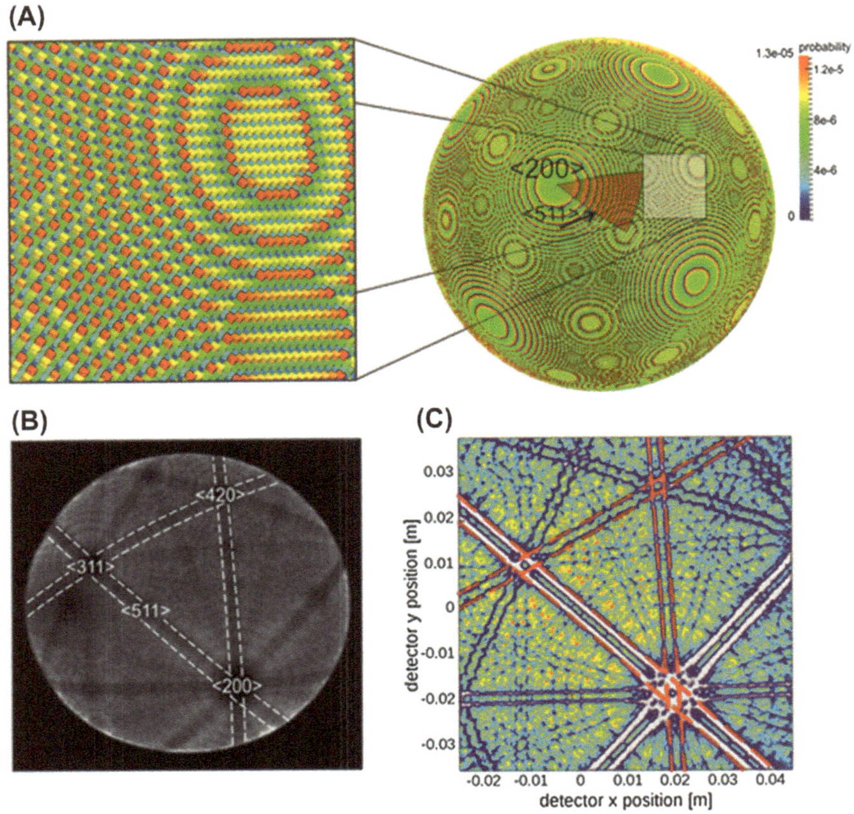

Figure 7.28 Comparison of experimental (B) and simulated field evaporation maps (C) obtained by modeling ion trajectories of evaporated atoms from the tip surface to the detector for an aluminum sample. (A)The structure of the simulated tip corresponds to an Al tip oriented along the direction <511>. Local density variations in (C) are similar to experimental observations in (B) and are explained by systematic deviations of trajectories in the first steps of flight.

time consuming and cumbersome, since the real structure of millions of atoms in the tip must be modeled and the calculation of millions of ion trajectories must be performed at the vicinity of an atomically defined tip.

Another way to improve the image is to assume that the atomic density inside the tip should be constant. Image distortions in Fig. 7.28 induce mostly compression or dilatation of the impact density on the detector. Density correction algorithms were developed to restore a quasi-constant density of atoms or impacts onto the detector (Fig. 7.30). This method was first proposed by De Geuser et al. [82] and is implemented in commercial

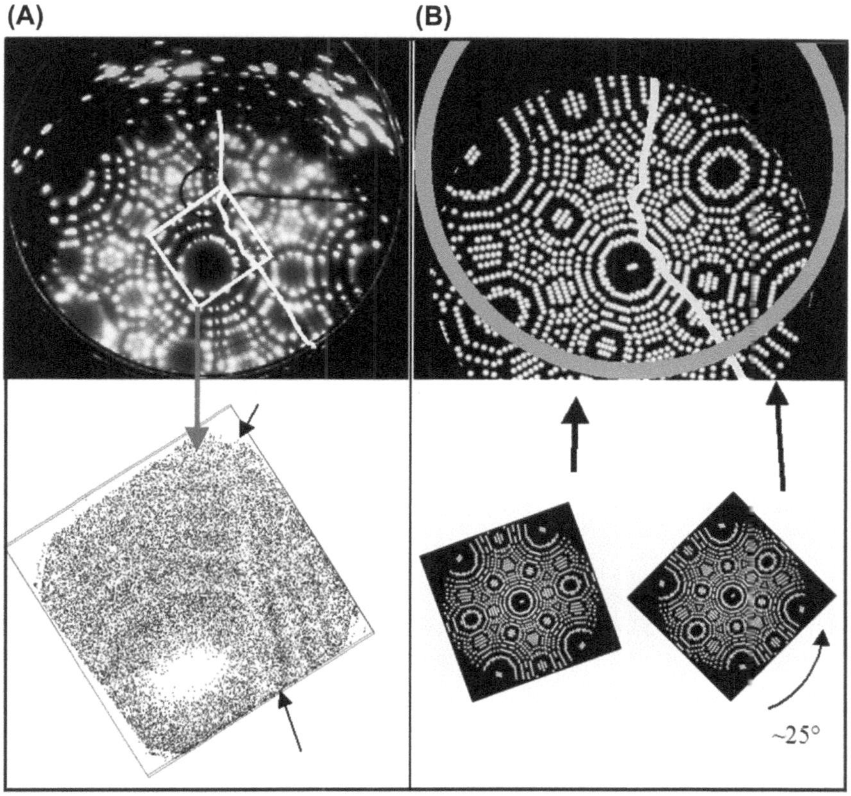

Figure 7.29 Density artifacts observed at the grain boundary of a *W* specimen. Note also the low density of the (011) pole center (A). The grain boundary has a low tilt angle, and the tip is oriented along the <011> common direction of both grains as modeled in (B) to match the field ion microscopy image of the tip surface.

reconstruction software. A grid is superimposed to the ion impacts, and the positions of impacts are evaluated with reference to this grid. Inside each cell of the grid composed of four nodes, a number of atoms can be calculated. The density of impacts or atoms is simply this number divided by the surface of a cell. By moving step by step the nodes of the grid to increase or decrease the surface of each cell, the density can be homogenized. Nodes are simply displaced following the density gradient calculated inside the grid. The last step consists in replacing atoms with reference to the nodes.

Although the obtained image appears much more homogeneous, these kinds of corrections are practically not often used. The reason is first the correction is generally bad close to the boundary of the volume (possible deformations exist outside the probed area and are not taken into account).

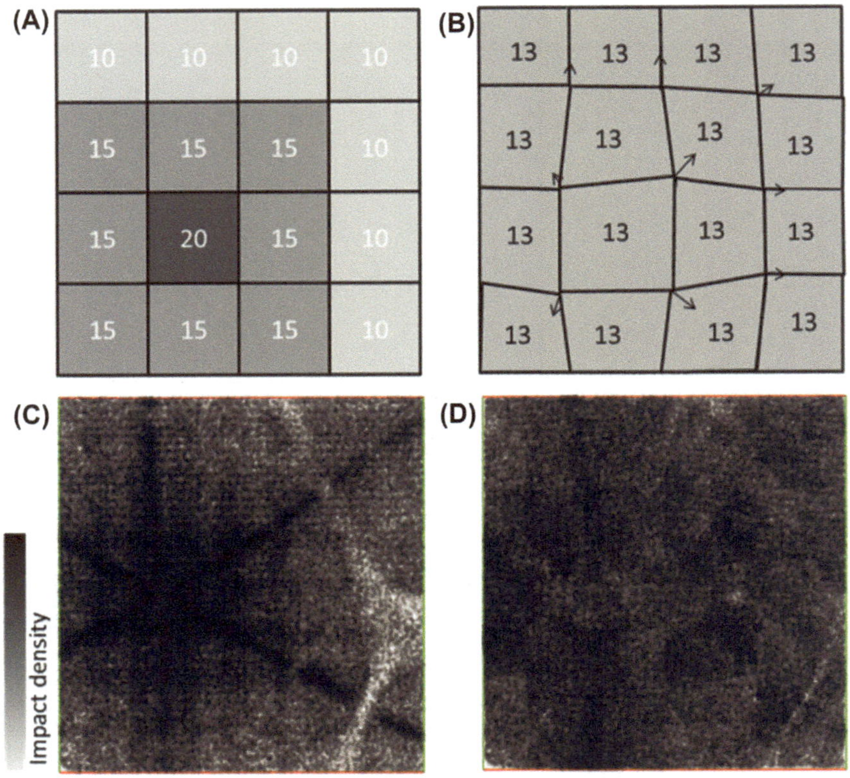

Figure 7.30 Lateral relaxation method and application to correction of zone lines contracts aberration. The impact map is restored to a uniform detector illumination. Local structure is preserved. (A) Density is calculated on a grid; (B) grid is deformed to achieve homogeneous density; (C) application to an experiment done with an face centered cubic pure gold sample, presenting strong zone line aberrations as seen on the density map (128 × 128 nodes); (D) after relaxation the density is much more homogeneous.

Second, zone lines are useful to have accessible crystallographic information directly in the dataset and users prefer to keep them visible.

Spatial Distortions in Multiphase Materials: In-Depth Degradation

In homogeneous materials, evaporation field of the materials of interest can smoothly vary over 20% at the tip surface. When analyzing materials composed of different atoms, and following field evaporation theory, this variation can reach 300% in exotic cases, and very abruptly if no interdiffusion exists between adjacent phases, for instance. Let us consider a material composed of two (A and B) or more phases (C, D, etc.) stacked perpendicularly to the tip axis. Each phase is defined by a unique evaporation field

(F_A, F_B, etc.). At the interface between phases, we define a field ratio $f = F_A/F_B$. The emitter mean radius of curvature R is related to the surface field F by the relationship $F = V/k_F R$, where V is the electric potential applied to the tip and k_F the field factor (note this formula is only valid for R positive). We assume that this formula can be applied locally over any surface point, with a constant factor k_F. From this consideration, one can infer several rules governing the emitter shape evolution.

1. There is tangential continuity all over the emitter surface. Otherwise R would be locally null and F infinite: the surface residence time of such a point (ie, atom) is null.
2. Each phase of the sample develops a constant mean curvature surface (eg, the hemispherical cap when the sample is homogeneous). Indeed, the progressive removal of atoms with high evaporation probability results in a surface with a constant evaporation probability, and thus a constant electric field.
3. For the same reason, the mean radii of curvature of two different phases A and B compensates the evaporation field differences such as $R_A/R_B = F_B/F_A$.

As a result, in multiphase cases, the tip will develop a complex shape that diverges from the perfect sphere [83–88]. This divergence increases with the difference in evaporation field and is nonstationary (see Fig. 7.20, Chapter 2: Field Ion Emission Mechanisms). Note that the presence of two adjacent phases at the tip apex breaks the spherical symmetry of the tip shape. In the simple case of a bilayer stack along z, the lower phase develops different radii of curvature along the interface (R_N) and perpendicularly to the interface (R_C). The global curvature radius R is defined as $1/R = (1/2) \times (1/R_N + 1/R_C)$.

Using standard reconstruct algorithm on these cases is therefore incorrect, since the tip apex cannot be approximated with a single curvature radius. An example of degradation of the image induced by this effect can be observed in Fig. 7.31. The specimens consists of several Cu layers of ~2 nm thickness separated by $Co_{90}Fe_{10}$ layers (~3 nm thick) grown on a seed layer which ensures the films have the <111> direction perpendicular to the surface. The first one is perpendicular to the tip axis, whereas the second one is slightly tilted from this direction. Individual (111) planes in FeCo regions can be resolved. This indicates that the precision achieved in the depth direction is better than 0.2 nm. However, some unexpected density variations are observed at the Cu/CoFe interfaces. Depleted zones are exhibited below the Cu layer in both cases. In addition, the planar spacing

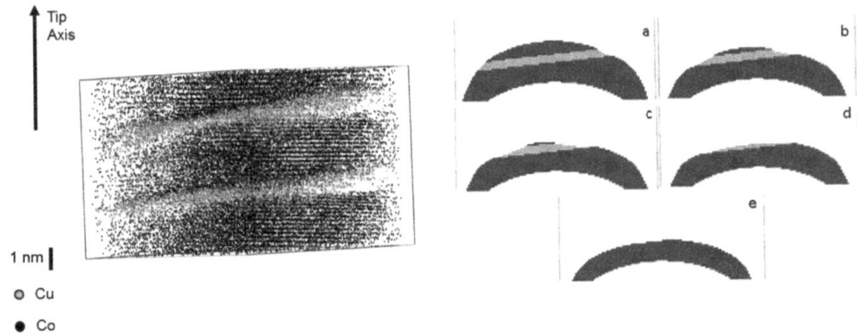

Figure 7.31 Reconstruction of a FeCo/Cu multilayer stack slightly tilted from the tip axis. Note that the upper parts of the volumes were the first to be evaporated. Evaporation field ratio $f = F_{Cu}/F_{Co}$ between Co and Fe is about 1.2. Deformation in z of the image (depleted zones) is induced by nonstationary evolution of the tip shape under field evaporation as seen on the simulated image (A–E).

is strongly perturbed in the close vicinity of the Cu layer. When the layer is strictly perpendicular to the tip axis, the reconstructed atomic density of the copper layer is strongly increased and the Cu planar spacing is simply compressed in the depth direction. When the layer is slightly tilted, additional deformations of the planes are observed. Close to the borders of the reconstructed volume, an increase of the planar spacing is observed within the copper layer.

The gradual change of the tip morphology can be simulated. Fig. 7.31A–E shows a series of tip profile snapshots for the tilted layer sample, in which 20 atomic layers have been evaporated between the first snapshot and the last one. Strong deformations of the tip morphology are exhibited. The copper layer is preferentially evaporated and it leads to a drastic modification of the initial hemispherical surface. The preferential evaporation not only flattens the copper layer surface but also leads to an apparent negative curvature at the tip surface in the plane of the figure. All nonhomogeneous evaporation from the surface may therefore lead to reconstruction errors of the z direction. Such artifacts were observed, for instance, in disordered alloys. Because of the preferential retention or evaporation, iron atoms in amorphous FeB alloy were found to cluster in unexpected plate shapes with high atomic density, and the depth resolution is made worse by this effect [89].

Two simple assumptions govern the depth reconstruction procedure. The tip surface is assumed to be hemispherical and atoms are assumed to be evaporated from this defined surface in a homogeneous way. This suggests that the hemispherical geometry is considered as a steady-state shape.

The depth direction is reconstructed from the order in which the ions strike the single atom detector. The snapshots presented in Fig. 7.31 clearly show that the tip shape is no more a hemispherical steady-state shape. The depth difference, between the mean hemispherical morphology of analyzed surface and the actual deformed surface, amounts to two to three atomic diameters. This difference between the real shape and the theoretical shape of the tip is not taken into account by the reconstruction algorithm and is sufficient to produce the observed deformation images.

Note that postreconstruction algorithms can be applied to correct these distortions. Since the evaporated atoms are received over the relatively large area of the detector, strong differences of the impact density are observed at its surface. Several layers could be evaporated locally, whereas on average only one atomic layer has been removed from the surface. Since the z direction is calculated from this mean evaporation from the surface, the increase in atomic density is simply due to artificial compression or expansion in the z direction. A correction method would only modify the positions in depth of each atom to restore a constant atomic density within the reconstructed volume. However, this kind of density correction neglects the magnification artifacts induced by the tip shape modification during the nonstationary evaporation process as it will be discussed hereafter.

Spatial Distortions in Multiphase Materials: Local Magnification Effects

The deformation of the tip shape in the analysis of multiphase sample induces also variations of the image projection, degrading lateral accuracy. The strong differences in physical properties induce strong distortions of the images that were also observed at the early stages of the technique. Solid-state precipitation of a given phase within a matrix is a key metallurgical phenomenon that is often used as a lever to optimize a material's properties and has hence been a topic of intense research using APT. Accurately reconstructing volumes containing precipitates can be challenging due to localized concentrations of atoms with a different evaporation field, which, as pointed out previously, lead to the development of local nonspherical curvature. Focusing or defocusing effects, named local magnification effects, were observed in multiphase materials when the phases are significantly different in terms of threshold field of evaporation F_{EV}. Using a phenomenological description of the field evaporation of the phases, and assuming a surface field proportional to the surface curvature ($F_{EV} \sim V/R$, with R the local radius of curvature at the specimen surface), these artifacts were explained qualitatively but the detailed nature of the imaging process remained unclear [43,44,90,91].

Indeed, high F_{EV} phases tend to develop high surface curvature compared to neighbor low F_{EV} phases. Since the magnification is proportional to the local curvature, a significant difference in magnification between phases is expected.

In a first approach, we may consider the magnification defined as $\boldsymbol{M} = L/\xi R$, so that each phase induces its own magnification $\boldsymbol{M}_A = L/\xi R_A = L \times F_A/\xi$ and $\boldsymbol{M}_B = L/\xi R_B = L \times F_B/\xi$. In this case, the magnification ratio between phases would be equal to the ratio of the evaporation field. We may also observe an overlap, or a strong depletion of ion trajectories at the interface (Fig. 7.32A). However, this schematic approach finds some limits at the interface. Indeed, considering that the electrostatic surface should not present sharp roughness (otherwise the local curvature would be null, and therefore the local field infinite), a continuity of the surface at the interface, and a tangential continuity should exist between phases. Initial trajectories of ions from both sides of the interface should be theoretically parallel. In this case, no discontinuity of the projection law should exist (Fig. 7.32B).

Nevertheless, size effects in the extent of local magnification, the anisotropy of the deformation of the surface, and the presence of trajectory overlaps in the vicinity of interfaces of real 3D images limit the practical use of this simple approach. The impact of the atomic nature of the surface is, for instance, evident. Understanding the imaging process in APT requires not only determining the consequence of mesoscopic features on the image formation. A microscopic to macroscopic inspection of the field evaporation mechanism, taking into account at the same time the crystallographic nature of the sample, the presence of different phases or atoms, the initial shape of the sample, and its dynamic evolution is therefore mandatory.

Field evaporation models were used to understand the nature of the distortions that are induced by the presence of several phases with different evaporation fields. Local magnification and trajectory overlaps were studied for small spherical precipitates with a small diameter embedded in a pure crystalline matrix [92]. The evaporation field of the precipitate was varied in the range 0.85−1.15. Snapshots of the tip shape evolution for two particles with a difference in evaporation field of 15% evaporation field precipitate ($f = 0.85$ and $f = 1.15$) are shown in Fig. 7.33. A few atomic layers were evaporated until the particle emerged. Note the slight deformation of terraces intersecting the particle. The existence of protrusions or cusps

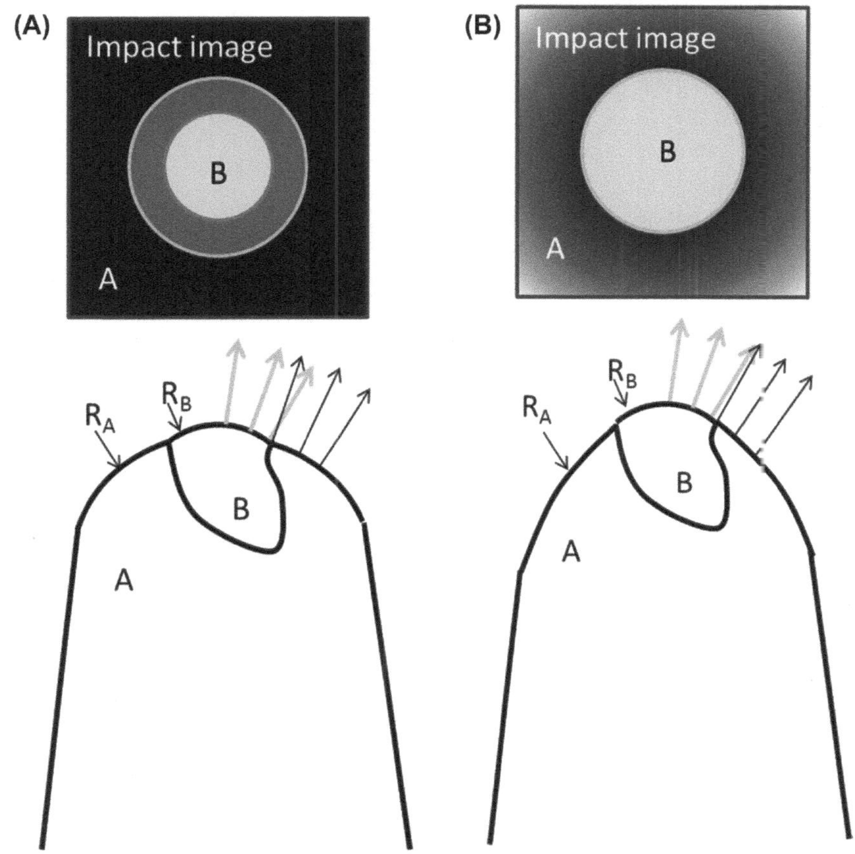

Figure 7.32 Schematic image of the local magnification effects in multiphase analyses. Each phase develops a curvature radius that depends on its evaporation field (A) classical interpretation that predicts defocusing effect of the high-field phase (B) and trajectory overlap at the interphase (B) interpretation taking into account tangential continuity, magnification is anisotropic and no trajectory overlaps can occur.

in the particle region, however, appears to be difficult to ascertain. The notion of local curvature is in fact difficult if not impossible to define at this scale. The ion trajectories from such an emitter were calculated for a few tens of atomic layers evaporated until the particle was removed. Simulated positions of impacts on the detector were then used to reconstruct the volume that was analyzed in the simulation. The 3D reconstruction of a low or high evaporation field precipitate illustrates in a spectacular way the image distortions introduced by the so-called local magnification effects. The standard procedure, which assumes a homogeneous magnification at the tip

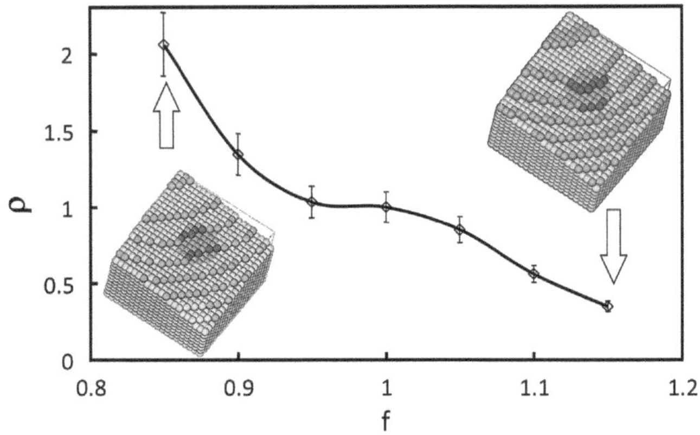

Figure 7.33 Variation of the impact density ρ on the detector inside a particle 2 nm in diameter, having a field evaporation different from the matrix. Density ρ is calculated from the impacts inside the particle as a function of the ratio of evaporation field (*f*) between the precipitate and the matrix. Low-field (respectively, high-field) particle emerges at the surface with a significant surface curvature radius that gives rise to focusing effects (respectively, defocusing).

surface, reconstructs the sphere as an ellipsoid. The observed deformation is caused by the enhanced focalization of ion trajectories originating from the precipitate increasing the observed atomic density.

This effect is well known experimentally and observed in numerous materials. Fig. 7.34 shows some examples of different materials. The extent of density variations can be extremely large, especially at the interfaces between phases. When varying the field ratio f in a larger window, it is predicted that the atomic density can present a singular point for a field ratio of 0.5–0.7 (Fig. 7.35) [44]. This point indicates that the tip surface is electrostatically flat, so that all ion trajectories are parallel for ions of the particle. For lower value of f, ion trajectories from the particle are focused to a point in front of the tip, and defocused afterward, giving rise to a negative magnification, and to a decrease of the atomic density. This effect was observed experimentally in oxide dispersion strengthened steel, where small yttrium oxide particles are dispersed in a Fe-based matrix [94]. Note that the magnitude of the effects varies also significantly with the ratio of the size of the precipitate relative to the specimen radius. Smaller precipitates are subjected to stronger focusing or defocusing effects for the same field.

Note that negative magnification exists also locally when high-field precipitates are embedded in a low-field matrix. At the interface between the Al matrix and the Al_3Zr precipitate (Fig. 7.34C), the interface spike in density is

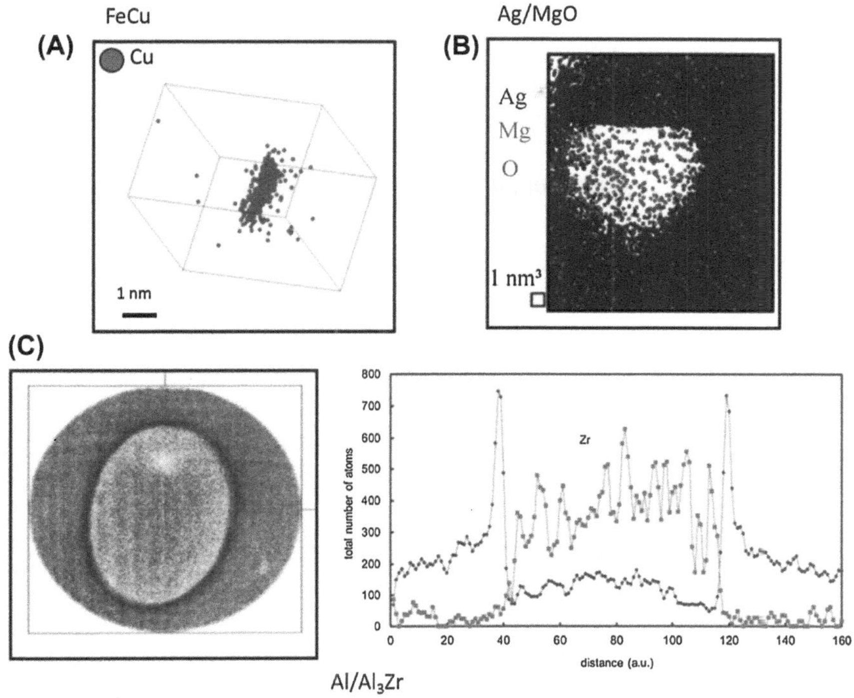

Figure 7.34 Three experimental examples showing local magnification effects. (A) A FeCu alloy showing small Cu particles (low field) artificially deformed into ellipsoidal shapes. (B) MgO particle (high field) embedded in pure Ag. MgO particles are experimentally reconstructed with a low atomic density [93]. (C) AlZr alloy with large Al_3Zr particles (high field). Note the huge drop of impact density inside the particle. *Courtesy of Williams Lefebvre.*

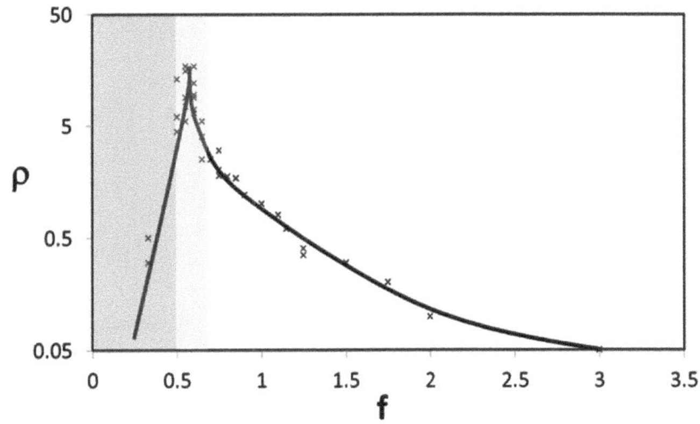

Figure 7.35 Density variation for large difference in evaporation field between phases. For $f \sim 0.5$–0.7 the magnification is close to 0 in the particle giving rise to a spike in ion impact density. Below this value the magnification is negative for the particle (absolute density then decreases for lower value of f).

induced by a negative local magnification effect. The Al phase is folded on itself at this location. This artifact induces local intermixing at the interface. This phenomenon is named trajectory overlap and can be severe for a large difference in evaporation fields ($f > 1.5$ or $f < 0.6$). In the general case (moderated difference in evaporation field), only the first atomic shell around the particle or phase intermixes with its neighborhood. In particular geometries (spherical, lamellar, etc.) the model can be used to evaluate the amount of intermixing between phases. This information is then used to correct significantly the measured composition [92,95].

Density Correction Applied to Local Magnification Effects

Since most of the aberrations introduced by the presence of different phases in the material are distortions of the density of impacts, it has been proposed to adjust the position of impacts laterally to achieve an equal density in the volume. The method is based on relaxation method, identical to the method presented in section Lateral Distortions in Homogeneous Materials (Constant Evaporation Field). The method proposed by F. De Geuser adjusts

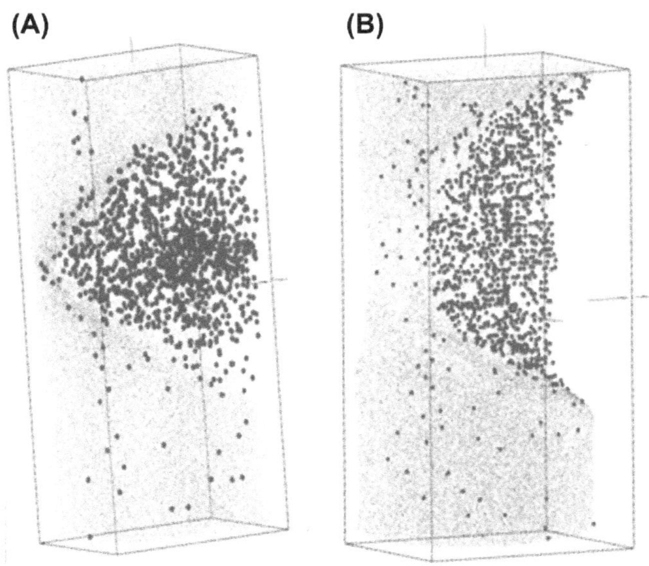

Figure 7.36 Local magnification effects corrected for an AlZrSc alloy. (A) Initial three-dimensional volume containing a high-field particle (Al_3Zr) presenting an internal low density of atoms due to local magnification effects; (B) after the density relaxation method, the density is homogenized in the whole volume. Note the presence of atomic planes inside the particle. *Courtesy of F. De Gueser.*

iteratively the position of atoms laterally in the volume in a 3D grid [20]. Not that no abrupt rupture of the volume is created in z by the method using linear interpolation between slices. Convergence criteria based on the mean quadratic difference between theoretical and measured densities are applied to stop the iteration of the algorithm. An example is presented in Fig. 7.36.

CONCLUSIONS AND PERSPECTIVES

Getting a 3D image in APT is extremely easy. After the generation of the dataset, you can virtually push a button, and easily obtain a fancy and colored 3D image of your sample that can give you lots of information. Using APT as a scientific microscope with robust estimation of metrological parameters is far more difficult. The bottleneck of the technique is certainly its spatial accuracy that is strongly specimen dependent. The first implementations of atomistic simulations of the imaging process have been a crucial reality check that the effects induced by the combination of atomic roughness and local curvature across the specimen surface are extremely complex. At least we are aware of the distortions possibly existing in the image. A step further would be to recompute the exact trajectory of each ion based on an atomically defined specimen inserted within a simulation model, and hence allow overcoming the distortions induced by the simplicity of the projection used in the current algorithm.

Haley et al. and Larson et al. proposed such methods [82,96]. Nevertheless, the computational cost of large-scale atomic-scale simulations, as well as the set of assumptions required to construct the simulation, precludes the implementation of such methods for the time being. However, simulations have demonstrated their merit beyond facilitating the interpretation of complex data and validating hypotheses with regard to fundamental aspects of field evaporation.

Possible improvements of the methods include the use of simulations to inform the development of algorithms yielding more accurate reconstruction by better modeling experimental data. Nevertheless, correlative microscopy (see Chapter 10: Correlative Microscopy by (Scanning) Transmission Electron Microscopy and Atom Probe Tomography) will be required to introduce in the models sufficient information necessary to describe at least at different depth inside the sample, the nature of the lens of our microscope, which will always be our specimen itself.

REFERENCES

[1] Müller EW, Panitz JA, McLane SB. Rev Sci Instrum 1968;39:83.
[2] Muller EW. Z Phys 1951;131:13.
[3] Panitz JA. Imaging atom probe microscopy. Prog Surf Sci 1978;8:219–62.
[4] Miller MK, Cerezo A, Hetherington MG, Smith GDW. Atom probe field ion microscopy, monograph on the physics and chemistry of materials. 52, Oxford science Publication. Oxford, UK: Clarendon Press; 1996.
[5] Wagner A, Seidman DN. Phys Rev Lett 1979;42(515).
[6] Blavette D, Bostel A, sarrau JM, Gallot J. In: Proceeding of the 29th IFES; 1982. p. 497.
[7] Goodman SR, Brenner SS, Low Jr JR. Metall Trans 1972;4:p2371–2378.
[8] Cerezo A, Godfrey TJ, Grovenor CRM, Heteherington MG, Hoyle RM, Jakubovics JP, et al. J Microsc 1989;184:215.
[9] Boll T, Al-Kassab T, Yuan Y, Liu ZG. Ultramicroscopy 2007;107:796–801.
[10] Geiser BP, Kelly TF, Larson DJ, Schneir J, Roberts JP. Microsc Microanal 2007;13: 437–47.
[11] Moody MP, Gault B, Stephenson LT, Haley D, Ringer SP. Ultramicroscopy 2009; 109(7):815–24.
[12] Cadel E, Vurpillot F, Lardé R, Duguay S, Deconihout B. J Appl Phys 2009;106(4): 044908.
[13] Vurpillot F, da Costa G, Menand A, Blavette D. J Microsc 2001;203:295–302.
[14] Vurpillot F, Renaud L, Blavette D. A new step towards the lattice reconstruction in 3DAP. Ultramicroscopy May–June 2003;95:223–9.
[15] Moody MP, Gault B, Stephenson LT, Marceau RKW, Powles RC, Ceguerra AV, et al. Microsc Microanal March 2011;17(02):226–39.
[16] Moody MP, Ceguerra AV, Breen AJ, Cui XY, Gault B, Stephenson LT, et al. Atomically resolved tomography to directly inform simulations for structure–property relationships. Nat Commun 2014;5:5501.
[17] Gruber M, Vurpillot F, Bostel A, Deconihout B. Field evaporation: a kinetic Monte Carlo approach on the influence of temperature. Surf Sci December 2011; 605(23–24):2025–31.
[18] Cerezo A, Clifton P, Gomberg A, Smith G. Ultramicroscopy 2007;107:720.
[19] Gault B, Moody MP, de Geuser F, Haley D, Stephenson LT, Ringer SP. Appl Phys Lett 2009;95:034103.
[20] Vurpillot F, Oberdorfer C. Modeling atom probe tomography: a review. Ultramicroscopy December 2015;159(Part 2):202–16.
[21] Vurpillot F, Bostel A, Cadel E, Blavette D. The spatial resolution of 3D atom probe in the investigation of single-phase materials. Ultramicroscopy 2000;84(3):213–24.
[22] Vurpillot F, Bostel A, Blavette D. A new approach to the interpretation of atom probe field-ion microscopy images. Ultramicroscopy 2001;89(1):137–44.
[23] Blakely J. Surface physics of materials. 1st ed., Materials Science and Technology; 2013. ISBN:9781483191478.
[24] Gault B, Müller M, La Fontaine A, Moody MP, Shariq A, Cerezo A, et al. Influence of surface migration on the spatial resolution of pulsed laser atom probe tomography. J Appl Phys 2010;108:044904.
[25] Wada M. On the thermally activated field evaporation of surface atoms. J De Physique, Colloque C9, supplément au n012, Tome décembre 1984;45:C9–89.
[26] Waugh AR, Boyes ED, Southon MJ. Surf Sci 1976;61:109.
[27] Sanchez CG, Lozovoi AY, Alavi A. Field-evaporation from first-principles. Mol Phys 2004;102(9–10):1045–55.
[28] Karahka M1, Kreuzer HJ. Ultramicroscopy September 2013;132:54–9.

[29] Parviainen S, Djurabekova F, P Fitzgerald S, Ruzibaev A, Nordlund K. Atomistic simulations of field assisted evaporation in atom probe tomography. J Phys D Appl Phys 2016;49(4):045302.
[30] Gault B, Danoix F, Hoummada K, Mangelinck D, Leitner H. Impact of directional walk on atom probe microanalysis. Ultramicroscopy February 2012;113:182–91.
[31] Moore AJ. Philos Mag 1981;A 43:803.
[32] Bas P, Bostel A, Deconihout B, Blavette D. A general protocol for the reconstruction of 3D atom probe data. Appl Surf Sci 1995;87–88:298–304.
[33] Cerezo A, Hetherington MG, M.Hyde J, Miller MK, Smith GDW, Underkoffler JS. Visualization of 3-dimensional microstructures. Surf Sci 1992;266:471–80.
[34] Blavette D, Sarrau JM, Bostel A, Gallot J. Direction and depth of atom probe analysis. Rev De Phys Appl 1982;17:435–40.
[35] Blavette D, Sarrau JM, Bostel A, Gallot J. Direction et distance d'analyse a la sonde atomique. Rev Phys Appl 1982;17:435–40.
[36] Walls JM, Southworth HN. Magnification in the field-ion microscope. J Phys D Appl Phys 1979;12:657–67.
[37] Wilkes TJ, Smith GDW, Smith DA. On the quantitative analysis of field ion micrographs. Metallography 1974;7:403–30.
[38] Cerezo A, Warren PJ, Smith GDW. Some aspects of image projection in the field-ion microscope. Ultramicroscopy 1999;79:251–7.
[39] Brandon DG. Accurate determination of crystal orientation from field ion micrographs. J Sci Instrum 1964;41:373–5.
[40] Newman RW, Sanwald RC, Hren JJ. A method for indexing field ion micrographs. J Sci Instrum 1967;44:828–30.
[41] Southworth HN, Walls JM. Projection geometry of field-ion image. Surf Sci 1978;75:129–40.
[42] Gault B, Haley D, de Gueser F, Moody MP, Marquis EA, Larson DJ, et al. Advances in the reconstruction of atom probe tomography data. Ultramicroscopy 2011;111:448–57.
[43] Geiser BP, Larson DJ, Oltman E, Gerstl SSA, Reinhard DA, Kelly TF, et al. Wide-field-of-view atom probe reconstruction. Microsc Microanal 2009;15:292–3.
[44] Gault B, de Geuser F, Stephenson LT, Moody MP, Muddle BC, Ringer SP. Estimation of the reconstruction parameters for atom probe tomography. Microsc Microanal 2008;14:296–305.
[45] Vurpillot F, Gault B, Geiser BP, Larson DJ. Reconstructing atom probe data: a review. Ultramicroscopy September 2013;132:19–30. http://dx.doi.org/10.1016/j.ultramic.2013.03.010.
[46] Larson DJ, Gault B, Geiser BP, De Geuser F, Vurpillot F. Atom probe tomography spatial reconstruction: status and directions. Curr Opin Solid State Mater Sci 2013;17(5):236–47.
[47] Walck SD, Buyuklimanli T, Hren JJ. Extended depth profiling with the IAP. J De Physique 1986;47:451–8.
[48] Gipson GS, Eaton HC. The electric field distribution in the field ion microscope as a function of specimen shank,. J Appl Phys 1980;51:5537–9.
[49] Gipson GS. An improved empirical formula for the electric field near the surface of field emitters. J Appl Phys 1980;51:3884–9.
[50] Gipson GS, Yannitell DW, Eaton HC. On the electric field distribution within the field ion microscope and near the surface of field emitters. J Phys D Appl Phys 1979;12:987–96.
[51] Bajikar SS, Kelly TF, Camus PP. Electrostatic analysis of local-electrode atom probes. Appl Surf Sci 1996;94–95:464–71.

[52] Rose DJ. On the magnification and resolution of the field emission electron microscope. J Appl Phys 1956;27:215–20.
[53] Haley D, Petersen T, Ringer SP, Smith GDW. Atom probe trajectory mapping using experimental tip shape measurements. J Microsc 2011;244:170–80.
[54] Vurpillot F, Gaillard A, Da Costa G, Deconihout B. A model to predict image formation in atom probe tomography. Ultramicroscopy September 2013;132:152–7.
[55] de Castilho CMC. Ion trajectories in atom probe field ion microscopy and gas field ion sources. J Phys D Appl Phys 1999;32:2261–5.
[56] Loi ST, Gault B, Ringer SP, Larson DJ, Geiser BP. Electrostatic simulations of a local electrode atom probe: the dependence of tomographic reconstruction parameters on specimen and microscope geometry. Ultramicroscopy 2013;132:107–13.
[57] Vurpillot F, Gruber M, Da Costa G, Martin I, Renaud L, Bostel A. Pragmatic reconstruction methods in atom probe tomography. Ultramicroscopy 2011;111(8): 1286–94.
[58] Vurpillot F, Bostel A, Blavette D. The shape of field emitters and the ion trajectories in three-dimensional atom probes. J Microsc 1999;196:332–6.
[59] Müller EW, Bahadur K. Phys Rev 1956;102:624.
[60] Dreschler M, Wolf P. Zur Analyse von Fieldionenmikroscop-Aufnamen mit atomer Auflösung. Proc 4th Int Congr Electron Microsc 1958;1:835.
[61] Bolin PL, Ranganathan BN, Bayuzick RJ. J Phys E-Sci Instrum 1976;9:363–5.
[62] Webber RD, Smith R, Walls JM. J Phys D Appl Phys 1979;12(9):1589–95.
[63] Larson D, Geiser B, Prosa T, Ulfig R, Kelly T. Non-tangential continuity reconstruction in atom probe tomography data. Microsc Microanal 2011;17:740–1.
[64] Arslan I, Marquis EA, Homer M, Hekmaty MA, Bartelt NC. Towards better 3-D reconstructions by combining electron tomography and atom-probe tomography. Ultramicroscopy 2008;108:1579–85.
[65] Petersen TC, Ringer SP. Electron tomography using a geometric surface-tangent algorithm: application to atom probe specimen morphology. J Appl Phys 2009;105.
[66] Gorman BP, Diercks D, Salmon N, Stach E, Amador G, Hartfield C. Hardware and techniques for cross-correlative TEM and atom probe analysis. Microsc Today 2008; 16:42–7.
[67] Grenier A, Duguay S, Barnes JP, Serra R, Rolland N, Audoit G, et al. Three dimensional imaging and analysis of a single nano-device at the ultimate scale using correlative microscopy techniques. Appl Phys Lett 2015;106(21):213102.
[68] Gault B, Loi ST, Araullo-Peters VJ, Stephenson LT, Moody MP, Shrestha SL, et al. Dynamic reconstruction for atom probe tomography. Ultramicroscopy 2011; 111(11):1619–24.
[69] Bolin PL, Ranganathan BN, Bayuzick RJ. Determination of field ion tip shapes. J Phys E Sci Instrum 1976;9:363–5.
[70] Fortes MA. Shape of field-evaporated metal tips. Surf Sci 1971;28:95–116.
[71] Moore AJW. Structure of atomically smooth spherical surfaces. J Phys Chem Solids 1962;23:907–12.
[72] Sakurai et Muller 1977 JAP.
[73] Webber RD, Smith R, Walls JM. Shape of field-ion emitters. J Phys D Appl Phys 1979;12:1589–95.
[74] Shariq A, Mutas S, Wedderhoff K, Klein C, Hortenbach H, Teichert S, et al. Investigations of field-evaporated end forms in voltage- and laser-pulsed atom probe tomography. Ultramicroscopy April 2009;109(5):472–9.
[75] Geiser BP, Larson DJ, Prosa TJ, Kelly TF. Microsc Microanal August 2013;19(S2): 936–7.
[76] Panitz JA, McLane SB, Muller EW. Rev Sci Instrum 1969;40:1321.
[77] Brenner SS, McKinney JT. Surf Sci 1970;23:88–111.

[78] Waugh AR, Boyes ED, Southon MJ. Field-desorption microscopy and the atom probe. Nature 1975;253:342–3.
[79] Waugh AR, Southon MJ. Surface studies with an imaging atom-probe. Surf Sci November 1977;68:79–85.
[80] Moore AJW. The simulation of FIM desorption patterns. Philos Mag A 1981;43:803–14.
[81] Vurpillot F, Bostel A, Menand A, Blavette D. Trajectories of field emitted ions in 3D atom-probe. Eur Phys J Appl Phys 1999;6:217–21.
[82] De Geuser F, Lefebvre W, Danoix F, Vurpillot F, Forbord B, Blavette D. An improved reconstruction procedure for the correction of local magnification effects in three-dimensional atom-probe. Surf Interface Anal 2007;39(2–3):268–72.
[83] Larson DJ, Geiser BP, Prosa TJ, Gerstl SSA, Reinhard DA, Kelly TF. Improvements in planar feature reconstructions in atom probe tomography. J Microsc 2011;243:15–30.
[84] Larson DJ, Geiser BP, Prosa TJ, Kelly TF. On the use of simulated field-evaporated specimen apex shapes in atom probe tomography data reconstruction. Microsc Microanal 2012;18:953–63.
[85] Oberdorfer C, Eich SM, Schmitz G. A full-scale simulation approach for atom probe tomography. Ultramicroscopy 2013;128:55–67.
[86] Rolland N, Vurpillot F, Duguay S, Blavette D. Dynamic evolution and fracture of multilayer field emitters in atom probe tomography: a new interpretation. Eur Phys J Appl Phys 2015;72(2):21001.
[87] Vurpillot F, Cerezo A, Blavette D, Larson DJ. Modeling image distortions in 3DAP. Microsc Microanal 2004;10(03):384–90.
[88] Vurpillot F, Larson D, Cerezo A. Improvement of multilayer analyses with a three-dimensional atom probe. Surf Interface Anal 2004;36(5–6):552–8.
[89] Vurpillot F, Bostel A, Blavette D. Ultramicroscopy 2001;89:137–44.
[90] Miller K. The effects of local magnification and trajectory aberrations on atom probe analysis. J De Physique 1987;48:565–70.
[91] Miller MK, Hetherington MG. Local magnification effects in the atom probe. Surf Sci 1991;246:442–9.
[92] Vurpillot F, Bostel A, Blavette D. Trajectory overlaps and local magnification in three-dimensional atom probe. Appl Phys Lett 2000;76(21):3127–9.
[93] Kluthe C, Al-Kassab T, Kirchheim R. Investigation of the internal oxidation in Ag-1at %Mg by 3D-TAP microscopy. Mater Sci Eng A 2002;327:70–5.
[94] Hatzoglou C. Quantification et correction des biais inhérents à l'analyse par sonde atomique tomographique des nanoparticules d'un acier ODS : application à l'étude de leur formation et à leur comportement sous irradiation [Ph.D. thesis]. University of Rouen; 2015.
[95] Blavette D, Vurpillot F, Pareige P, Menand A. A model accounting for spatial overlaps in 3D atom-probe microscopy. Ultramicroscopy 2001;89(1):145–53.
[96] Haley D, Petersen TC, Ringer SP, Smith GDW. J Microsc 2011;244(2):170–80.

CHAPTER EIGHT

Laser-Assisted Field Evaporation

A. Vella, J. Houard
Groupe de Physique des Matériaux, University and INSA of Rouen, Normandie University, France

INTRODUCTION

The field evaporation process can be triggered by electrical pulses or laser pulses.

Laser pulses are used to reduce the risk of rupture of fragile materials such as multilayers [1,2]. The use of electrical pulses induces a repeated mechanical stress applied to the sample, which can be at the origin of its rupture. Laser pulses, however, are directly focalized at the apex of the samples, keeping the mechanical stress due to the electric field almost constant. Moreover, thanks to the laser the voltage applied to the sample is reduced, decreasing the electrostatic stress applied to the sample.

Laser pulses are also used to analyze nonconductive materials (semiconductors, oxides, and polymers). If electrical pulses are used to field evaporate such kind of materials, due to their very high resistivity, the electrical pulse will be strongly distorted during the propagation from the base to the apex of the field emitter (or from the counterelectrode to the tip apex). An electrical pulse of 1 ns will be elongated up to hundreds of nanoseconds and its amplitude will decrease becoming too low to trigger the evaporation process. Hence, the analysis of nonmetallic field emitters is possible today only thanks to the use of laser pulses [2,3].

In the case of metallic samples, the use of the laser pulses can strongly improve the performances of atom probe tomography (APT), in terms of mass resolution or sensitivity [in terms of signal to noise (S/N) ratio], if the illumination conditions are well chosen [4,5]. The improvement in the mass resolution is related to a decrease of the evaporation time and of the energy spread of the evaporated ions, as we will show all along this chapter.

The use of laser pulses allows also the increase of the angle of ion collection and hence the increase of the analyzed volume, simply reducing the distance between the field emitter and the detector, as shown in Fig. 8.1 [3,6]. In fact, if the evaporation time decreases when laser pulses are used, then

Atom Probe Tomography
ISBN: 9780128046470
http://dx.doi.org/10.1016/B978-0-12-804647-0.00008-5

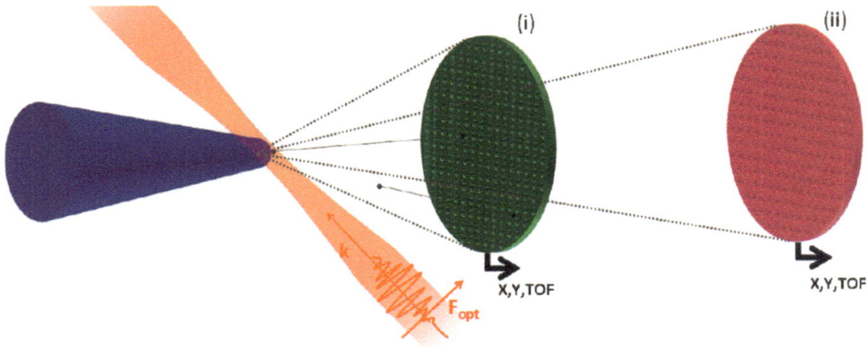

Figure 8.1 Schematic diagrams of the laser-assisted atom probe in large angle (i) and in small angle (ii). Laser propagates along k and its electric field is along F_{opt}. In this chapter, we only consider the case where k is perpendicular to tip axis and F_{opt} parallel to tip axis. *TOF*, time of flight.

shorter flight time can be used without affecting the precision on time measurement and the related mass resolution.

In this chapter, after a short historical background on field evaporation assisted by laser pulses, the optical properties of the nanometric field emitters will be studied as a first step for the discussion on the laser-assisted evaporation mechanisms. Then the consequences of the laser—tip interaction on the APT performances will be discussed, in the last section.

HISTORICAL BACKGROUND OF FIELD EVAPORATION ASSISTED BY LASER PULSES

Around the mid-1960s, the first attempts to evaporate a filed emitter using laser pulses started in the group of E. W. Muller. However, due to the instability of the laser, the first experiments were unsuccessful. The first laser-assisted atom probe analyses were reported around the end of the 1970s by Block and Tsong [7—9]. Using nanosecond laser or hundreds of picosecond laser, metallic and nonmetallic samples were analyzed such as silicon, germanium, and silica, SiC, and III-V semiconductors (more details are reported in Chapter 7 of [3]). However, in these early experiments, the focalization conditions were usually very constraining as reported by Tsong [10]. Due to the low energy per pulse of the laser system, the laser spot was strongly focalized on the sample and it was very difficult to reproduce the beam positioning on the sample from one experiment to the others. Hence, the repeatability of the analysis was not assured, even though similar laser illumination conditions were used. However, these first analyses opened the door to the study of the physical mechanisms of laser-assisted field evaporation. At first, the purely quantic ionization process by resonant absorption of one or several photons was considered. The direct interaction between the photons and one adatom can theoretically lead to its own ionization: one electron of the external layer of the adatom absorbs a photon and transmits to the material. However, Viswanathan et al. [8] estimated that the lifetime of the holes in the energy levels of the metal or the adatom are of about 10^{-15} s, which is much faster than the evaporation time. This mechanism was thus quickly considered irrelevant in the case of metals, but it was reported in the case of silicon tips by Tsong [11]. Indeed, in the case of semiconductors, due to the presence of a band gap, the lifetime of holes is longer than the evaporation time; this will be discussed in section "Resonant Excitation: Photoionization."

At the same time, Kellogg and Tsong [12], Kellogg [13], and Lee and Robins [14] showed experimentally that the temperature of the field emitter can increase a few hundred Kelvin after the interaction with the laser beam. Hence, they considered the possibility of a thermally assisted evaporation by a thermal pulse. To represent this thermal pulse, numerical and analytical models were developed to describe the tip's heating and cooling dynamics. These models consider homogeneous heating in depth on all the illuminated area, of several microns. Accounting for the cone-shaped geometry of the tip, the calculated cooling time was of several nanoseconds for tungsten [15,16] and longer for metals with poor conductivity such as steels. Slow cooling of the tip will lead to uncertainty in ion's evaporation time and thereby uncertainty in flight time resulting in poor mass resolution, as explained later in section "Nonresonant Excitation: Thermal Pulsing."

The poor mass resolution, coupled with the difficulties of obtaining reproducible results reduced considerably the use of the laser-assisted APT (La-APT) around the end of the 1980s. At this time, Tsong [10] and Cerezo [3] proposed to use ultrashort laser pulses, to improve the performances of La-APT. According to them, the electrical field of the electromagnetic wave for subpicosecond laser pulses could trigger the evaporation by acting as an ultrashort high-voltage pulse. In fact, the laser field is of the same magnitude as the evaporation field of atoms.

At the beginning of 2000, new ultrafast lasers with high energy per pulse (amplified laser system in femtosecond range) became available. The strong focalization is not anymore necessary and the first results obtained using ultrashort laser pulses are exciting [2,6,17]. The results obtained on metallic

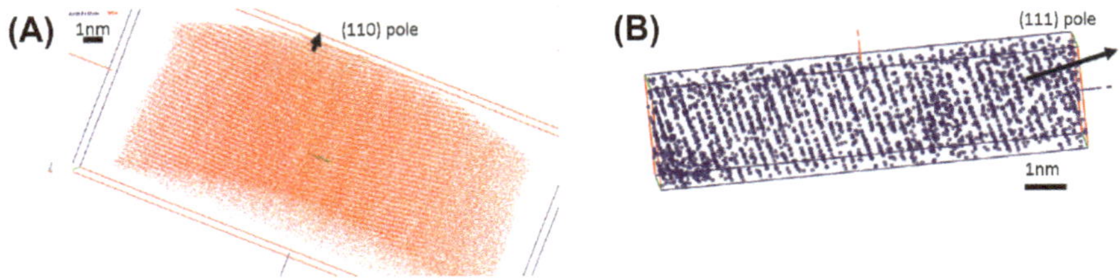

Figure 8.2 Volume reconstruction of tungsten (A) and silicon (B). [110] planes of tungsten are well reconstructed, for silicon [111] double planes (due to the diamond cell of the crystal) are seen as a single plane.

and semiconducting samples gave good spatial resolution images, as shown in Fig. 8.2, where the crystallographic structure of tungsten and silicon can clearly be seen. On the same setup and sample, compared to electrical pulses atom probe, in La-APT mass resolution is greatly increased, and was mostly limited by the detector's time response, as shown in Figs. 8.2 and 8.3.

In the case of nonmetallic sample, first results were also encouraging [2,18,19]. These results showed that the evaporation of the surface atoms was fully controlled and that the use of laser pulses did not lead to the tip's ablation, corresponding to the evaporation of several ions or clusters of ions for each laser pulse. Moreover, the clear observation of the crystallographic structure was correlated to the absence of any important vibration

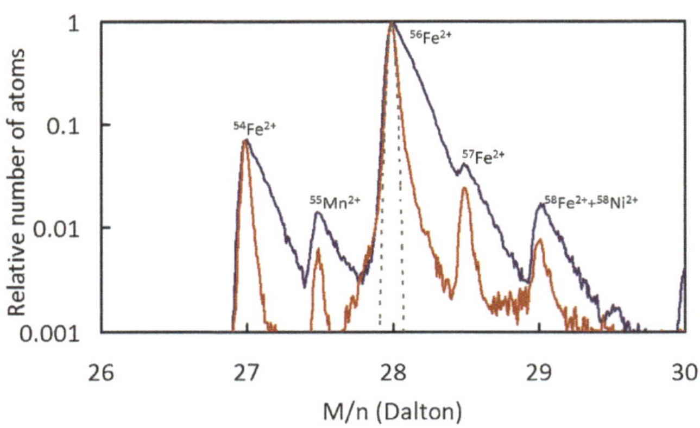

Figure 8.3 Mass spectra of iron alloy tips collected on a 45-cm free flight length atom probe. Spectrum in blue was obtained in electrical mode (corresponding mass resolution: full width at half maximum (FWHM) = 200; full width at tenth maximum (FWTM) = 78; full width at 1% of maximum (FW1%M) = 35). Spectrum in red was obtained in laser mode (corresponding mass resolution: FWHM = 560; FWTM = 235; FW1%M = 90). In this case, mass resolution is mostly limited by time resolving power of detector (drawn as the gray dashed line, a Gaussian function with a standard deviation of 0.7 ns).

of the crystal lattice such as the one expected for nanosecond pulses [14] due to the sample's heating. Moreover, good mass resolution proves that, if the evaporation is thermally assisted, the cooling process is very fast. It means that the temperature decreases almost a factor of two in less than 1 ns, in contradiction with predictions of Lee [16] and Liu et al. [20].

Hence, to understand the possible mechanisms of the laser-assisted field evaporation in ultrafast regime, it is necessary to understand the interaction of the laser beam with the nanometric field emitter.

OPTICAL PROPERTIES OF NANO TIPS

The interaction of light with matter is well described when the optical properties of the materials are known, in terms of its complex refractive index. Hence, the reflected, transmitted, and absorbed energy can be calculated, according to the thickness of the sample and the properties of the incident light [21]. However, when the sample is a nanometric tip, as in the case of field emitter samples, the nanometric dimension of the apex and the elongated shape of the tip can strongly influence the laser—sample interaction due to several physical effects such as diffraction, field enhancement, and plasmonic excitation. These contributions will be studied separately in the case of metallic and band-gap materials.

Metals

In the case of metals, the laser's photons interact with electrons in the conduction band, which is partially filled. After these interactions, electrons are promoted in the higher energy levels of the conduction band (hot electrons) and then relax their energy via electron—electron collisions (in femtosecond time scale) and electron—phonon collision (in picosecond time scale) [22]. Therefore, the lattice temperature increases.

The tip shape can strongly influence the distribution of the laser's photons at the surface of the field emitter [4,5,23—25]. In La-APT, the field emitter is illuminated from one side, which means that the propagation direction of the laser beam k is perpendicular to the axis of symmetry of the field emitter (see Fig. 8.3A). Moreover, the polarization of the laser pulse (ie, the direction of laser light intrinsic electric field F_{opt}) is in most of the case parallel to the tip axis. Effects of polarization of laser light can play a large role, but it will not be discussed in the present book. More information on this subject can be seen in Ref. [23,24]. Considering that the laser intensity (W/cm^2 s) is constant on several microns, a uniform laser energy

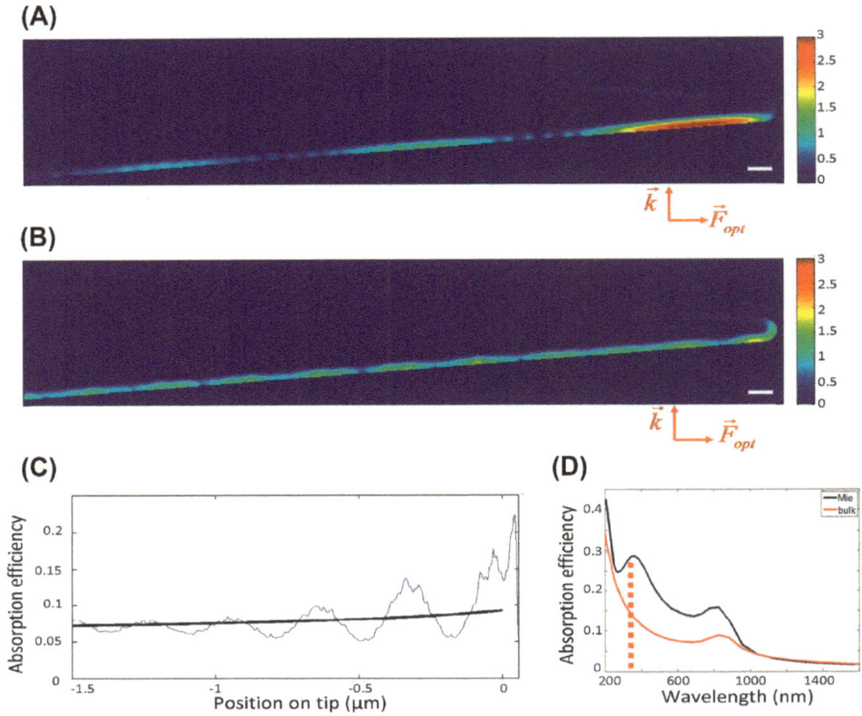

Figure 8.4 Finite-difference time domain computed absorption map in the incident plane (k, F_{opt}) (in $10^6\ m^{-1}$) of Al tip in axial polarization at (A) $\lambda = 1200$ nm, (B) $\lambda = 360$ nm. The white scale bar size is 100 nm, and laser is coming from the bottom following vector k and its polarization F_{opt} is parallel to the tip axis. (C) Absorption efficiency along tip axis. $x = 0$ corresponds to the beginning of the half hemispherical cap of the tip. Blue curve is computed from map (B), black curve comes from infinite cylindrical model (see Ref. [16]), (D) is the computing of absorption, on bulk aluminum [red] and on a 55-nm-radius aluminum particle (corresponding to tip apex).

absorption by the electrons along all the illuminated area is expected. However, if numerical software (such as *Lumerical* [26]) are used to solve Maxwell's equations for the real three-dimensional (3D) shape of the field emitter by finite-difference time domain (FDTD) method [26], nonuniform absorption maps are obtained, as reported in Fig. 8.4, in the case of an aluminum tip.

These maps show that, at the surface of the specimen, there are some maxima of absorption that are separated by a distance equal to the laser wavelength. Moreover, the intensity of these maxima decreases from the tip apex to its base. The apex of the tip can thus strongly absorb the light or be totally transparent depending on the ratio between the laser wavelength and the tip end radius.

These numerical results are due to several physical effects.

The first one is the diffraction of the incoming light at the discontinuity between the metallic tip and the vacuum. The diffraction effects were studied by Sommerfeld in 1896 [27], in the case of a half infinite plate, perfectly conductive, infinitely thin, and enlightened perpendicularly to the half plate. He predicted the presence of a spatially modulated electronic current density along the plane with a periodicity equal to the laser wavelength, due to the interference of photons scattered from the apex of the specimen with the primary laser beam. On real metallic materials, the absorption properties are linked to the surface current density through their conductivities. In the case of a tip, the apex acts as a diffraction source emitting in all directions. This source interacts with the incident and reflected beams giving rise to oscillating surface currents on the shank surface. It leads, in the case of nonperfectly conductive materials, to absorption maxima seen in Fig. 8.4C.

The second effect to consider is the collective excitation of surface electrons at the tip apex [28]. In fact, the apex of the field emitter can be approximated by a half-sphere. The optical properties of nanosphere have been widely studied. The absorption is shown to increase due to the collective excitation of electrons (surface plasmonic excitation) for resonant excitation wavelength obtained by the equation [29]:

$$\left(\frac{48\pi^2 R^2}{5\lambda^2} + \Re(\varepsilon(\lambda)) + 2\right) = 0 \tag{8.1}$$

where $\Re(\varepsilon(\lambda))$ is the real part of the dielectric constant of the material. In Fig. 8.4D, this resonance can be seen on a nanoparticle with a radius of 55 nm at $\lambda = 360$ nm (absorption efficiency of nanoparticle is twice higher than for bulk). This explains why in Fig. 8.4C the absorption efficiency is much more higher at the tip apex (ie, for position >0). In the case of iron nanoparticles, taking into account the variation of the dielectric constant with the laser wavelength [30], the excitation of the plasmonic resonance with green light is, for example, possible for a radius of 60−70 nm [31].

At last, we have to take into account the cone shape of the sample. For an infinite cylinder, the solution of Maxwell's equations shows that, for an axial polarization of the incident light, the absorption decreases as a function of the ratio $R_{\text{cylinder}}/\lambda$ [32]. If we consider the shank of the tip as an assembly of small cylinders, as schematically represented in Fig. 8.5, their radius increases from the apex to the base of the tip. In this case, Maxwell equations can be solved on each part of the tip, considering each part as an infinite cylinder. Hence, the decrease of the absorption maxima in

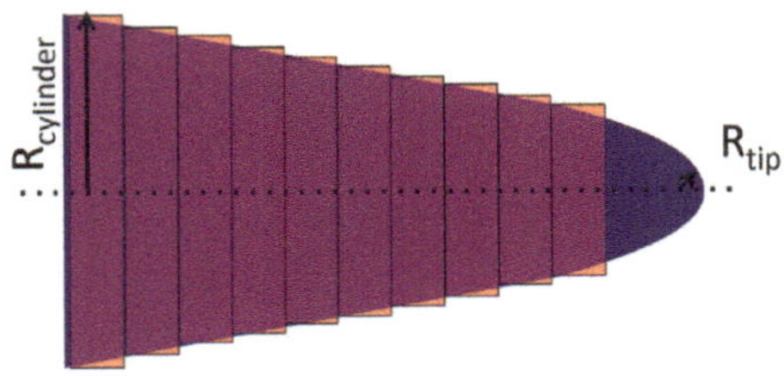

Figure 8.5 Schematic representation of the tip as an assembly of cylinder with different radii $R_{cylinder}$.

Fig. 8.4C is related to decrease of the cylinder absorption when their radius increases.

Band-Gap Materials

In the case of band-gap materials, as in the case of metals, the interaction with light can be completely described if the optical properties, in terms of refractive index, are well known [21]. These optical properties depend on the band structure (related to doping and defects). At low temperature, the valence band is generally filled and the conduction band empty. Hence, laser photons interact with electrons in the valence band, which can be promoted in the conduction band or not, depending on the photon energy [22]. In the first case, the material absorbs the light; in the second case it is transparent to the light. To explain the process of light absorption, we can distinguish the case of direct band-gap materials and indirect ones.

In the case of direct band-gap materials, the absorption of photons with energy (E_{ph}) lower than the band gap $\left(E_{ph} < E_g^{direct}\right)$ is negligible and it becomes very strong for $\left(E_{ph} > E_g^{direct}\right)$, see Fig. 8.6A. In this latter case, the penetration depth of the light is a few nanometers and all the photon energy is absorbed at the surface by valence electrons promoted to the conduction band creating an electron–hole pair [33,34].

For indirect band gap semiconductors, as shown in Fig. 8.6B, the absorption of photons with $E_{ph} > E_g^{indirect}$ is assisted by phonons and it is very low when the photon energy is close to the band gap energy. In this case the penetration depth can be large, up to 1 mm [33].

As in the case of metallic samples, the tip shape can strongly influence the distribution of the laser photons at the surface of the field emitter and also in its volume, when the penetration depth is of several hundreds of nanometers.

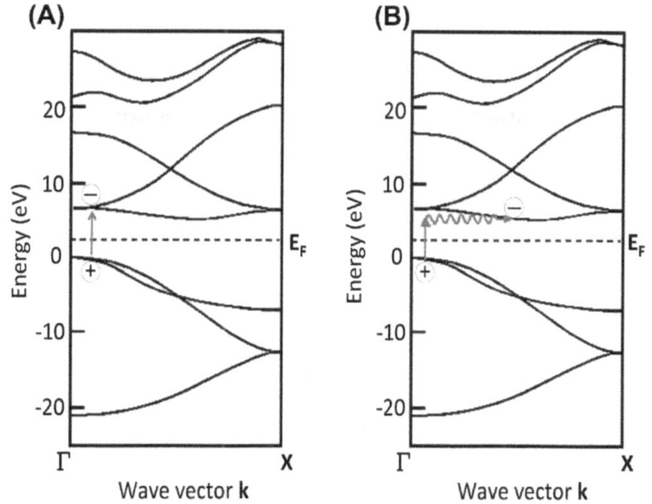

Figure 8.6 Schematic representation of the absorption process for (A) $E_{ph} > E_g^{direct}$; (B) $E_{ph} > E_g^{indirect}$.

Using numerical software (such as Lumerical [26]) Maxwell's equations can be solved for the real 3D shape of the field emitter by the FDTD method, for band-gap materials. In Fig. 8.7, we report the absorption maps obtained in the case of a silicon tip for (A)$E_{ph} > E_g^{direct}$, (B)$E_{ph} > E_g^{indirect}$, (C)$E_{ph} = E_g^{indirect}$. These absorption maps represent the energy absorbed in the incident plane (x,z), considering that the laser is propagating along the x axis, from negative to positive values (as represented in Fig. 8.7 by the vector k), and it is linearly polarized along the z axis (vector F_{opt} in Fig. 8.7).

When the photon's energy is higher than the direct band-gap energy, the absorption depth is a few nanometers, hence all the energy is absorbed at the surface and the absorption map shows the spatial modulation (with maxima along the tip axis) as already reported for metals. This behavior is due to the interaction between the incident and the partially reflected beams giving rise to spatial modulation of the electric field.

For $E_{ph} > E_g^{indirect}$ and $E_{ph} = E_g^{indirect}$, the absorption maps reported in Fig. 8.7B and C, show new features. The absorption maxima are no longer located at the surface of the field emitter but in its volume. Moreover, even though the maps are very complex, the absorption maxima seem to be separated by a distance depending on the laser wavelength, as visible by comparison of Fig. 8.7B and C.

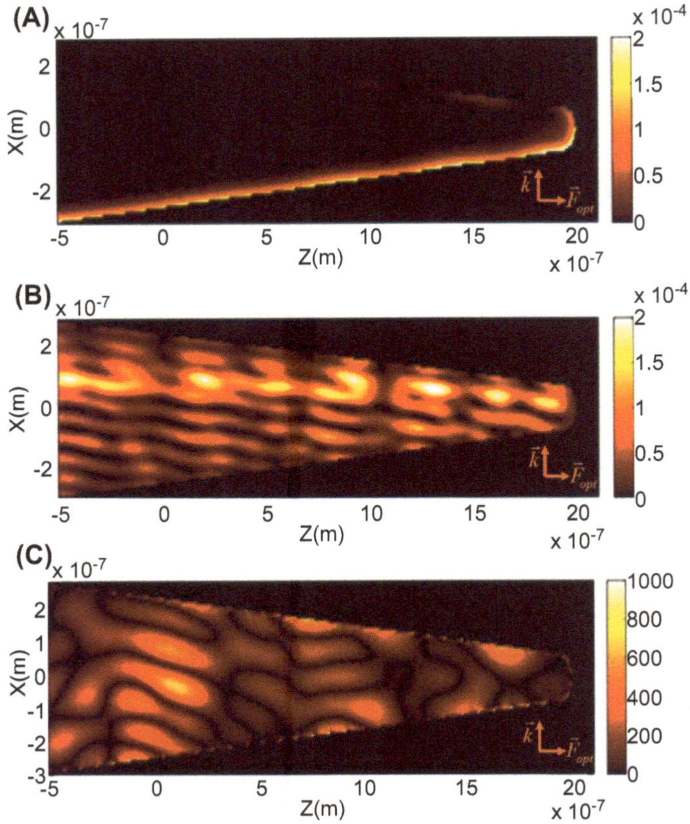

Figure 8.7 Absorption maps computing by finite-difference time domain of a silicon tip ($R = 50$ nm) at (A) $\lambda = 343$ nm, (B) $\lambda = 515$ nm, (C) $\lambda = 1030$ nm (colorbar unit is in m^{-1}). In each case, the laser is coming from the bottom following vector k and is polarized parallel to the tip axis following vector F_{opt}.

This spatial distribution of absorption is due to multiple reflection and refraction of the laser beams propagating inside the tip, which can interfere. The presence of absorption maxima can be explained considering the absorption properties of nanowires. In the geometry of an infinite cylinder the Maxwell's equation can be solved analytically and the existence of resonant modes, for given ratios of $R_{cylinder}/\lambda$, can be proved. These resonant modes have been reported both experimentally and theoretically in the case of semiconductor nanowires [35,36]. The maxima of absorption reported in Fig. 8.7B and C correspond to the positions along the tip axis for which the ratio $R_{cylinder}/\lambda$ (with $R_{cylinder}$ the local "cylindrical" radius) is close to the resonant values. That is why, when the wavelength increases, from green light to IR light, as in Fig. 8.7B and C, the resonant values of $R_{cylinder}$ increase; hence, the spacing between maxima of absorption also increases.

Furthermore, the optical properties of bulk semiconductors can also be affected by the presence of a strong static electric field [37]. Recently Sileava et al. [38] reported that, when a high voltage is applied to high resistivity materials (such as band-gap materials at cryogenic temperature), a strong DC field is generated inside the sample. As shown in Fig. 8.8, this strong DC field then generates, by field-assisted ionization processes (such as impact ionization [39]), free carriers inside the samples, which, driven by the field, move to the sample's surface in a few nanoseconds and screen the field.

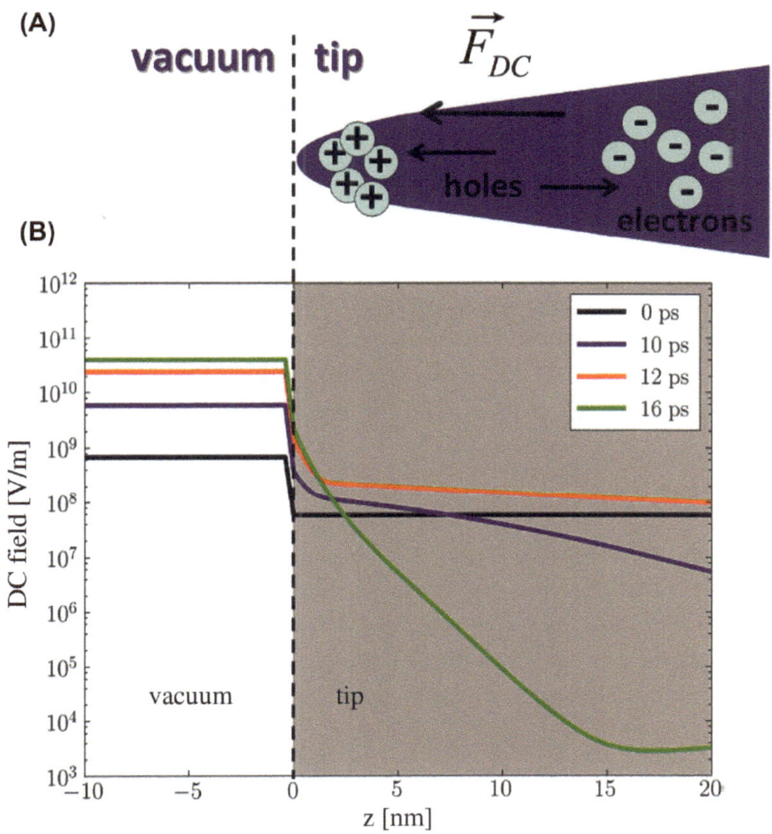

Figure 8.8 (A) Schematic representation of the charges motion inside the tip driven by the standing field F_{DC}, after the application of the high standing voltage to the tip base; (B) computing by impact ionization of DC field distribution along the tip and at its surface at different instants after the application of the high DC voltage to the tip base [38].

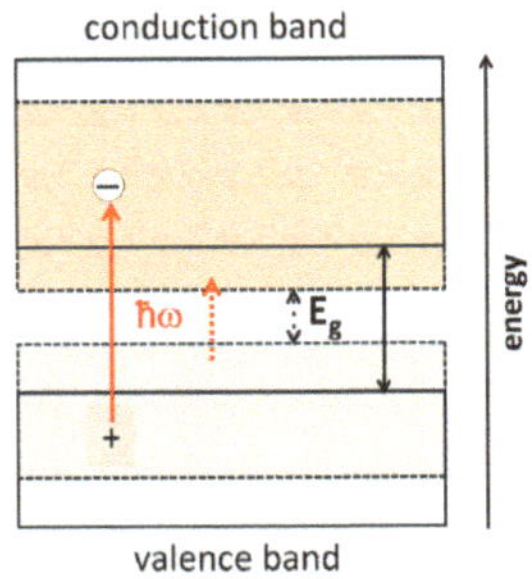

Figure 8.9 Schematic representation of the band-gap reduction under high DC field. Dotted lines correspond to the new band structure under the DC field, *dotted red and black arrows* correspond to the new photon energy and the band gap, respectively.

After the screening process, the strong DC field affects only a few atomic layers at the surface of the sample. This field modifies the band structure of the materials, changing the band gap [38,40] or bending the bands [41]. By increasing the value of the external electric field, subbandgap absorption ($E_{ph} < E_g$) becomes possible, as shown in Fig. 8.9.

PHYSICAL MECHANISMS OF FIELD EVAPORATION ASSISTED BY LASER PULSES

As explained in previous chapters, two basic theories were developed to describe field evaporation mechanisms: the force image model [42] and the intersection model [3,43]. Both theories predicted a change of the field evaporation activation energy Q_n, in the presence of an external field F_{DC}:

$$Q_n = \Lambda + \sum_n I_n - n\phi_e - f(F_{DC}) \tag{8.2}$$

where Λ, I_n, and ϕ_e are, respectively, the heat of sublimation, the nth ionization energy, and the work function of the emitting surface.

The expression of the function $f(F_{DC})$ depends on the model but, around the threshold of evaporation, a linear behavior of $f(F_{DC})$ was experimentally observed [44]. The rate of process is given by the Arrhenius equation:

$$\phi = \nu \exp(-Q_n/k_B T) \tag{8.3}$$

where ν is the vibration frequency of the surface atoms, k_B the Boltzmann constant, and T the emitter temperature. The typical time τ_{evap} required for evaporating is given by:

$$\tau_{evap} = \tau_0 \exp(Q_n/k_B T) \tag{8.4}$$

where τ_0 is the vibration time of surface atoms, typically a few picoseconds [45,46]. When a laser pulse is used to trigger the evaporation, different evaporation mechanisms can occur:

1. the photoionization, corresponding to resonant ionization of surface atoms and its following ejection as ion;
2. the absorption of laser energy and the following heating of the specimen, which allows thermal-assisted evaporation;
3. the increase of the surface field, which allows the evaporation by energy barrier reduction.

These mechanisms are presented in the following sections and their experimental observation and the consequences on the evaporation behavior will be discussed in the next section.

Resonant Excitation: Photoionization

The interaction of a photon with an electron of the surface atom (adatom) could lead to the transition of an electron of the adatom to a vacant state of the band structure and thus to the ionization of the adatom as proposed by Tsong et al. [47].

The presence of an electric field at the surface induces band-bending; hence, the excited state (hole) of the adatoms can fall in the band gap of the surface (Tsong [41], Tsong [11]). The ion lifetime is, then, long enough to allow the evaporation.

The photoexcitation probability is equal to the number of absorbed photons at the tip surface: $N_{\text{abs}}^{\text{tip}}$. If we assume that all the ionized atoms are field evaporated, the evaporation probability (or rate) is expected to be equal to $N_{\text{abs}}^{\text{tip}}$. However, an activation energy Q_n^{ionic}, as the difference between the minimum and the maximum of the ionic potential, exists, decreasing the photoionization probability of at least a factor of 10 [48].

Following this model, the atoms are first ionized to the 1+ state and then postionized near the tip surface, as predicted by the Kingham theory [49]. Although the photoionization mechanism is a quasi-instantaneous process, the field evaporation takes some time, about 0.5 ps [48]. This time can be estimated as the lifetime of the photoionized atoms on the field emitter surface. The resonant photoionization can be observed only at very low laser energy, where the thermal effects are drastically reduced. However, the evaporation rate is generally lower than in the case of thermal-assisted evaporation.

Nonresonant Excitation: Thermal Pulsing

As explained in the previous section, the nanometric tip can absorb the energy of the light. This absorption process induces a heating of the sample and the increase of the temperature triggers the evaporation process by "a thermal pulse." In this case, the evaporation rate follows the temperature dynamics of the apex of the specimen, as predicted by Eq. (8.3) replacing T by $T(t)$ the temperature evolution of the apex after the interaction with the laser beam. If the light is absorbed at the tip apex, the heating process can be really fast (of a few picoseconds); hence, the dynamics of the evaporation rate depends only on the cooling dynamic of the field emitter. An analytic expression of the cooling dynamic can be obtained in the case of a wire (one-dimensional system), studying the thermal diffusion behavior as predicted by Fourier's law with a thermal diffusivity α [50]:

$$T(t, z = 0) = T_0 + \frac{T_{\text{rise}}}{\sqrt{1 + 2\frac{t}{\tau_{\text{cooling}}}}} \tag{8.5}$$

where

$$\tau_{\text{cooling}} = w^2/\alpha \tag{8.6}$$

is the tip's cooling time value and w is the waist of the heated zone, supposed to have a Gaussian shape. A schematic representation of the temporal evolution of $T(t,z = 0)$ is reported in Fig. 8.10. At time $t = 0$, corresponding to the end of thermalization of the electrons with the lattice (a few picoseconds after the pulse [51]), the temperature profile along the wire is

$$T(t = 0, z) = T_0 + T_{\text{rise}}\exp\left(-z/2w^2\right) \tag{8.7}$$

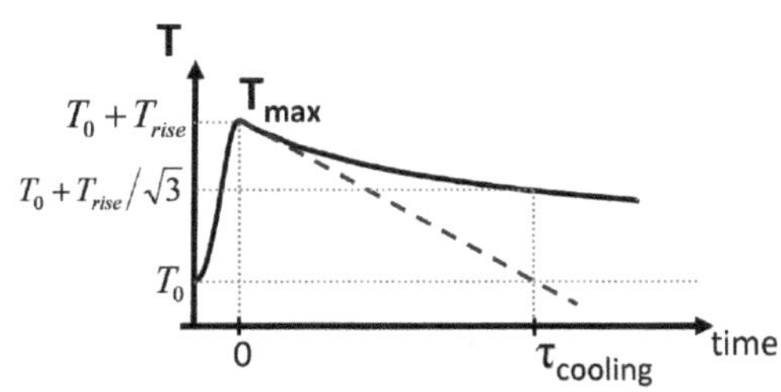

Figure 8.10 Schematic representation of the temperature evolution at the tip apex after the interaction with the laser beam.

with T_0 the base temperature of the sample usually ranging from 20 to 80K. Following Eqs. (8.5) and (8.6), the thermal pulse is a function of the thermal properties of the samples (via its thermal diffusivity α) and of the spatial distribution of the absorption along the sample (via the term w), which, as reported in section "Optical Properties of Nano Tips," depends on the illumination conditions and tip shape. Hence, for the same illumination condition, the duration of the thermal pulse can change from one sample to another, depending on its geometry.

Nonresonant Excitation: Optical Pulsing

The interaction of the laser beam with the field emitter can induce a change of the evaporation barrier. The electric field of the laser beam (optical F_{laser}) can be of a few volts per nanometer, thanks to the enhancement of the field due to the tip shape of the emitter [52,53]. This optical field will affect the activation barrier, however, due to the optical oscillations frequency (10^{15} Hz), the mean value of the barrier, on typical evaporation time (of a few picoseconds) will be unchanged.

In the case of metallic field emitters, a new physical mechanism was proposed, based on the nonlinear optical properties of surfaces [54]: the optical rectification effect at metal surfaces [55,56] which corresponds to the generation of an ultrafast electrical pulse at the surface of the sample when it is illuminated by an ultrafast laser [57,58]. The optical rectification is based on nonlinear second-order processes that occur in all noncentrosymmetric materials and at the surface of metals where the electronic density of states exhibits a discontinuity that yields strong anharmonic and nonlinear effects [59,60]. Following this model, the energy barrier will, in the first approximation, decrease linearly with the laser intensity (I):

$$Q_n(t) = \Lambda + \sum_n I_n - n\phi_e - \gamma(F_{\text{DC}} + F_{\text{OR}}) \tag{8.8}$$

with $F_{\text{OR}} \propto I$ the electric field generated by optical rectification at the surface of the sample.

In the case of nonmetallic samples, Tamura et al. [61] suggest an evaporation mechanism based on a change of evaporation energy barrier Q_n by hole accumulation near surface atoms. Holes would be created during the laser pulse by photoabsorption and the energetic electrons would be driven away by the electric field present in the band-bending region, at the surface. Using electric field modeling by the boundary element method with a mesoscopic model, and atomistic ab initio calculations, it has been shown that a certain amount of hole accumulation reduces the activation barrier for field evaporation substantially, triggering the evaporation process.

Discussion on the Evaporation Mechanisms

All the physical mechanisms of laser-assisted field evaporation presented here can happen; however, not all these mechanisms have been experimentally reported.

In the case of metallic field emitters, the nonresonant excitation was never observed and the optical pulsing, by the reduction of the activation energy barrier, was proved to be negligible. Hence, the thermal pulsing is now considered as the dominant evaporation process for all metallic field emitters [4,5].

In the case of band-gap materials, for low laser intensities and really low evaporation rate (less than 0.001 atoms/pulse), the field evaporation was proved to be assisted by a single-photon absorption mechanism, in the case of Si and MgO samples [48,61]. As shown in Fig. 8.11, the number of evaporated ions/pulse starts to vary exponentially when the laser energy is higher than 400 nJ (about 500 J m^{-2}). In fact, at high laser intensity and higher evaporation rate, a thermal contribution is reported which clearly accounts for the long evaporation process. Moreover, resonant and nonresonant mechanisms are still under discussion to explain the fast evaporation process [4,5], reported at high laser intensity.

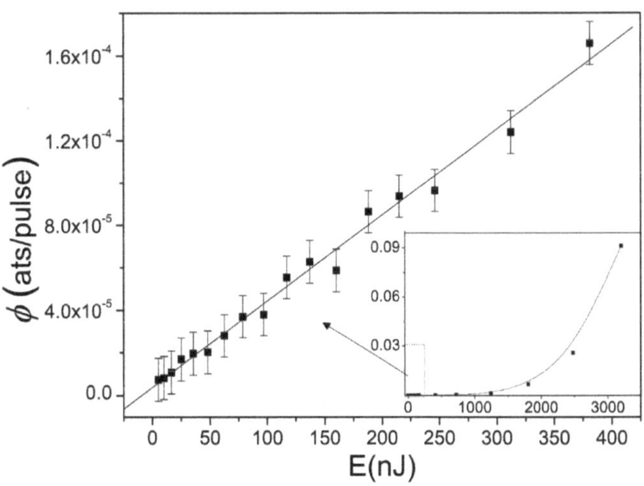

Figure 8.11 Number of evaporated ions/pulse for a Si tip analyzed using IR laser ($\lambda = 1030$ nm) with a spot size of 30 μm diameter.

CONSEQUENCES ON THE La-APT PERFORMANCES

We will now discuss the performances of La-APT analyses in terms of mass resolution and image reconstruction and their dependences on the experimental conditions, to give the reader the tools to optimize the experimental conditions before starting La-APT analysis. Concerning the decrease of the mass resolution due to the energy spread of evaporated ions, as reported in the previous chapter for electrical pulses, it was shown recently, by Sevelin-Radiguet et al. [62] and Arnoldi et al. [63,64], that this energy spread is only a few electronvolts in the case of laser pulsing for metallic and nonmetallic samples. Hence, in the following discussion, we neglect the contribution of the energy spread to the widening of mass peaks.

Metallic Samples

In the case of metallic samples, the evaporation process is triggered by thermal pulses; hence, to get the best experimental conditions, this thermal pulse should be as short as possible. To reduce the duration of the thermal pulse two conditions have to be fulfilled:

1. minimize the heating time of the tip apex, allowing the direct absorption of the laser energy at the apex of the field emitter, playing with the tip geometry and the laser wavelength;
2. minimize the cooling time (reported in Eq. (8.5)), by the reduction of the heated zone or the increasing of the cone angle of the field emitter.

By the analysis of the same steel tip at different laser wavelengths, it was shown by Houard et al. [24] that the mass resolution strongly depends on the illumination conditions, as reported in Fig. 8.12A.

As explained in section "Optical Properties of Nano Tips," changing the laser wavelength changes the absorption map, and as reported in Fig. 8.4, the larger the laser wavelength the larger the size of the heated zone. Hence, the bad mass resolution at FWHM reported for large laser wavelength is due to larger absorption regions, w. Moreover, we can also remark that, for large laser wavelengths, the first maximum on the absorption map is further away from the apex of the field emitter. Therefore, the heating time increases, increasing the width of mass spectra peaks and reducing the mass resolution. Similar behavior was also reported keeping the laser wavelength constant but changing the tip dimensions, especially the end radius; see Fig. 8.12B [31]. As explained before, for values of the ratio R/λ close to the resonance of the end half-sphere at the tip apex,

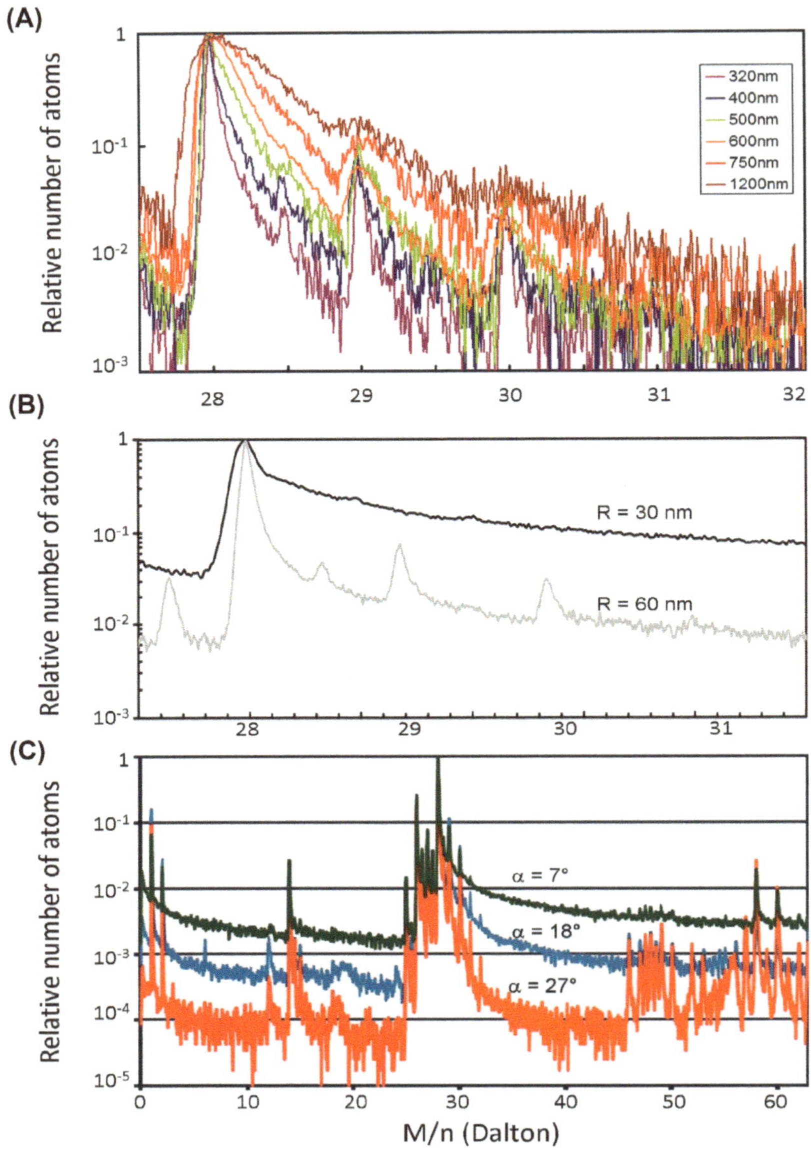

Figure 8.12 Mass spectra of steel tip at (A) different wavelengths, (B) $\lambda = 515$ nm, constant cone angle $\alpha = 5°$, different radius (black $R = 30$ nm, gray $R = 60$ nm), (C) $\lambda = 515$ nm, $R = 60$ nm and different cone angles [red $\alpha = 27°$, blue $\alpha = 18°$, green $\alpha = 7°$].

the absorption is confined at the apex reducing the heating time and the size of the heated zone (and hence the cooling time). For values of the tip radius far from the resonance, the absorption is not so well confined at the apex, increasing the size of the heated zone and reducing the mass resolution.

As discussed in several papers, the cooling time also depends on the cone angle of the field emitter: the larger the cone angle, the shorter the cooling time. This geometrical parameter strongly affects the S/N ratio and hence the sensitivity of La-APT analyses. As reported in Fig. 8.12C, changing the cone angle from 2° to 15° can improve the S/N ratio by almost two orders of magnitude [31].

For materials with poor thermal conductivity, such as steels, the S/N ratio depends also on the repetition rate of the laser, as proved by Cerezo et al. [65]. The higher the repetition rate, the shorter the time between two laser pulses (τ_{trains}). Hence, if the cooling time is longer than τ_{trains}, the effects of the pulses will cumulate and the sample will never cool back to the base temperature. This higher temperature between laser pulses is at the origin of a degradation of S/N ratio.

Still in the case of metals with poor thermal diffusivity, it was reported that the tip shape can be affected by the illumination conditions [66]. Because the laser energy is absorbed at the surface by electrons, the transfer of the energy to the lattice and the following heating of the tip apex depends on the energy transfer from electrons to lattice. In metals with poor thermal conductivity, the ballistic motions of electrons after the absorption of laser energy are a few nanometers; hence, the transfer of the energy to the lattice happens at the surface. This heating process was studied by Houard et al. [67], solving the heat equations for electrons and lattice, taking into account the real 3D geometry of the tip. It was shown that the illuminated side reaches in a very short time (less than 1 ps) a high temperature of 280K; however, the opposite side reaches the maximum temperature of 250K only 100 ps after the laser—tip interaction. This delay is due to the 3D heat propagation from the illuminated side to the hidden one. The evaporation rate can be 10 times higher for the illuminated side; hence, this side of the tip ejects more ions (preferential evaporation) and the local curvature radius increases, as reported in Ref. [66]. This effect is known as the shadow effect. This change of the tip shape can have strong consequences on the quality of the reconstructed images, if the hypothesis of a homogeneous radius of curvature is done for the reconstruction method.

Nonmetallic Samples

When nonmetallic samples are analyzed by La-APT, two evaporation processes are visible: a fast evaporation process and a slow one as sketched in Fig. 8.13. Their relative contribution strongly depends on the experimental conditions, in terms of laser illumination conditions and sample shape [5,68].

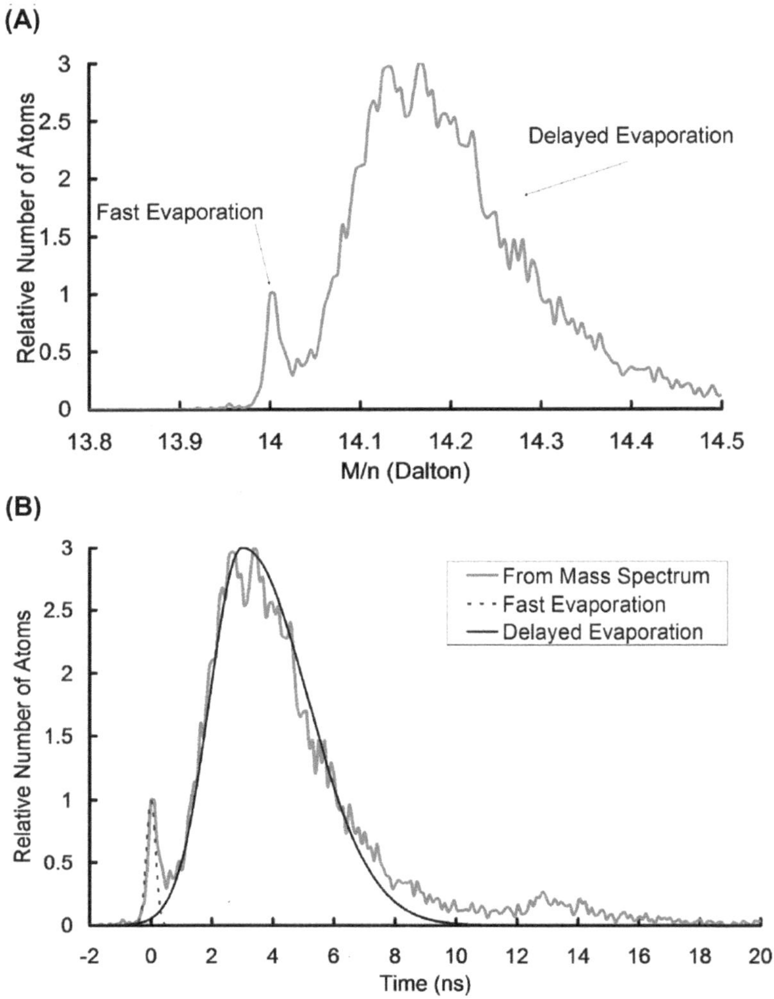

Figure 8.13 (A) Mass spectrum (gray curve) of a silicon tip analyzed at $\lambda = 650$ nm in a 20-cm atom probe. It is centered on the Si_{28}^{2+} peak, the main one. (B) Conversion of the mass spectrum showed in (A) in time, with $t = 0$ corresponding to the maximum of emission of the Si_{28}^{2+} peak. The fast component (black dots) corresponds to a Gaussian function with a standard deviation of 150 ps (detector limitation). The slow component (black curve) is delayed by 3 ns, and has a rise time of 1.5 ns and a decay time of 3 ns.

The slow evaporation process is related to a "thermal pulsing" that occurs a few nanoseconds after laser irradiation of the tip. Fig. 8.13 shows that on silicon, a thermal pulse takes 1.5 ns to reach its maximum temperature. This maximum occurs 3 ns after the fast evaporation, and then decays with a characteristic time of 3 ns. However, the physical origin of the fast evaporation process is under discussion. For this reason, it is difficult as of now to argue on the best experimental conditions for the analysis of nonmetallic samples.

Experimental results on silicon show an increase of the contribution of the slow evaporation process with the laser wavelength and laser power as reported in Figs. 8.14 and 8.16. Those results have also been reported for other semiconducting materials like MgO or SiC [68,69].

As reported in section "Optical Properties of Nano Tips," the absorption map of band-gap materials depends on the laser wavelength and tip shape. Hence, the spatial distribution of free carriers generated by the light will follow the distribution of the absorbed energy. These free carriers then transfer their energy to the lattice by electron—electron (hole—hole) collisions (in a few picoseconds) and electron—hole recombination (>100 ps). These processes take place during the drift-diffusion dynamics of the carriers. These dynamics can be studied, as in the case of metals, by heat equation for carriers and lattice, coupled with drift diffusion continuity equations [5].

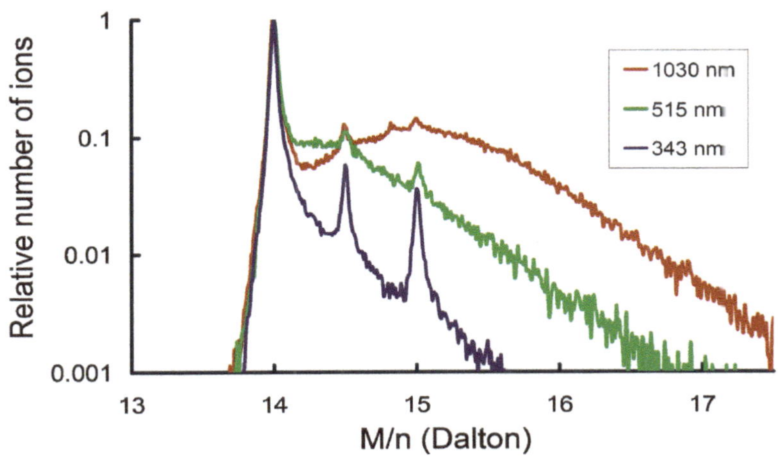

Figure 8.14 Mass spectrum of a silicon tip analyzed at different wavelengths in a 10-cm atom probe centered on the Si^{2+}_{28}, the main peak. Evaporation rate ≈0.01 at/pulse and DC field (fixed around 21 V/nm, deduced from ratio between Si^{1+} and Si^{2+}) were fixed parameters.

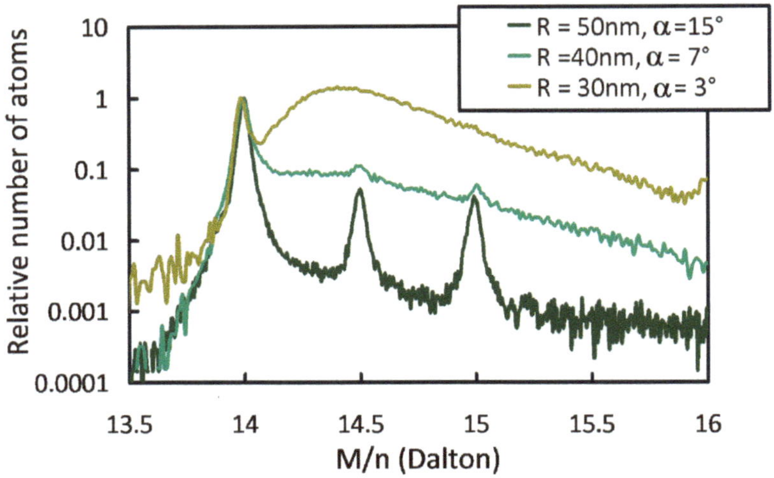

Figure 8.15 Mass spectra of silicon tips of different shapes obtained in 10-cm atom probe using $\lambda = 515$ nm wavelength.

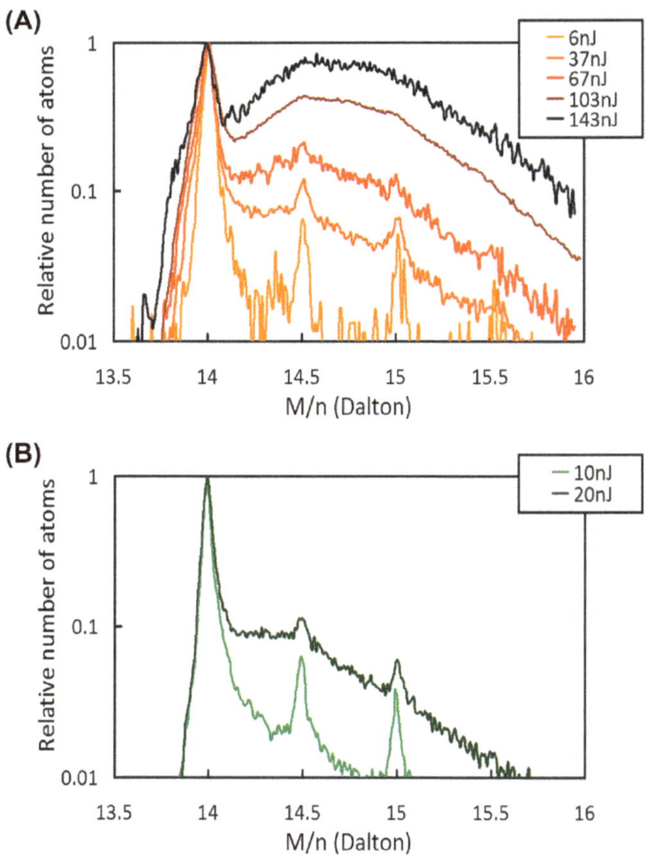

Figure 8.16 Mass spectra of silicon tip at different laser energies in 10-cm atom probe using (A) $\lambda = 1030$ nm and (B) $\lambda = 515$ nm wavelength.

For example, in the case of silicon tips, when high-energy photons are used (this is the case of UV light), the energy is absorbed at the surface and the thermal and drift-diffusion equations predict a heating localized at the surface, which will induce the shadow effect, as in the case of metallic samples with poor thermal diffusivity, discussed in the previous section. The change of the tip shape with the consequences on the image reconstruction was reported experimentally by Koelling et al. [70].

When photons with energy lower than the direct band-gap are used (this is the case of green and infrared light for silicon samples), the light energy is absorbed in the volume of the tip along its shank as shown in Fig. 8.7B and C. For green light, these maxima of absorption can occur close to the tip apex, if the radius is larger than 50 nm. However, for infrared light, we can note the absence of maxima of absorption close to the apex. For this reason, in the case of green light, the contribution of the slow evaporation process strongly depends on the tip geometry (radius and cone angle) as shown in Fig. 8.15, changing from a long tail to a second broad peak after each narrow peak in the mass spectra. However, in the case of infrared light, a second broad peak is always observable at middle and high laser intensity, as reported in Fig. 8.16A.

This slow evaporation process, assisted by thermal pulses, can be minimized in the case of green light, adjusting the tip shape. However, in the case of infrared light, to obtain absorption maxima close to the tip apex, very large tip radius, of around 100 nm [35], should be used, which cannot be easily analyzed in standard La-APT, due to the need of very high voltages.

Concerning the fast evaporating process, it can be associated with a thermal ejection, only for materials that show strong absorption confined at the tip apex. This confined heating will give rise to a fast heating and cooling process of the sample.

However, in the case of silicon analyzed using infrared light, or in the case of large band-gap materials analyzed using UV light, it means when $E_{ph} < E_{gap}$, any absorption is expected close to the apex. In this case, other evaporation mechanisms should be taken into account such as the lowering of the activation barrier, by hole accumulation due to the absorption of the laser beam at the surface [61]. Furthermore, the absorption of the surface of band-gap materials was proved to strongly change under high electric DC field [38]. Hence, the enhancement of the surface absorption under the DC field can probably allow new fast ejection mechanisms that have to be demonstrated unequivocally to be completely accepted. Only the full understanding of the physical origin of this fast evaporation will allow the optimization of the analysis conditions, by the maximization of this fast evaporation process.

As a last point, we want to discuss the performances of La-APT on the composition measurements. Recently, compositions far from the stoichiometry have been reported on several semiconductors and oxides. Moreover, Mancini et al. [71] have shown that the composition measurements are strongly dependent on the DC field more than on the laser intensity. These experimental observations suggest the presence of new process at the surface of the sample related to the domain of high field chemistry, which can be assisted by the laser heating. Ab initio calculations are a useful help in the understanding of these complex phenomena, as reported recently by Xia et al. [72]. More details on these analyses are reported in "Chapter 9, Data mining." Today, more theoretical and experimental studies are necessary to elucidate these processes and determine the optimal experimental condition to get good composition measurements in La-APT on band-gap materials.

CONCLUSIONS AND PERSPECTIVES

The use of laser pulses to trigger the evaporation of surface atoms in APT opened the technique to many new fields for the analysis of materials from micro- and nanoelectronics and photonics to biology and geology. Even though APT was proved to be very powerful to provide the detailed structure of materials with truly quantitative elemental measurement, these results are far from being straightforward. APT analysis of nanostructures can simply fail or their composition measurements can strongly depend on the experimental conditions, as discussed in this chapter. These difficulties arise from the complexity of the laser-induced field evaporation process.

As explained in this chapter, the interaction between a nanometric sample (shaped as a tip) and an ultrashort laser pulse (up to a few picoseconds) is strongly dependent not only on the material but also on the shape of the sample and on the illumination conditions (in terms of laser wavelength and intensity).

In the case of a metallic sample, the evaporation mechanism was proved to be assisted by a thermal process, at nanoscale, due to the heating of the sample after the interaction with the laser beam. The heating and cooling processes of the tip can be modeled considering the absorption properties of nanomaterials and the thermal properties of the sample. Hence, the optimal experimental parameters can be determined to achieve better mass resolution, S/N ratio, and 3D reconstruction. Usually a compromise can be found, depending to the information required from the analysis.

In the case of semiconductors and dielectrics, the exact timescale of the evaporation process and the evaporation mechanisms are still not known. Several mechanisms have been proposed, from thermal processes to field-assisted processes. New progress has been made in the understanding of the interaction between the nonmetallic sample with the laser beam, taking into account the effects of the strong DC field at the surface of the samples. A strong DC field changes the band structure of materials, changing their optical properties and allowing new laser-assisted processes. Because the physical mechanisms of the evaporation of nonmetallic samples are not well understood, the experimental parameters to optimize APT analysis of these materials cannot be determined by simple calculations. More experimental and theoretical work is necessary to better understand the behavior of nonmetallic samples under high DC field and ultrafast pulses and hence be able to find the optimal experimental conditions, to achieve good composition measurements and 3D reconstruction.

REFERENCES

[1] Grenier A, Lardé R, Cadel E, Vurpillot F, Juraszek J, Teillet J, et al. Atomic-scale study of $TbCo_2$. 5/Fe multilayers by laser-assisted tomographic atom probe. J Appl Phys 2007;102(3):033912.

[2] Kelly TF, Miller MK. Invited review article: atom probe tomography. Rev Sci Instrum 2007;78:31101.

[3] Miller MK, Cerezo A, Hetherington MG, Smith GDW. Atom probe field ion microscopy. Oxford Science Publications – Clarendon Press; 1996.

[4] Vella A. On the interaction of an ultra-fast laser with a nanometric tip by laser assisted atom probe tomography: a review. Ultramicroscopy September 2013;132 5–18.

[5] Kelly TF, Vella A, Bunton JH, Houard J, Silaeva EP, Bogdanowicz J, et al. Laser pulsing of field evaporation in atom probe tomography. Curr Opin Solid St M 2014;18(2): 81–9.

[6] Gault B, Vurpillot F, Vella A, Bostel A, Menand A, Deconihout B. Design of a femtosecond laser assisted tomographic atom probe. Rev Sci Instrum 2006;77:043705.

[7] Tsong MNTT, Block JH, Viswanatha B. Photon stimulated field ionization. J Chem Phys 1976;65(6):2469.

[8] Viswanathan B, Drachsel W, Block J, Tsong T. J Chem Phys 1979;70:2582.

[9] Nishigaki S, Drachsel W, Block J. Photon-induced field ionization mass spectrometry of ethylene on silver. Surf Sci 1979;87(2):389–409.

[10] Tsong T. Atom-probe field ion microscopy : field ion emission and surfaces and interfaces at atomic resolution. Cambridge University Press; 1990.

[11] Tsong TT. Pulsed-laser-stimulated field emission from metal and semiconductor surfaces. Phys Rev B 1984;30(9):4946.

[12] Kellogg G, Tsong T. J Appl Phys 1980;51:1184.

[13] Kellogg G. Determining the field emitter temperature during laser irradiation in the pulsed laser atom probe. J Appl Phys 1981;52:5320.

[14] Lee M, Reifenberger R, Robins E, Lindenmayr H. Thermally enhanced field emission from a laser-illuminated tungsten tip: temperature rise of tip. J Appl Phys 1980;51(9): 4996–5006.

[15] Liu H, Tsong T. Numerical calculation of the temperature evolution and profile of the field-ion emitter in the pulsed-laser time of flight atom probe. Rev Sci Instrum 1984; 55(11):1779.

[16] Lee M, Robins E. Thermal relaxation of a laser illuminated field emitter. J Appl Phys 1989;65(4):1699–706.

[17] Deconihout B, Vurpillot F, Gault B, Costa GD, Bouet M, Bostel A, et al. Surf Interf Anal 2004. accepted.

[18] Gault B, Vurpillot F, Bostel A, Menand A, Deconihout B. Appl Phys Lett 2005;86: 094101.

[19] Gilbert M, Vurpillot F, Vella A, Bernas H, Deconihout B. Some aspects of the silicon behaviour under femtosecond pulsed laser field evaporation. Ultramicroscopy 2007; 107(9):767–72.

[20] Liu H, Liu H, Tsong T. Numerical calculation of the temperature distribution and evolution of the field-ion emitter under pulsed and continuous-wave laser irradiation. J Appl Phys 1986;59(4):1334.

[21] Born M, Wolf E. Principles of optics. Cambridge University Press; 1983.

[22] Ashcroft NW, Mermin ND. Solid state physics. Saunders College; 1976.

[23] Houard J, Vurpillot F, Vella A, Deconihout B. Conditions to cancel the laser polarization dependence of a subwavelength tip. Appl Phys Lett 2009;94:121905.

[24] Houard J, Vella A, Vurpillot F, Deconihout B. Phys Rev B March 2010;81:125411.

[25] Miller MK, Forbes R. Atom probe tomography : the local electrode atom probe. US: Springer; 2014.

[26] L. S. Inc. FDTD solution release 6.5. 2009.

[27] Sommerfeld A. Mathematische theorie der diffraction. Math Ann 1896;47:317–74.

[28] Novotny L. Near-field optics and surface plasmon polaritons, ch. forces in optical near-fields. Berlin: Springer Verlag; 2000. p. 123–41.

[29] Blaber M, Arnold M, Harris N, Ford M, Cortie M. Plasmon absorption in nanospheres: a comparison of sodium, potassium, aluminium, silver and gold. Phys B 2007;394: 184–7.

[30] Palik E, Ghosh G. Handbook of optical constants of solids. Academic Press; 1985.

[31] Arnoldi L, Vella A, Houard J, Deconihout B. Antenna effect in laser assisted atom probe tomography : how the field emitter aspect ratio can enhance atomic scale imaging. Appl Phys Lett 2012;101(15):153101.

[32] Robins ES, Lee MJG, Langlois P. Can J Phys 1986;64:111.

[33] Dash WC, Newman R. Intrinsic optical absorption in single-crystal germanium and silicon at 77K and 300K. Phys Rev August 1955;99:1151–5.

[34] Kittel C. Introduction to solid state physics. 8th ed. Hoboken (NJ): John Wiley and Sons; 2004.

[35] Ding G, Chan C, Zhang Z, Sheng P. Resonance-enhanced optical annealing of silicon nanowires. Phys Rev B 2005;71(20):205302.

[36] Bogdanowicz J, Gilbert M, Innocenti N, Koelling S, Vanderheyden B, Vandervorst W. Light absorption in conical silicon particles. Opt Express 2013; 21(3):3891–6.

[37] Silaeva EP, Karahka M, Kreuzer H. Atom probe tomography and field evaporation of insulators and semiconductors: theoretical issues. Curr Opin Solid St M Oct. 2013;17: 211–6.

[38] Silaeva EP, Arnoldi L, Karahka ML, Deconihout B, Menand A, Kreuzer HJ, et al. Do dielectric nanostructures turn metallic in high-electric DC fields? Nano Lett Nov. 2014;14:6066–72.

[39] RamÃ-rez R, GonzÃ¡lez R, Pareja R, Chen Y. Semiconducting property of a wide-band-gap oxide crystal: impact ionization and avalanche breakdown. Phys Rev B Jan. 1997;55:2413–6.
[40] Karahka M, Kreuzer H. Field evaporation of oxides: a theoretical study. Ultramicroscopy 2013;132.
[41] Tsong T. Field penetration and band bending near semiconductor surfaces in high electric fields. Surf Sci 1979;81(1):28–42.
[42] Mueller E. Phys Rev 1956;103:618.
[43] Gomer R. J Chem Phys 1951;31:341.
[44] Tsong TT. J Phys F Metal Phys 1978;8:1349.
[45] Rusina GG, Ereneev S, Borisdava SB, Sklyadnea F, Chulkov E. Surface phonon on Al(111) surface. Phys Rev B 2005;71:245401.
[46] Kellog G. Phys Rev B 1984;29(8):4304.
[47] Tsong T, Kinkus T. Energy distribution of pulsed-laser field-desorbed gaseous ions and field-evaporated metal ions: a direct time-of-flight measurement. Phys Rev B 1984; 29(2):529.
[48] Mazumder B, Vella A, Gilbert M, Deconihout B, Schmitz G. Reneutralization time of surface silicon ions on a field emitter. New J Phys 2010;12:113029.
[49] Kingham D. J Phys D Appl Phys 1982;15:2537.
[50] Vurpillot F, Gault B, Vella A, Bouet M, Deconihout B. Estimation of the cooling times for a metallic tip under laser illumination. Appl Phys Lett 2006;88:094105
[51] Hohlfeld J, Wellershoff SS, Gudde J, Conrad U, Jahnke V, Matthias E. Chem Phys 2000;251(1–3):237–58.
[52] Bachelot R, H'Dhili F, Barchiesi D, Lerondel G, Fikri R, Royer P, et al. Apertureless near-field optical microscopy: a study of the local tip field enhancement using photosensitive azobenzene-containing films. J Appl Phys 2003;94(3):2060–72.
[53] Bouhelier A, Beversluis M, Hartschuh A, Novotny L. Near-field second-harmonic generation induced by local field enhancement. Phys Rev Lett 2003;90(1):13903.
[54] Shen YR. The principles of nonlinear optic. Wiley-Interscience Publication; 1984.
[55] Vella A, Vurpillot F, Gault B, Menand A, Deconihout B. Evidence of field evaporation assisted by nonlinear optical rectification induced by ultrafast laser. Phys Rev B 2006; 73:165416.
[56] Vella A, Deconihout B, Marrucci L, Santamato E. Femtosecond field ion emission by surface optical rectification. Phys Rev Lett 2007;99:046103.
[57] Bragas AV, Landi SM, Martinez OE. Laser field enhancement at the scanning tunneling microscope junction measured by optical rectification. Appl Phys Lett 1998;72:2075–7.
[58] Sipe JE, So VCY, Fukui M, Stegeman GI. Analysis of second-harmonic generation at metal surfaces. Phys Rev B 1980;21:4389–402.
[59] Bass M, Franken P, Ward J, Weinreich G. Optical rectification. Phys Rev Lett 1962; 9(11):446.
[60] Rudnick J, Stern EA. Second-harmonic radiation from metal surfaces. Phys Rev B 1971;4(12):4274–90.
[61] Tamura H, Tsukada M, McKenna KP, Shluger AL, Ohkubo T, Hono K. Laser-assisted field evaporation from insulators triggered by photoinduced hole accumulation. Phys Rev B 2012;86:195430.
[62] Sévelin-Radiguet N, Arnoldi L, Vurpillot F, Normand A, Deconihout B, Vella A. Ion energy spread in laser-assisted atom probe tomography. EPL Europhys Lett 2015; 109(3):37009.
[63] Arnoldi L, Silaeva E, Gaillard A, Vurpillot F, Blum I, Rigutti L, et al. Energy deficit of pulsed-laser field-ionized and field-emitted ions from non-metallic nano-tips. J Appl Phys 2014;115(20):203705.

[64] Arnoldi L, Silaeva E, Vurpillot F, Deconihout B, Cadel E, Blum I, et al. Role of the resistivity of insulating field emitters on the energy of field-ionised and field-evaporated atoms. Ultramicroscopy 2014.
[65] Cerezo A, Clifton P, Gomberg A, Smith G. Aspects of the performance of a femtosecond laser-pulsed 3-dimensional atom probe. Ultramicroscopy 2007;107(9):720–5.
[66] Sha G, Cerezo A, Smith G. Field evaporation behavior during irradiation with picosecond laser pulses. Appl Phys Lett 2008;92:043503.
[67] Houard J, Vella A, Vurpillot F, Deconihout B. Three-dimensional thermal response of a metal subwavelength tip under femtosecond laser illumination. Phys Rev B 2011; 84(3):033405.
[68] Mazumder B, Vella A, Vurpillot F, Martel G, Deconihout B. Surface carrier recombination of a silicon tip under high electric field. Appl Phys Lett 2010;97:073104.
[69] Vella A, Mazumder B, Da Costa G, Deconihout B. Field evaporation mechanism of bulk oxides under ultrafast laser illumination. J Appl Phys 2011;110:044321–044321–7.
[70] Koelling S, Innocenti N, Schulze A, Gilbert M, Kambham A, Vandervorst W. In-situ observation of non-hemispherical tip shape formation during laser-assisted atom probe tomography. J Appl Phys 2011;109:104909.
[71] Mancini L, Amirifar N, Shinde D, Blum I, Gilbert M, Vella A, et al. Composition of wide bandgap semiconductor materials and nanostructures measured by atom probe tomography and its dependence on the surface electric field. J Phys Chem C 2014; 118(41):24136–51.
[72] Xia Y, Karahka M, Kreuzer HJ. Field evaporation of ZnO: a first-principles study. J Appl Phys 2015;118(2).

CHAPTER NINE

Data Mining

W. Lefebvre-Ulrikson, G. Da Costa, L. Rigutti, I. Blum
Groupe de Physique des Matériaux, University and INSA of Rouen, Normandie University, France

Data mining can basically be defined as an approach consisting in extracting quantitative information from a data set. In atom probe tomography (APT), this information may include morphologies, size of objects, volume fractions, compositions, number densities, interface positions, and levels of segregations. One rapidly understands from this nonexhaustive list that a single algorithmic tool might not be able to answer all the questions the user may have about a data set. It becomes implicit that, depending on the feature of interest, a category of tools will be of relevance. The choice of the relevant tool is hence the first step of the data processing. The second step deals with the smart definition of processing parameters. Indeed, most algorithmic tools require a quantity of free parameters, the choice of which has often a critical influence on the measured quantities. Defining processing parameters remain consequently as crucial as the definition of experimental conditions. Once (or before) a result is provided by the data processing, one still has to wonder whether any artifact or intrinsic feature of the instrument may decrease the degree of confidence in this result.

The imperfection of an APT data set must, however, be evaluated before any quantitative information is extracted. For instance, an intrinsic limitation of any APT data set relies on the fact that only a portion of the evaporated atoms hitting the detector are actually detected. The average ratio between the evaporated atoms hitting the detector and the amount of atoms available in the reconstruction has been defined as the detection efficiency Q (see Chapters 5: Time-of-Flight Mass Spectrometry and Composition Measurements and 7: Three-Dimensional Reconstruction in Atom Probe Tomography: Basics and Advanced Approaches). The detection efficiency might have less obvious impacts on the measurements. Let us consider a cluster made of 20 atoms that a given algorithmic tool is capable of distinguishing in an APT data set. Considering this cluster inevitably leads to the following question: how many atoms were actually contained in this cluster inside the material? The answer is 20 or more. For a detection efficiency of 50%, an amount of 40 will get the largest probability (ie, the mean value of the

Atom Probe Tomography
ISBN: 9780128046470
http://dx.doi.org/10.1016/B978-0-12-804647-0.00009-7

negative binomial distribution function introduced in Chapter 5: Time-of-Flight Mass Spectrometry and Composition Measurements). However, an amount of 60 would also be possible, although with a much smaller probability. This example illustrates that the detection efficiency systematically underestimates the cluster size distributions. This phenomenon has been studied and modeled by Stephenson et al. [1]. The imperfect character of data sets also relies in the resolution of the instrument. It has been explained earlier (see Chapter 7: Three-Dimensional Reconstruction in Atom Probe Tomography: Basics and Advanced Approaches) that the spatial resolution is anisotropic and that, unless for some very rare exceptions, the true atomic positions are not accessible. Furthermore, for a crystal, pole figures display low densities at the center of poles. It is hence important to avoid extracting quantitative data along these zones. More generally, simple potential artifacts have to be considered before processing the data and once a result is obtained.

This chapter provides an overview of the methods available for the data mining of APT analyses. Since the literature about data mining in APT is very abundant, the purpose is not to provide an exhaustive list of the algorithms developed in this field. The reading of this chapter may hence be accompanied by the reading of some published reviews or book chapters to get a complete overview of the methods proposed in the APT community [2–6]. It will be considered here that all possible mistakes in the prior processing of the APT data set have been avoided. This means that the mass spectrum, mass ranges, experimental conditions, and reconstruction parameters are optimal. Insofar as the tool depends on the feature to investigate, some preliminary definitions of the categories of features present in an APT data set are given. Then, the various ways APT data sets can be processed are presented, followed by a classification of different methodologies available as a function of the quantities to be extracted. In the last section, the various ways to reveal experimental artifacts and to correct them are presented.

PRELIMINARY DEFINITIONS

Thanks to the use of ultrafast laser pulse and also to the better ergonomics (see Chapters 1: Early Developments and Basic Concepts, 2: Field Ion Emission Mechanisms, and 8: Laser-Assisted Field Evaporation), increased acquisition rates of APT, and the drastic improvement of specimen preparation methods (see Chapter 4: Atom Probe Sample Preparation), the use of APT has spread to many fields of materials science and beyond. Nevertheless, the data sets still look the same insofar as they display atomic positions in a three-dimensional (3D) space, the chemical nature of atoms being known. In these volumes, several categories of features might be distinguished and it is relevant to unambiguously define them before further considering their analysis.

Solid Solution

The term "solid solution," mostly associated to the field of physical metallurgy, is also relevant to describe doped crystals in semiconductors. A solid solution is composed of two elements or more. It has the crystalline structure of the major element (solvent) and the minor elements (solutes) occupy positions on this lattice (in a substitutional solid solution) or on the interstitial sites of the lattice (in an interstitial solid solution). Although interstitial elements positions can be identified in field ion microscopy, the spatial resolution of APT does not allow making the distinction between interstitial and substitutional sites. It must be noticed that the term "matrix" is commonly used to designate the solid solution. This will often be the same in this chapter. When the solid solution is in a metastable state (eg, the case of a supersaturated solid solution), fluctuations of concentrations might exist. The statistical analysis of APT data sets has to reveal whether or not these fluctuations are larger than those induced by the APT analysis itself. Fluctuations may correspond to very early stages of phase decomposition. Quantities like short-range order parameter might hence be considered. As a consequence, a category of tools has been built by the APT community to assess the detailed analysis of solid solutions [2,3,7−15]. Most of these tools also apply for other cases and are definitely not dedicated only to alloys or semiconductors.

A solid solution is rarely free of crystalline defects which can interact with other defects such as vacancies or solute elements. The detailed study of crystalline defects structure (eg, dislocations and stacking faults) is impossible in APT owing to the spatial resolution of this instrument. This justifies why a correlative approach combining APT with another high-resolution technique might be relevant [see Chapter 10: Correlative Microscopy by (Scanning) Transmission Electron Microscopy and Atom Probe Tomography].

Clusters, Precipitates, and Particles

Clusters and precipitates are 3D defects which currently are formed by the decomposition of a solid solution. Establishing a distinction between clusters and precipitates is not straightforward. For the reader familiar to the field of physical metallurgy, a reference to the classical theory on nucleation can be helpful (see, for instance, [16]). In this theory, not free of underlying hypotheses, the formation of a new phase from a matrix involves the creation of particles of radius r. The classical theory of nucleation assumes that, from the beginning, particles of the new phase have a defined composition and share an interface with the surrounding matrix. One can, however, wonder about the significance of a particle composition or an interface when the particle is only constituted of several atoms. According to the classical nucleation theory, the variations of free enthalpy associated with the precipitation of a second phase in the form of spherical particles follow the graph shown in Fig. 9.1. Above a critical radius $r^\star$, the growth of particles, which are identified as the nuclei of a second phase, decreases the free enthalpy of the system. Hence, one can propose a distinction between clusters and precipitates. The term cluster can be attributed to the small particles forming in the very early stages of the precipitation process and precipitates can be identified as supercritical clusters. The former contain small quantities of atoms, which do not justify the definition of a clear interface, whereas the latter have a given composition and share an interface with the surrounding matrix.

Solute-rich clusters may also form when complex precipitation sequences are involved in an alloy. Seminal illustrations are provided by age hardenable aluminum-based alloys [17], for which clusters (often identified or designated as Guinier–Preston zones [18,19]) precede a sequence of

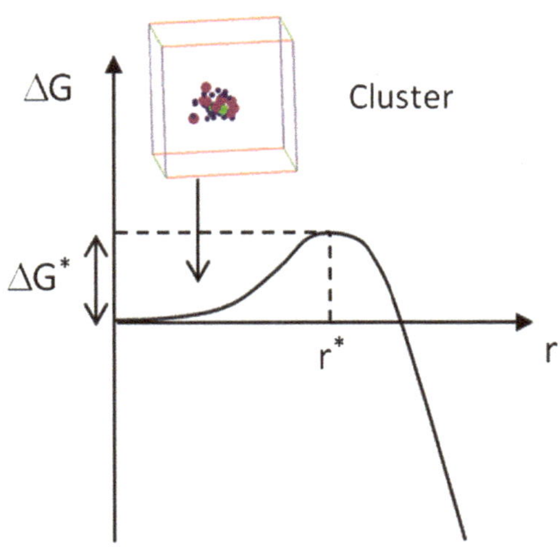

Figure 9.1 Variation ΔG of the free enthalpy associated with the nucleation of a spherical nucleus of radius *r*, according to the classical theory of nucleation [16]. r^* is the critical radius of nucleation and is the energy barrier calculated for the critical nucleus.

metastable phases. Clustering can therefore be simply defined as a phenomenon leading to local enrichments in solute elements inside a solid solution. Clusters have typically the same crystalline lattice of the solid solution and do not share any defined interface in opposition to precipitates. As a direct consequence, methodologies may often differ to quantify clustering and precipitation in APT. It must be noticed that precipitates are sometimes referred to as particles in alloys. Particles commonly designate any kind of 3D defect exhibiting a different composition (and often structure) than the surrounding medium with which an interface is shared. This generic definition is also relevant for the designation of 3D defects in most materials including biomaterials, composites, and geomaterials. Their study by APT will therefore require the same tools as those useful for the investigation of precipitates.

Interfaces

An interface is a two-dimensional (2D) defect separating two media. An interface may separate two regions of the same phase (eg, grain boundaries in crystalline materials) or of different structures having similar or distinct compositions. In such a case, interaction with other defects (eg, solutes or precipitates) may reveal the presence of the interface in APT. Without interactions of that kind, revealing the presence of such interfaces will require dedicated tools. When interfaces separate two media of different compositions, they can be identified by composition-based methodologies. In the next section, the bases of algorithmic methods of data mining are presented.

BUILDING THE TOOLS

One of the main features of APT relies on its ability to provide both a global and a local view of the material at the atomic scale. The global view of the atom distribution is useful to define the different phases included in the material for example, whereas the local view gives the possibility to count the number of atoms included in each specific phase with a resolution close to the angstrom and to perform statistic treatments by considering each atom and its neighborhood. The aim of data mining is to study the global and local organizations of atoms to define the different properties of the analyzed material.

Structure of the Data Set

At the end of the analysis, two complementary files are available for the user. The acquisition data file contains the different state variation of the instrument and for each detected atom, a list is available of relative data measured during the analysis. Several versions of acquisition data files are available. The minimum definition of a data file used for basic data mining is composed of a set of four components (X, Y, Z position on the specimen and M/n ratio) measured for each detected atom. Even if these parameters are sufficient, more components [eg, flight time, direct current (DC) voltage, XY position on the detection system, cluster id] are sometimes needed to increase the possibilities of investigation of the reconstruction. In addition, many of these other parameters are from the beginning required in any cases to calculate the 3D reconstruction.

As an example, for atom probes available at the University of Rouen (Fig. 9.2), the acquisition data file is composed of a header that gives all the information concerning the analysis and a list of components available for each detected atoms. As the number of atoms is quite large in wide-angle configuration, most of the components are compressed to decrease the size of the data file.

An additional file, designated as "environment file," contains the definition of the chemical nature and the mass range of the studied material, the

File Header

AP information

Analysis main parameters

Specimen temperature
Pulse fraction / Laser power
Vacuum values

Atom i

Atom Id
Number of pulse since the previous atom
XYZ atom position on the tip
Measured time of flight
Calculated m/n value
Cluster Id
Measured DC voltage
XY position on the detection system
Amplitude of the signal measured on MCP
Propagation times measured on XY delay lines

Figure 9.2 Definition of the structure of the acquisition data file.

main parameters linked to the atom probe, and the different parameters used to calculate the 3D reconstruction.

Grids Versus Atomic Positions

Different algorithms based on 3D investigation of atom composition or densities inside the volume are useful during data mining. Indeed, these algorithms give access to global and local fluctuations of each element composition (or densities), and allow the representation of these fluctuations and the assignment of atoms to distinct phases. The estimation of the morphology of these phases and their related composition can be then extracted.

Since considering the x,y,z position of each atom for calculations requires long processing, a 3D composition grid for each defined element is calculated while the atom file is loading. The two main parameters used to build this composition grid are the size of sampling volume in x,y,z dimension and the delocalization value.

The size of the sampling volume is very important because it defines the resolution of the grid versus the accuracy of the composition calculation. For example, Fig. 9.3A shows a very small sampling volume in 2D. In this case, the resolution of the grid is good. However, only few atoms are present inside each sampling box. This produces a huge dispersion over the composition because of statistical fluctuations. On the contrary, Fig. 9.3B shows a quite high sampling volume in 2D. In this case, the resolution of the grid becomes poor but the statistical significance of the measured composition

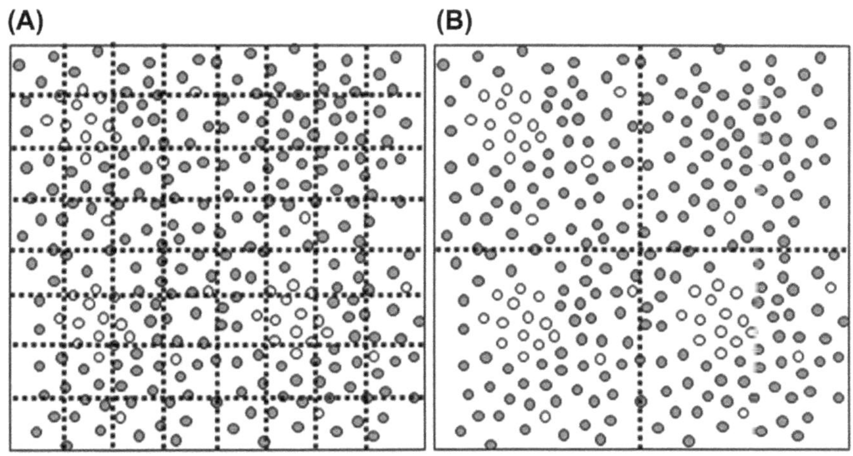

Figure 9.3 (A,B) Schematic illustration of the influence of sampling size over the statistical significance of compositions measured with a grid.

is increased because of the larger number of atoms inside each sampling volume.

The easiest way to evaluate the related composition of each sampling box is to take into account all the atoms inside the box (ie, N atoms) to calculate the composition values for each defined element ($C_i = N_i/N$) and assign each composition value to the center of the sampling box (Fig. 9.4). Based on these calculations, deriving the local composition for each atom position in the volume is not straightforward. At least, it requires consideration of the compositions calculated in sampling volumes in the vicinity of the atom, as illustrated in Fig. 9.4B.

Delocalization

To reduce the effect of sampling, a delocalization function based on a Gaussian weighting correction is applied to each atom to calculate each value of the composition grid. This Gaussian weighting was initially introduced by Hellman et al. [20,21]. Fig. 9.5 illustrates the principle of the

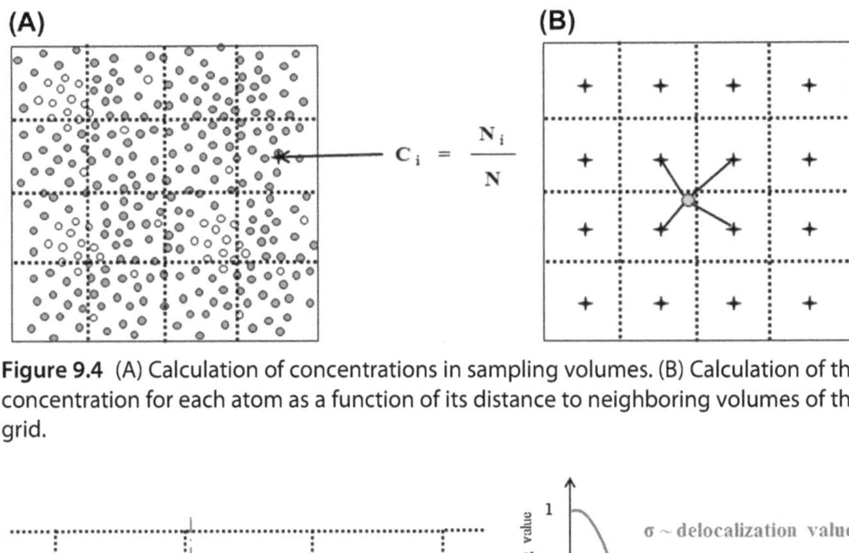

Figure 9.4 (A) Calculation of concentrations in sampling volumes. (B) Calculation of the concentration for each atom as a function of its distance to neighboring volumes of the grid.

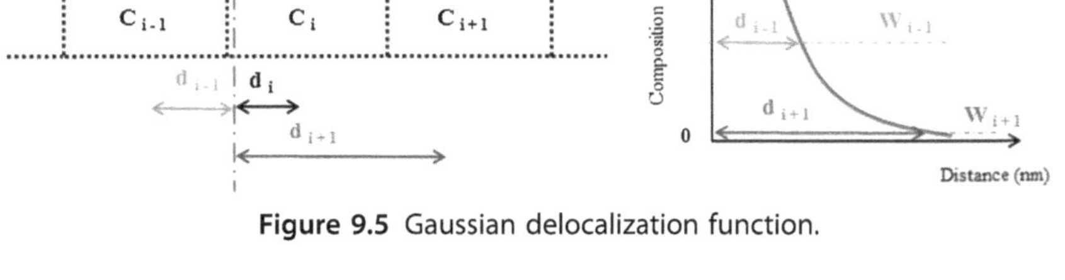

Figure 9.5 Gaussian delocalization function.

delocalization in 1D. Each center of the composition grid sample is superimposed on a Gaussian function with a defined standard deviation. The local concentration on each atom in the data set is a weighted average of the concentrations calculated in the neighboring sampling boxes, each concentration being weighted by a factor W_i. The distance d_i from an atom position to the center of the sampling box i defines the weight W_i of the average.

The number of sampling boxes involved in the calculation for each atom position directly depends on the standard deviation σ of the Gaussian (ie, the delocalization value) and the original size of the sampling grid. For instance, if the size of the sampling grid is 1 nm and the delocalization value is 0.5 nm, the calculation will involve only the first neighboring sampling volumes of each atom position, insofar as the weight values are negligible beyond 3σ. Gaussian smoothing is commonly used in the field of signal or image processing. In APT reconstructions, it is very useful to decrease the roughness of isosurfaces and was introduced by Hellman et al. as a mandatory step for the calculation of proximity histograms, better known as "proxigrams" [20,21].

Interpolation

If the knowledge of very local variations of concentrations is needed, to study clustering, for instance, delocalization is not always applied to APT data sets. When the 3D grid of composition values is available, the relative composition of a defined element can be calculated at any position inside the volume. Fig. 9.6 shows the principle of the linear interpolation in 1D,

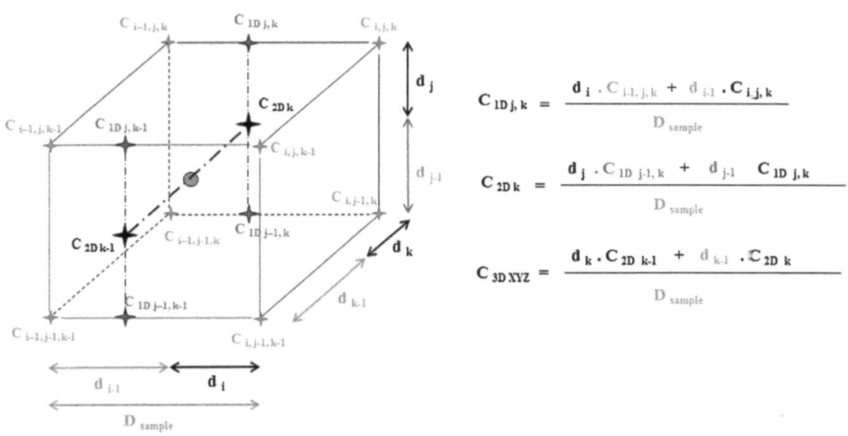

Figure 9.6 Principle of the 1D, 2D, and 3D interpolations.

2D, and 3D, used to extract the composition value associated with a defined position from the original composition grid.

This way, the fluctuation of the composition or the density of atoms in the 3D volume can be represented (eg, iso composition surfaces or curves, iso density). Any atom of the volume can then be filtered as a function of the composition calculated from its position. This is particularly useful for the identification of clusters or precipitates, when a significant difference of composition exists between these objects and the surrounding matrix.

RELEVANCE OF THE APPROACH AS A FUNCTION OF THE FEATURE OF INTEREST

In the following, we detail the methodologies as a function of the feature of which they enable the investigation. It should, however, be noticed that this classification, although useful from a didactical point of view, is not unique and that some methodologies have a larger spectrum of application.

Solid Solution—Clusters

Careful Selection of a Region of Interest

The degree of complexity of the solid solution analysis might vary as a function of the actual purpose of the study. But whatever the purpose is, attention must first be paid to the selection of the volume investigated. Indeed, the careful selection of a region of interest is not trivial. This is still true when the APT data set displays a region of apparently homogeneous composition. As stated earlier, we consider here that the mass spectrum, mass intervals, experimental conditions, and reconstruction parameters are optimal. Despite this, some zones of the volume might disrupt its investigation. This is typically true when a crystalline phase is analyzed. This phenomenon has been deeply investigated and considered by Stephenson et al. [22] and is illustrated in Fig. 9.7. Lower densities of atoms in the APT volume are closely related to the symmetry of the crystal and are easy to distinguish by eye. Let us consider a simple algorithmic tool which would count the amount of B atoms (minor element) in sampling volumes of a given size (eg, boxes of $1 \times 1 \times 1\ \mathrm{nm}^3$). This tool enables one to plot a histogram displaying the frequencies of boxes containing n atoms B. It is clear that frequencies will be artificially increased at low values n along

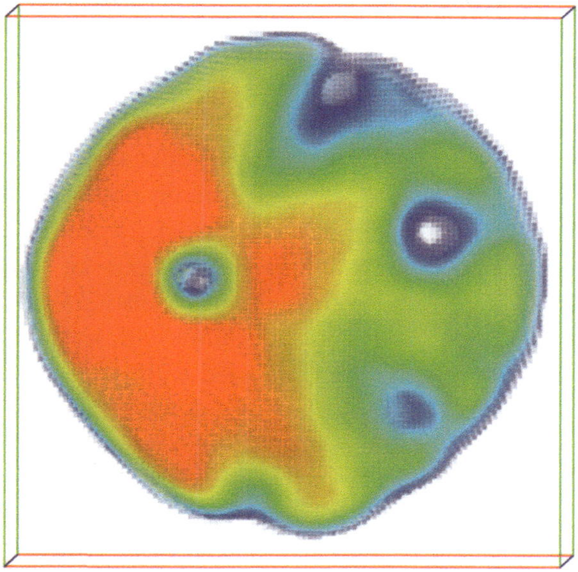

Figure 9.7 Atom density map calculated on each atom position in the volume, for a Gaussian delocalization performed with a standard deviation of 2 nm and a grid size of 1 nm. The material is an AA-6101 alloy (Al-0.88 at.%Si-0.68 at.%Mg). The volume of the box is 37 × 37 × 37 nm^3. Low-density regions, which appear as bright disks, are located at the center of low index poles.

regions of lower densities. On the contrary, if an artifact-like surface migration of B atoms (or "directional walk") toward the poles occurs during the APT analysis [23], larger frequencies will be estimated for larger values of *n* along poles. Consequently, the pole figures must be considered prior to any further investigation. Based on this consideration, Stephenson et al. [22] have proposed an algorithmic tool based on the distance to the *K*th nearest neighbor (KNN) distances and enabling the visualization and filtering of data in the reconstruction as a function of local density of atoms. Nevertheless, it must be kept in mind that such a filtering may also remove the signal we are interested in. Parameters of filtering have hence to be set with care, as explained in Ref. [22]. Regions identified as low indexes crystallographic zones or poles can be filtered by density thresholding and their composition compared to other regions, surface migration of a minor element would result in a larger (diffusion to the poles) or weaker (diffusion from the poles) concentrations. If the latter method is not available, it is still possible to localize regions of interest manually, based on the visualization of a density map similar to the one displayed in Fig. 9.7.

Randomness of Atom Distribution

In some cases, it might be of interest to determine whether or not the volume is a random solid solution/matrix. In alloys, for instance, it can be important to evaluate if early stages of phase decomposition are involved. But more simply, since the eye is not a perfect tool, it is always relevant to check whether a volume which looks apparently homogeneous is actually composed of a single phase. For this purpose, an experimental frequency distribution can be calculated by counting the number of atoms of a given species B in sampling volumes containing a fixed number N of atoms. The observed distribution is designated as $O(n)$ with $n = 0,1,\ldots,N$. Then, the observed distribution has to be compared with an expected theoretical distribution $e(n)$ (Fig. 9.8). If B atoms are expected to be randomly distributed in the reconstructed volume, $e(n)$ can be estimated by using the binomial distribution $B(n)$ as follows:

$$e(n) = N \cdot B(n) = N \cdot \frac{N!}{n!(N-n)!} X_B^n (1 - X_B)^{(N-n)} \tag{9.1}$$

This can be simply done by a χ^2 test, which has been applied for long to atom probe and APT data [24]. The statistical distance between the

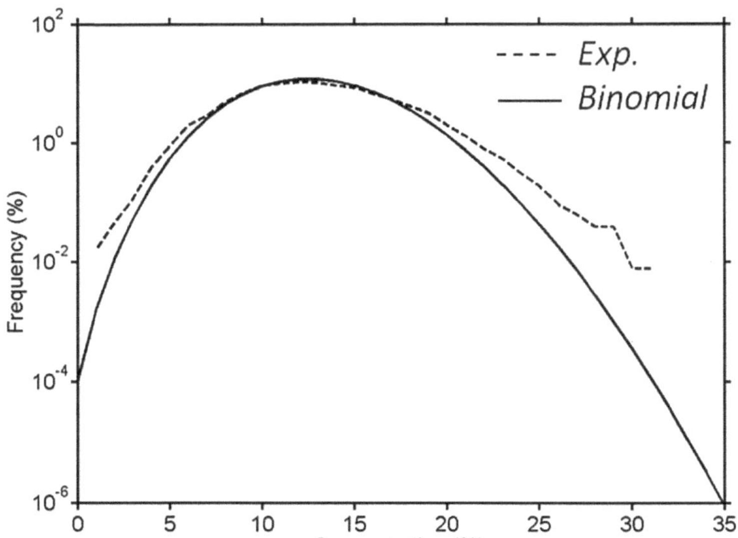

Figure 9.8 Frequency distribution of the concentration of solute elements B in a simulated volume (25 × 25 × 45 nm^3) containing clusters enriched in B ($C_c = 70$ at.%) in a matrix of concentration $C_m = 10$ at.% of B.

observed and expected distributions is evaluated by the parameter χ_m as follows:

$$\chi_m^2 = \sum_{n=0}^{N} \frac{(e(n) - O(n))^2}{O(n)} \tag{9.2}$$

Depending on the value of this parameter, the random character of the region of interest will be considered as relevant or not. More precisely, a null hypothesis has to be evaluated which is: the B atoms are randomly distributed across the volume investigated. Then, to estimate if χ_m is lying on a χ^2 distribution with N degrees of freedom (ie, the $N+1$ possible values -1), with a level of tolerance $P = 0.05$ (common values for this statistical test), one has to refer to a calculated χ^2 table and check if the value at the column P and row N is less than χ_m. If this value is more than χ_m, the null hypothesis is rejected with a probability or error equal to P. Doing this test, the user has to pay attention that the minimum amount of observations per class is five. This means that too low values of N would be detrimental to the quality of the result. Consequently, for very small concentrations of B (ie, <1 at.%), large values of N would have to be chosen. It is hence unlikely that very small clusters of B are detected with this statistical method, for which nearest neighbor analysis would be more helpful.

Much details about the way to use the comparison to a binomial distribution for APT data can be found in the work by Moody et al. [7], who discuss the applicability of the method for early stages of phase decomposition and highlight the influence of the detector efficiency on the segregations measured thanks to this statistical analysis.

Neighborhood-Based Methodologies (KNN, Radial Distribution Functions, Pair Correlations)

As stated in the preliminary definitions of this chapter, clusters are tiny objects constituted of very small amount of atoms for which no clear interface can be defined. Deciding from the visualization of a reconstruction whether clusters are present is hence a very challenging issue. Among the dedicated tools relevant for the investigation of clustering, several allow plotting graphical data, whereas some others also enable displaying clusters. However, when opting for the latter possibility, it is always mandatory to establish if the clusters shown actually correspond to reality or, in other words, if the signal to noise ratio of the filtering algorithm is large enough to trust the results.

The tools dedicated to clustering rely on the investigation of atoms neighborhood. A very straightforward method consists in using the radial distribution function (RDF) or the pair correlation function [14,26–29]. As illustrated in Fig. 9.9A, the RDF calculates the number of atoms $n_{\mathrm{RDF}}(r)$ in a spherical shell $(r,2\Delta r)$ of radius r and thickness $2\Delta r$ around atoms of a selected element i. This calculation is performed in the whole volume (ie, for all "i" atoms in this volume) for increasing values of r and fixed value of Δr. The normalized RDF is derived by dividing $n_{\mathrm{RDF}}(r)$ by the average number of atoms that should be present in each shell, as deduced from the average density $\overline{\rho}$ in the whole volume.

$$\mathrm{RDF}_i(r) = \frac{n_{\mathrm{RDF}}(r)}{\overline{\rho}\left(4/3\right)\pi\left((r+\Delta r)^3 - (r-\Delta r)^3\right)} \tag{9.3}$$

As shown in Fig. 9.9B, owing to a given detection efficiency and to an imperfect spatial resolution, coordination shells are not resolved by the APT analysis, although KNN distances can still be shown as an indication. However, statistical information about the very first neighbors is still accessible for most crystalline systems. By considering only the atoms of an

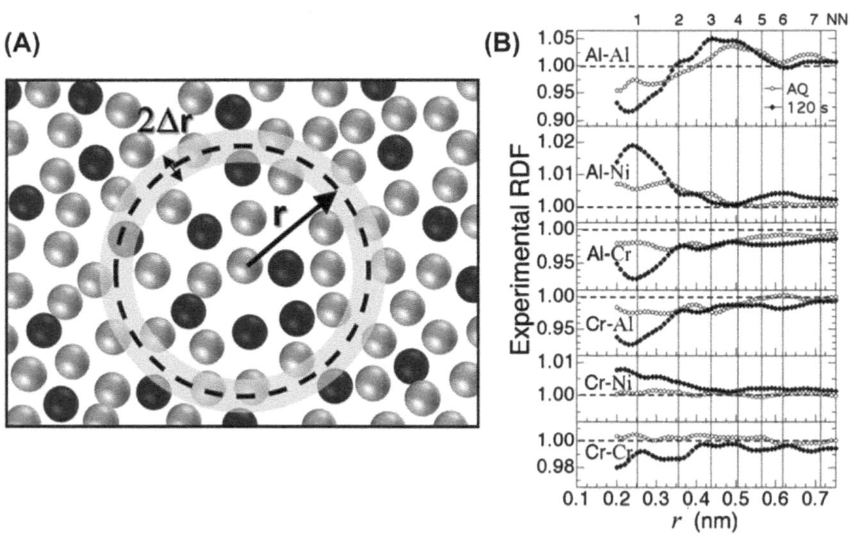

Figure 9.9 (A) Schematic illustration of the way radial distributions or pair correlation functions are calculated. (B) Radial distribution functions for an as-quenched (AQ) Ni-5.2 Al-14.2 Cr at.% alloy and after aging for 120 s at 873K. *Figure reproduced with permission from the work by Sudbrack CK, Noebe RD, Seidman DN. Phys Rev B 2006;73:212101.*

element j in the shell $(r,2\Delta r)$ around i atoms, one gets the expression of a partial RDF, which can be assimilated to a partial pair correlation function $g_{ij}(r)$ as introduced in Ref. [14]:

$$g_{ij}(r) = \frac{n_j(r)}{c_j\left(\frac{4}{3}\right)\pi\left((r+\Delta r)^3 - (r-\Delta r)^3\right)} \tag{9.4}$$

In the above-mentioned equation, n_j is the amount of j atoms in the shell $(r,\Delta r)$ and c_j is the average concentration (in at.%) of element j in the volume investigated. Values of the pair correlation function $g_{ij}(r)$ are very indicative of the clustering behavior of solute elements. Indeed, values of $g_{ii}(r) > 1$ indicate a coclustering behavior between species i and j, whereas values of $g_{ii}(r) < 1$ indicate a repulsive behavior between the two species. Obviously, values of $g_{ii}(r) = 1$ would be obtained when local and average concentrations are identical, which indicates an absence of clustering.

The Maximum Separation Distance Algorithm

Once the evidence of clustering is provided by a statistical analysis, one can attempt to identify clusters in the reconstruction. However, attention has to be paid to the way such quantifications are processed. As interfaces are not considered for clusters, filtering methods are mostly based on nearest neighbor distances. The most widely used method, the maximum separation distance (MSD) [29,30], uses the distance to the first nearest neighbor. Generally, a Kth nearest neighbor is written as KNN. The MSD algorithm is hence a 1 NN-based method. The algorithm works as follows. An element is specified. For all atoms of this element in the volume investigated, the 1 NN distances are calculated. Then, an MSD d_{max} is chosen and atoms closer than d_{max} are gathered in a same cluster. Finally, only clusters constituted by more than N_{min} atoms are kept. This latter parameter is used to remove artificial clusters generated by statistical fluctuations. This method, although extensively used, has a rather low sensitivity and often suffers from a large noise. This can be explained by considering Fig. 9.10, where the 1 NN distances have been plotted for solute atoms in a volume containing clusters embedded in a matrix. For the chosen values of concentrations, a clear overlap occurs between the two distributions. Applying a d_{max} distance hence results in the impossibility to detect some real clusters and in the detection of artificial clusters originating from statistical fluctuations in the matrix. These later are expected to be removed by imposing an N_{min} value.

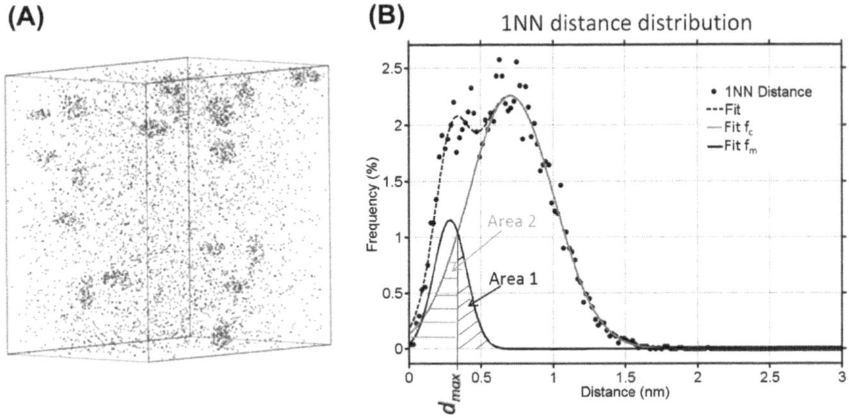

Figure 9.10 A 30 × 20 × 30 nm^3 simulated APT reconstruction of an A-B system. About 1% of atoms are in clusters. These clusters contain 20% of B in average, whereas the matrix contains 1% of B. (B) Corresponding 1 NN distance distribution. This distribution has been fitted with two Gaussian components. Each component is deduced from the fit of 1 NN data (f_c or f_m for the clusters and matrix, respectively). An arbitrary d_{max} is chosen. Area 1 indicates the fraction of B atoms in clusters which would not be selected by the maximum separation distance method. Area 2 corresponds to the fraction of B atoms present in the matrix, which are assigned to clusters by the maximum separation distance method. These latter fractions might be reduced by imposing a value N_{min} for the minimum amount of solutes per cluster. It might be noticed that the possibility of deriving quantitative information from the fitting of 1 NN distance distribution has been discussed in Ref. [13].

Let $f^{1\,NN}(d)$ designate the 1 NN distance distribution of a solute element in the volume investigated. The theoretical expression of $f^{1\,NN}(d)$ is composed of three contributions:

$$f^{1\,NN}(d) = (1 - x_S)\cdot\left[x\cdot f_c^{1\,NN}(d) + (1 - x)\cdot f_m^{1\,NN}(d)\right] + x_S\cdot f_s(d) \tag{9.5}$$

In the above-mentioned expression, x_S is the fraction of 1 NN distances involving mixed pairs (ie, clustered atoms and atoms from the matrix). Among other 1 NN distances, x is the fraction of 1 NN distances in clusters and $(1 - x)$ is the fraction of 1 NN distances in the matrix. $f_S^{1\,NN}(d)$ is the distribution of 1 NN distances in mixed pairs. $f_c^{1\,NN}(d)$ and $f_m^{1\,NN}(d)$ are the 1 NN distance distributions in the clusters and in the matrix, respectively. By neglecting the contribution of mixed pairs (ie, by assuming $x_S \ll 1$) and by imposing an MSD d_{max}, one can simply derive the relative proportion of "lost pairs" (corresponding to area 1 in Fig. 9.10), which are not considered as belonging to clusters:

$$\int_{d_{\max}}^{+\infty} f_c(d) \cdot dd \tag{9.6}$$

Similarly, the relative proportion of detected pairs which actually come from matrix atoms is written as:

$$\int_{0}^{d_{\max}} f_m(d) \cdot dd \tag{9.7}$$

The above-mentioned quantity represents the noise level of the MSD algorithm, which is expected to be reduced by imposing a minimum amount of atoms $N_{\min}$ per cluster. However, the choice of a relevant value for $N_{\min}$ is not straightforward and requires a heuristic approach [12,31] insofar as analytical determination of $N_{\min}$ is very complex [32,33].

To associate other species to clusters, the MSD algorithm is commonly associated to the envelope method [34], which consists in defining a grid to the reconstruction and to assign grid occupied by selected (Fig 9.11A). An alternative method relies on the rough association of other species to the cluster, followed by an erosion [35] (Fig 9.11B). Both methods require the user to define additional parameters which are susceptible to influence the reproducibility of results.

Kth Nearest Neighbors Distance Methodologies

From the earlier simple considerations, it appears obvious that a better sensitivity of the cluster selection algorithm would be reached by reducing the overlap between the respective NN distances distributions, ie, by reducing the areas shown in Fig. 9.10. This is what methods based on higher order nearest neighbor distances achieve [12,36,37]. The density-based spatial clustering of applications with noise (DBSCAN) [12,38] methodology introduced the possibility to impose a $d_{\max}$ distance for the KNN. As for the Delaunay cluster selection algorithm [39,40], its sensitivity relies on the use of 3 NN distances. In Fig. 9.12 plots of KNN distances are represented for a simulated data set. They illustrate how the overlap between contributions from the matrix and clusters are reduced as the order of the nearest neighbor is increased. More striking, the overlap between 1 NN distance distribution and the matrix is almost complete, whereas high orders enable distinguishing the two contributions.

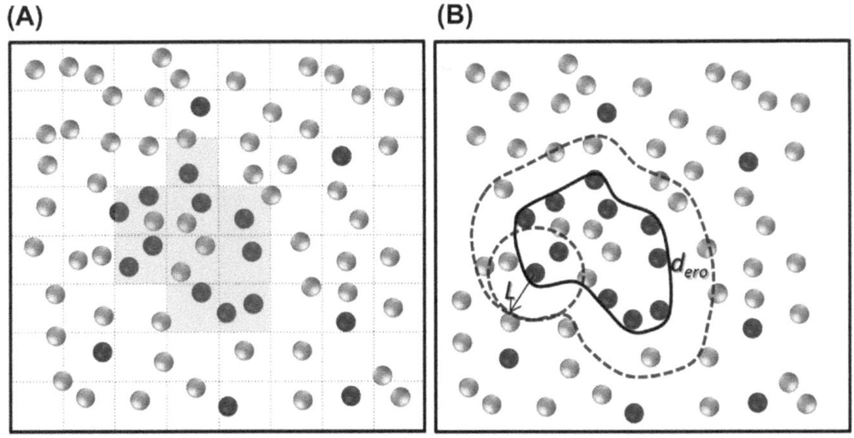

Figure 9.11 (A) Illustration of the envelope method [34]. Gray atoms from the matrix inside boxes containing cluster atoms are assigned to the cluster. Boxes surrounded by "cluster boxes" are assigned to the same cluster. This method requires the user to define a grid size. (B) Illustration of the association and erosion method [35]. In the first step, all atoms at a distance L from a cluster atom is associated to the cluster. In the second step, the cluster is eroded of a distance d_{ero}.

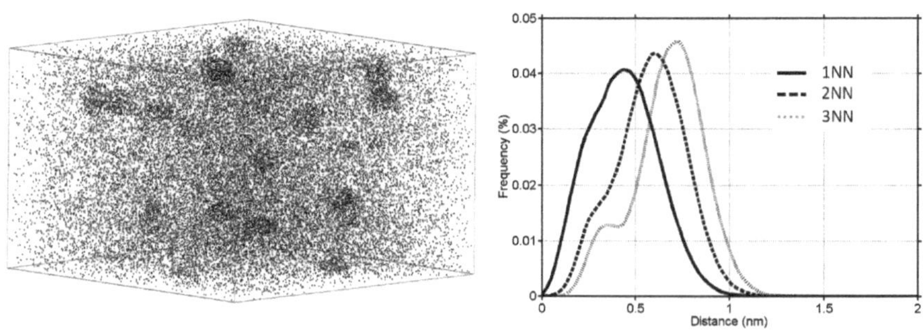

Figure 9.12 (*Left*) Simulated APT data sets containing ~800,000 atoms (30 × 30 × 20 nm^3). Parameters of the simulation are: $C_p = 50.54$ at.%; $C_m = 4.00$ at.%; 20 precipitates. The fraction of atoms in precipitates is 1.08%. Only B atoms are shown in the data set. (*Right*) Corresponding 1, 2, and 3 NN distance distributions for B atoms.

Based on the use of higher order neighbors, it is possible to conduct precise cluster search. However, the definition of threshold values ($d_{\min}$) as well as the order K of the neighbor is never trivial. It is hence mandatory to compare the experimental KNN distance distributions to those obtained from an equivalent randomized reconstruction. Doing this, clear trends are distinguishable when clustering is involved. Such a comparison helps defining the couple (K,d_{max}). Nevertheless, it must be kept in mind that, although a high signal over noise ratio can be reached, a portion of the signal (ie, real clusters) might have been missed and some noise (ie, fake clusters generated by the filtering method) is often present. Consequently, the use of an N_{min} value to remove part of the noise is useful.

Once the cluster search has been performed based on a KNN distribution and a d_{max}, association of other species has to be done. This can be performed owing to several methods: (1) the core linkage [12]; (2) the association and erosion [35]; (3) the Delaunay tessellation [40]. Among these methods, only the Delaunay tessellation does not require any additional parameter for the association of other species, as will be presented in the next section.

An interesting feature of nearest neighbor distances distribution calculated in APT reconstruction relies in their possible comparison of theoretical variations. Indeed, a KNN distribution can be calculated for a complete spatial random assumption (CSR) [36], as introduced for APT by Stephenson et al. [22]. The CSR assumption fits well to the description of most APT data sets, since the spatial resolution of this instrument blurs the respective positions of atoms. Nevertheless, in situations where the spatial resolution is high and for very small distances (<0.5 nm), the CSR assumption is not valid but this is not what is considered here. The generic probability $P_k(r,\rho)$ distribution function that the Kth nearest neighboring point to a central point of a CSR distribution lies in a spherical shell (r,dr) is given by:

$$P_k(r,\rho)\cdot dr = \frac{D}{(k-1)!}\lambda^k r^{kD-1} e^{-\lambda r^D}\cdot dr \tag{9.8}$$

where D is the number of dimensions, ρ is the density of points, and λ is calculated using the Gamma function:

$$\lambda = \rho\frac{\pi^{D/2}}{\Gamma\left(D/2+1\right)} \tag{9.9}$$

Quantification of Clustering

Quantification of clustering in alloys is a very challenging issue and APT is one of the very few techniques providing quantitative data. Very detailed studies are available in the literature. The reader is hence advised to read the following references for a more complete approach [2,9,15,41]. Although not much development will be done in this section, a very important point needs to be mentioned which is the following: as far as cluster size distribution is based on the number of atoms, the cluster size distribution measured in APT underestimates the physical cluster size distribution in the material. The reason for this is obvious and relies on the detection efficiency of the technique. Indeed, any selected cluster of *N* atoms in the reconstruction was initially made of *N* or more atoms in the material. Based on this knowledge, Stephenson et al. [1] have proposed a mathematical method which enables retrieving the physical cluster size distribution in the material. Although detection efficiency also induced a loss of information for precipitates, these larger entities are more generally described by their morphology and their estimated size distributions (in length) can directly be compared with those estimated by other techniques.

Precipitates—Particles

The definition of precipitates is an easier task than the identification of clusters. First, most of the tools described in the previous section are applicable to the detection of precipitates. In addition, other tools based on concentrations [42] and on the detection of interfaces like the proxigram [21], which have been introduced in section "Building the tools" of this chapter, are applicable. Among methods of cluster identification, the one based on Delaunay tessellation is rather straightforward and offers several advantages [40]. This method relies on the construction of a 3D tessellation from the position of a selected element in the reconstruction. The principle of construction of a Delaunay tessellation is illustrated in Fig. 9.13.

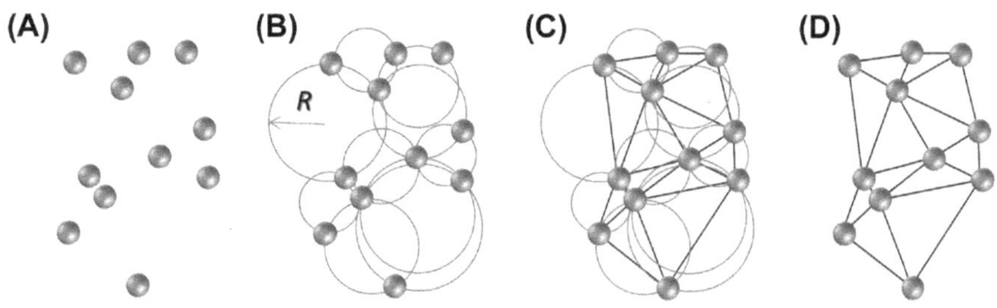

Figure 9.13 Schematic representation of the calculation of a Delaunay tessellation for a distribution of solute atoms in 2D. (A) Solute atoms. (B) Circumcircles are calculated for triangles joined by solute atoms, so that each circumcircle does not contain any atom other than the three that define it. (C) Each triangle is plotted with its circumcircle. (D) The Delaunay tessellation (or triangulation in 2D) is complete. *Figure adapted from Lefebvre W, Philippe T, Vurpillot F. Ultramicroscopy 2011;111:200–06.*

Once the tessellation (or "3D triangulation") is complete, the selection of precipitates is performed by fixing a threshold value R_T for the radius of the circumspheres, which corresponds to a maximum 3 NN distance. Indeed, for a 3 NN distribution, the development of Eq. (9.8) leads to:

$$P_3(r,\rho)\cdot dr = \frac{32\pi^3}{9}\rho^3 r^8 e^{-\frac{4\pi}{3}\rho r^3} \tag{9.10}$$

This expression corresponds exactly to the distribution of Delaunay cells circumspheres calculated by Muche [43]. Based on this statistic, it is possible to evaluate the reliability of the "Delaunay cluster identification," as explained in detail in Ref. [40]. Another advantage relies in the way the precipitates morphology is defined by the method. First, no user-defined parameter is required for associating other species inside the identified precipitates insofar as the volume of Delaunay cells for by clustered solute elements are defining the volume of the precipitates. Second, an expansion-erosion step allows by-passing the inherent limitation of standard d_{max} methodologies, which were illustrated in Fig. 9.14.

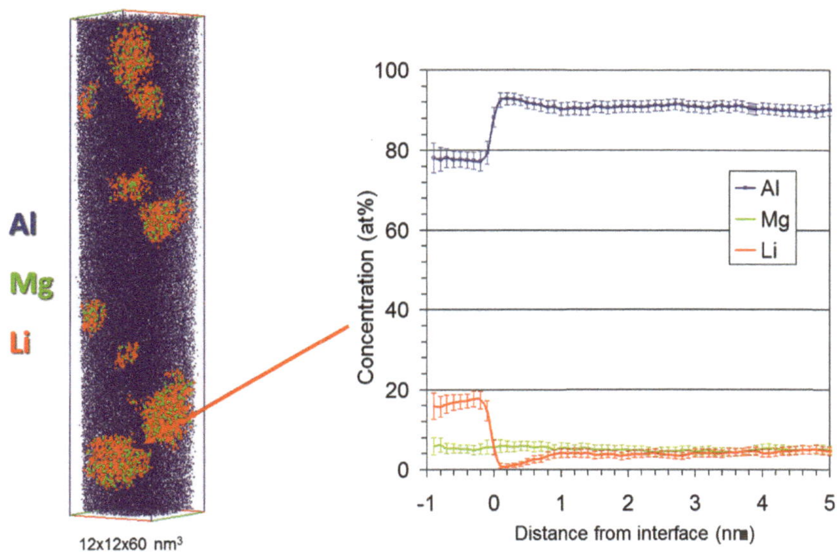

Figure 9.14 Al-Mg-Li alloy (Al−6.79 at.%Li−5.17 at.%Mg−0.074 wt.% Zr) aged for 8 h at 150°C and exhibiting Al_3Li-$L1_2$ precipitates, enriched in Mg. The erosion concentration profile on the right has been calculated once Delaunay cluster identification has been applied to Al_3Li. The Li depletion at the interface indicates a growth of precipitates assisted by the bulk diffusion of Li. *Images adapted from Deschamps A, Sigli C, Mourey T, de Geuser F, Lefebvre W, Davo B. Acta Mater 2012;60:1917–28.*

Metrics on Particle Dimensions and Orientations

We have seen previously that cluster size distribution is mainly displayed as a function of the number of atoms each cluster contains. To characterize a particle and a distribution of particles, other quantities are required such as, for instance: the average distance between particles, the mean radius (if spherical), or a morphological description. Such information may be simply deduced by the knowledge of atoms assigned to each identified particle.

First, the coordinates (x_0,y_0,z_0) of the center of mass of each particle can be calculated. Based on the coordinates of the center of mass, the distribution of interprecipitate distance can be computed. However, this requires defining the neighbors of each precipitate and, for this purpose, the use of Delaunay tessellation, as proposed by Karnesky [45] is extremely relevant. Then, an estimate of the particle dimension can be given by the calculation of its radius of gyration R_g as follows:

$$R_g = \sqrt{\frac{\sum_{i=1}^{n}\left\{(x_i - x_0)^2 + (y_i - y_0)^2 + (z_i - z_0)^2\right\}}{n}} \tag{9.11}$$

where n is the number of atoms in the particle and (x_i,y_i,z_i) are the coordinates of the ith atom in the particle. The radius of gyration is not a quantity which can be simply compared with radii measured by a projected image in transmission electron microscopy, for instance. It can, in specific cases, be compared with the Guinier radius determined by small-angle X-ray scattering [31]. Nevertheless, determining the radius of gyration allows estimating the evolution of a particle's size distribution in a given system.

Based on atom coordinates in each particle, much refined quantification can be assessed by the best fit ellipsoid method [46]. An illustration of this method is provided in Fig. 9.15 for an aluminum-based alloy displaying two particles size distribution, resulting from a complex heat treatment [47]. This method derives the three principle axes of an ellipsoid (fitting the spatial distribution of atoms in a particle) thanks to the diagonalization of the tensor of inertia of each particle. It allows performing a quantitative analysis of the particle orientation with respect to the crystal axis, for instance, as illustrated in the work by Karnesky et al. [46].

Determination of the Matrix Composition

The question related to the measurement of composition inside the main phase is often rapidly answered. Indeed, such a measurement can be performed by selecting a large area of interest away from any visible defect.

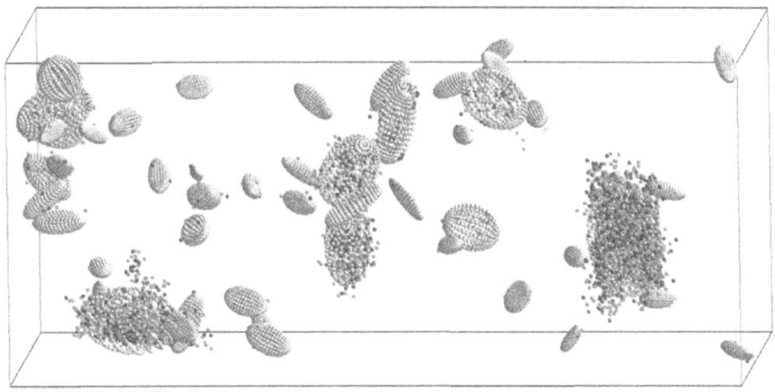

Figure 9.15 Application of the best fit ellipsoid method to an Al-Zn-Mg-Cu alloy submitted to retrogression and reaging treatment, as described in Ref. [47]. The volume dimension is $18 \times 18 \times 37$ nm^3. Only atoms belonging to particles (Mg, Zr, and Cu) are displayed. *Courtesy of Kopp V.*

Or if two phases are present (precipitates and major phase, or two large phases), the use of the proxigram [21] is very straightforward. Nevertheless, when the reconstruction displays a large density of intricate nanoobjects of different natures, it might become more complicated. For this particular case, a method based on the NN statistics has been developed [48]. This method, which is briefly described hereafter, only requires plotting the 1 NN distance distribution of a selected volume. This plot distance distribution is integrated so as to obtain the partition function. Then a variable is introduced:

$$n = 4/3\,\pi\, d^3 \rho_0 \tag{9.12}$$

where d is the distance and ρ_0 is the solute density in the whole reconstruction. Then the probability that a solute atom B is isolated from any other B atom at a distance d is deduced from the Poisson statistic and can be plotted as a function of n:

$$P(0, n) = n^0/0!\; e^{-n} = e^{-n} = e^{-4/3\;\;\pi^{d^3}\rho_0} \tag{9.13}$$

$P(0,n)$ is directly obtained from the experimental 1 NN distance distribution. Finally, the matrix concentration C_m is deduced from the slope of the following equation:

$$\mathrm{Ln}(P_{iso}(n)) \approx \mathrm{Ln}(1 - x) - C_m/C_0\; n \tag{9.14}$$

Where x is the atomic fraction of atoms in the second phase and C_0 is the average solute concentration in the whole reconstruction (an example is presented in Fig. 9.16).

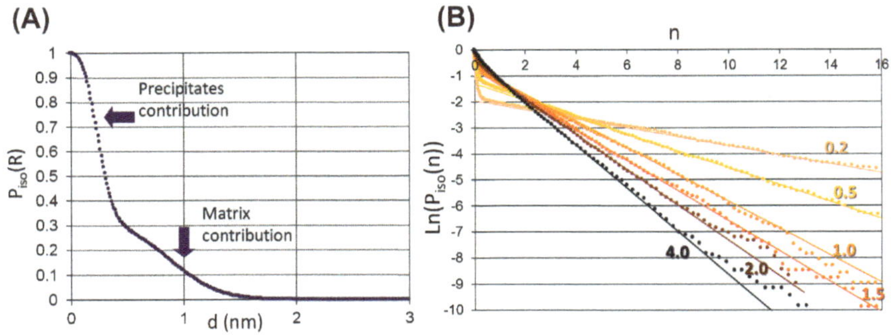

Figure 9.16 (A) Fraction of isolated solute atoms (ie, no solute atoms at a distance *d*) as a function of the distance in a simulated volume with $C_p = 20$ at.% and $C_m = 0.5$ at.%. (B) Plot showing how the C_m values can be obtained by a linear fit. The various curves correspond to plots obtained from various simulated APT reconstructions.

IDENTIFYING THE BIAS

If you don't promptly identify the bias, you might feel much less comfortable later. (old proverb of uncertain origin)

Compositional Biases

The measurement of composition by APT (laser or electric mode) is accurate only if all atomic species present in the analyzed specimen are field evaporated and detected with the same probability: unfortunately, this hypothesis does not generally hold. Due to different physical mechanisms, compound materials or alloys may exhibit complex field evaporation behaviors [49–53], which possibly translate into erroneous measurements of the elemental composition.

Identifying a Compositional Bias From the Mass Spectrum

The identification of a compositional bias might be straightforward, provided the composition of the system under study is a priori known. This situation is, for example, met in intrinsic GaN, a strictly stoichiometric material, where the fraction of detected Ga atoms should be 50% of the total number of atoms. Fig. 9.17A reports a mass spectrum acquired from intrinsic GaN in a laser assisted wide angle atom probe (LAWATAP) with the experimental parameters indicated in the figure. Besides the Ga^+ and the Ga^{2+} peaks, it is possible to recognize a peak related to $N_2{}^+$ and another peak at 14 atom mass unit (amu), which could be attributed either to N^+ or to $N_2{}^{2+}$. Several other molecular peaks are also present in the spectrum.

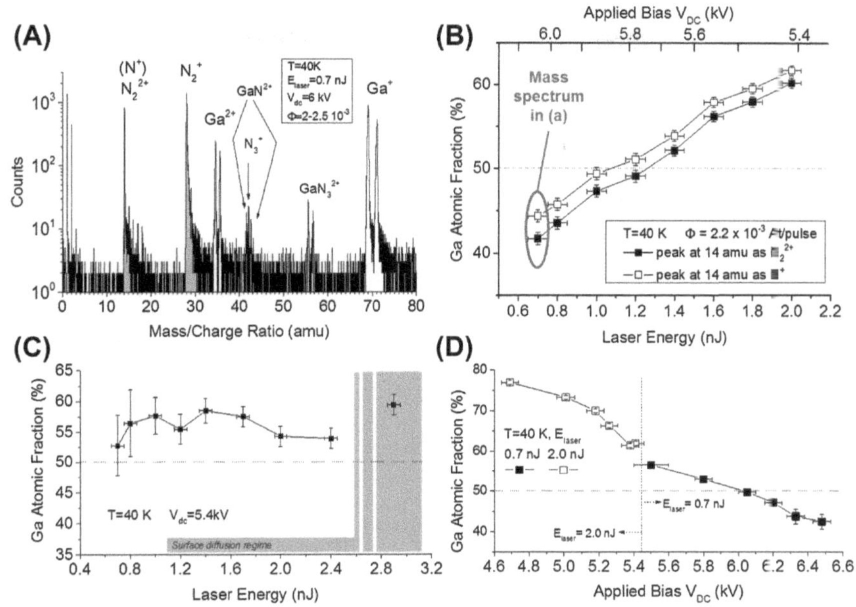

Figure 9.17 Compositional biases in intrinsic GaN: (A) mass spectrum issued from the collection of 2×10^5 ions in a LAWATAP at temperature $T = 40K$, UV laser energy $E_{las} = 0.7$ nJ, applied bias $V_{DC} = 6$ kV, and detection rate $\Phi = 2.2 \times 10^{-3}$ ions/pulse; (B) the measured atomic fraction of Ga plotted as a function of the laser energy (lower abscissa) and of the applied bias (upper abscissa) from a set of measurements performed at constant detection rate; filled (empty) squares correspond to the assumption of the mass/charge peak at 14 amu as N_2^{2+} (N^+), while the circle highlights the point corresponding to the mass spectrum in (A); (C) the atomic fraction of Ga from a set of measurements at constant applied bias; (D) the atomic fraction of Ga from two sets of measurements at two different constant laser energies.

The count of all atoms yields here a measured fraction of Ga equal to 42% if the peak at 14 amu is attributed to N_2^{2+} (and 44% if the peak at 14 amu is attributed to N^+), these results corresponding to the points highlighted with an ellipse in the plot of Fig. 9.16B. This clearly indicates that a compositional bias is present in the measurement, independently of the attribution of the peak at 14 amu to either N_2^{2+} or N^+.

If the composition of the material is not a priori known, it might be impossible to assess a compositional bias from the mass spectrum recorded in a single acquisition only. It is thus necessary to explore the space of experimental parameters to check if the measured composition is parameter dependent. The variation of laser energy per pulse, wavelength, frequency, applied bias, and base temperature may indeed influence the evaporation behavior and introduce artificial trends in the measured composition.

Dependence of the Compositional Bias on the Experimental Parameters

Varying the experimental parameters is not only a way (1) to assess a compositional bias, but also (2) to ascertain what are the parameters providing the most accurate measurement and (3) to study what are the physical factors driving the bias. In the following, the example of GaN will be developed.

In APT, it is customary to work at a constant detection rate throughout the analysis. This is why in most works the compositional biases are expressed as a function of the laser energy working at a given flux [2,54,55]. However, this approach has a major limitation: the simultaneous variation of laser pulse energy and DC applied bias makes it impossible to decide whether the measurement is influenced by one or both of these parameters. This is illustrated in Fig. 9.17B where the laser power and the DC applied bias vary to obtain a flux $\Phi = 0.0022$ ions/pulse. These results show that the composition of GaN measured at high bias and low laser energy is N-rich, and Ga-rich at low bias and high laser energy. The attribution of the peak at 14 amu does not influence the main trend of the composition, so in the following we will attribute this peak to N_2^{2+}.

To identify the parameter, and possibly the physical factor, driving the bias, the laser energy and applied bias should be varied independently, allowing the detection rate to vary.

Dependence on laser pulse energy. The plot reported in Fig. 9.17C refers to the composition measured under constant bias conditions. In this regime the laser does not significantly influence the compositional bias, all experimental points yielding consistent values within the error bars. When the laser begins to induce a departure from the composition of 55% Ga and 45% *N* in the series acquired at low DC bias, this is also accompanied by the appearance of an asymmetric pattern in the tip emission [54].

Dependence on applied bias. Fig. 9.17D reports the dependence of the measured Ga fraction in GaN in two series of measurements at constant laser energy. The first (second) series corresponds to a low (high) laser energy regime, with the laser set at 0.7 nJ (2.0 nJ). One can notice (1) that the behavior is similar to what is obtained during the measurements at constant detection rate plotted as a function of applied bias: Ga is overestimated at low DC bias, whereas *N* is slightly overestimated at high DC bias; (2) the laser has very little influence on the behavior: the two curves at different laser energies follow the same trend, and there is no clear discontinuity at the point where the laser energy passes from one value to the other. The reader should consider that in this point, which corresponds in the plot to the passage between 5.4 and 5.5 kV, the flux varies by a factor of ~15.

These data allow concluding that the main parameter driving the compositional bias in GaN is the applied bias V_{DC}, which points out to a primary role of the surface electric field in this issue.

Representation Through Charge State Statistics

A good representation of a compositional bias is the key to a correct interpretation and to the possibility of comparison with results obtained in other materials or by other experimentalists. Here, we will consider compositional biases that do not depend on laser energy (such a behavior is found in GaN, AlN, ZnO, and MgO [55]). A possible choice would be representing the data as a function of the pulse fraction, or field reduction, $PF = 1 - V_{DC}/V_0$, where V_0 is the evaporation potential found by extrapolating to zero laser power the nearly linear V_{DC}/E_{las} relationship that can be found in a series of measurements performed at constant detection rate. Furthermore, the pulse fraction is defined for a given flux, while the ensemble of the data might be collected at different detection rates. The pulse fraction is thus not a good parameter for a comparative representation of the whole data set [2].

Alternatively, the applied bias V_{DC} could be used. This is reasonable for bias-dependent data, but conical tips may exhibit either strong variations of the applied bias at constant flux, or of the surface field at constant bias due to tip evolution during the analysis. This also would not allow comparing the behaviors of tips with different sizes. The most convenient choice for the representation of composition measurements is as a function of a suitable E^{2+}/E^{1+} charge state ratio (CSR), where E is an element of the compound for which the CSR can be accurately defined throughout the whole parameter set [42,54,56,57].

Advantages of the Representation Through Charge State Statistics

The advantages of the representation of compositional biases through the charge state statistics are listed in the following:

1. The CSR is related to the surface field. This relationship might be made quantitative through the postionization statistics introduced by Kingham [56] (see Chapter 2: Field Ion Emission Mechanisms), as illustrated in Fig. 9.18A. In principle, the phenomenon of postionization takes place at a distance of several angstroms from the sample surface, and should be nearly independent of the material in which the atom was found if the

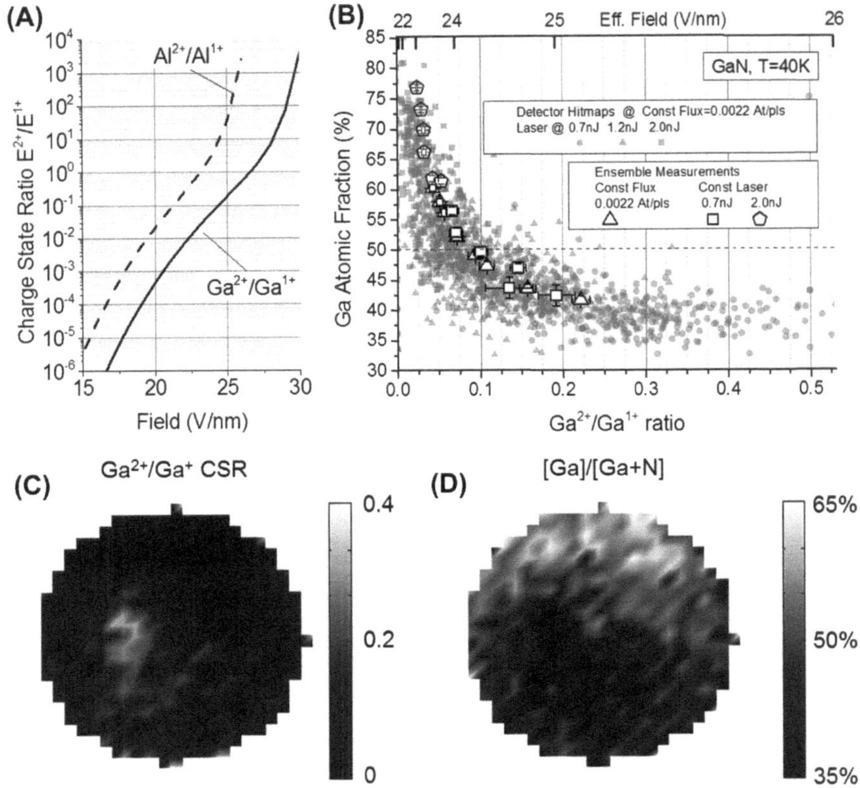

Figure 9.18 (A) Relationship between the surface field and the charge state ratios Ga^{2+}/Ga^{+} (*solid line*) and Al^{2+}/Al^{+} (*dashed line*); (B) Ga atomic fraction in pure GaN measured by La-APT at 40K plotted versus the $Ga^{2+}/Ga+$ charge state ratio. The plot contains both ensemble data issued by series of measurements performed at constant flux and constant laser power (large empty symbols) and microscopic data issued from the detector hitmaps. (C) and (D) show the spatial distribution (detector hitmaps) of the Ga^{2+}/Ga^{1+} charge state ratio and of the Ga atomic fraction on the detector issued from a data set containing 2×10^5 atoms collected at $T = 40K$ and with a UV laser energy of 1.2 nJ.

evaporation takes place in the singly charged state. The charge state abundances of the evaporated ions are thus a function of the surface field. Experimentally, the CSR is an increasing function of the applied DC bias, provided the tip shape remains nearly unchanged during the analysis. By inverting the CSR versus field relationship, it becomes even possible to calculate an effective surface field. However, this effective field should only be considered as an estimate of the actual surface field. First, it is known that the Kingham charge state statistics is based on approximations that make its application to 3D tips questionable [58,59]. Furthermore, data from nonmetallic compounds may show features which are not compatible with the Kingham statistics, such as, for instance, the simultaneous presence of the charge states Al^{1+}, Al^{2+}, and Al^{3+} in AlN [55].

2. For analyses performed on the same tip sample, the CSR allows for a robust evaluation of the composition trends even in the case where the tip shape significantly changes during the analysis.
3. For analyses performed on different tips of the same material, the CSR allows for a direct comparison, which would be extremely difficult to achieve by considering DC voltage, laser pulse energy or laser energy-related pulse fraction. The latter experimental parameters are indeed related to the electric surface field and to the absorbed energy per pulse through the geometrical shape of the tip, which is difficult to monitor during an APT experiment.
4. The CSR can be defined either as a global or as a local quantity, as it can be calculated either as an average or point by point by defining sampling pixels on the 2D detector. It thus becomes possible to establish microscopic relationships between the local surface field and the locally measured composition.
5. Finally, the CSR allows comparing analyses performed on different instruments and by different users.

Application of the Charge State Statistics to the Compositional Bias in GaN

The Ga fraction measured in the series at constant detection rate and constant laser pulse energy displayed in Fig. 9.18B and D is plotted versus the Ga^{2+}/Ga^{+} CSR in Fig. 9.18B. Remarkably, all points identify a well-defined trend, independently of the laser pulse energy used. The distribution of the Ga^{2+}/Ga^{+} CSR and of the Ga atomic fraction on the detector are displayed in Fig. 9.18C and D, respectively. These images show that the correlation between the CSR and the Ga fraction also holds at the microscopic scale, and that it follows the same trend as the ensemble data, as reported in Fig. 9.18B. This means that the nonuniform distribution of the surface electric field on the tip apex translates into a nonuniform measurement of the Ga fraction within the sampled volume.

The observed trends and the representation through charge state statistics allow drawing some partial conclusions about the physical mechanisms responsible for the compositional biases in atom probe measurements of GaN. These results are indeed consistent with the following proposed mechanism: at low field, *N* may form neutral N_2 molecules on the tip surface, which would easily evaporate and not be detected, leading to a Ga surplus. On the other hand, at high field, these molecules would undergo more and more postionization, making them detectable. In the same time, Ga would undergo preferential (ie, uncorrelated with the laser pulse) ionization, making it less and less identifiable in the mass spectrum [54–56]. A similar mechanism should be responsible for the compositional biases observed in oxide materials such as ZnO and MgO.

What Are the Most Convenient Parameters?

The answer to this question is left to the reader, as the most convenient set of parameters may depend on the material and on the features on which the analysis focuses. The choice of the parameters is particularly critical in materials or systems of unknown composition, especially in those containing more than two component atoms. It has been shown by Riley et al. [60] and by Mancini et al. [55] that the InN alloy fraction in InGaN can be accurately measured, as In and Ga have quite similar evaporation behavior (even if this is very different from that of N), but it may be much more difficult to determine the composition of other ternary materials (or doped binary materials). The experimentalist should also be aware that for some materials it could not be possible to determine the elemental composition by atom probe only.

Analysis of Multiple Events

Ideally, ions are evaporated one by one (ie, one ion per pulse), but it is often observed that multiple ions are detected together after the same laser or voltage pulse. Detection of such multiple events can be problematic, and partial detection of multiple events can sometimes cause significant biases in composition measurements depending on the type of detector and detection methods that are used (see Chapter 6: Atom Probe Tomography: Detector Issues and Technology). However, they can also contain data that can be used to gain more understanding about the physics of the evaporation and also information about the sample itself. Indeed, such multiple events are generally not the sum of random independent evaporation events. Their occurrence is generally higher than what Poisson's law would predict [61], indicating that actual physical phenomena must be involved in the production of a large fraction of such events. The study of the data contained in these events evidences two sources of multiple events: correlated evaporation [61], which means an evaporation event triggering the evaporation of a neighbor surface atom or molecule in a short time interval, and dissociation of molecular ions after evaporation [53,62–64] producing multiple molecule fragments. The study of such events can thus give information about the physics of surface evaporation during an atom probe analysis. One first interesting aspect of that correlation is that it should affect only ions actually originating from the tip surface, whereas detector noise events or noise events caused by residual gas atoms in the chamber should not be correlated. It follows that filtering a data set to contain only multiple events can significantly increase the signal to noise ratio to the point that peaks can be identified that were not visible in a conventional mass spectrum when single events are included [65]. Although this does not directly give quantitative information about the composition of the sample, it can improve the qualitative interpretation of the data.

Additional information about the processes taking place during evaporation can also be extracted from chemical correlations in multiple events themselves. We will present here two data treatment methods that have been developed to visualize such correlations and extract information from them. The first one, the correlation table [53], is a direct way to identify if the detection of two given ions in the same event is more likely than what would be expected in the case of independent evaporation. First, each possible combination of two ions is computed from each multiple event. For example, an event where ions A, B, and C are detected would give three pairs, AB, BC, and AC. Then the significance d_{ij} of the correlation of the evaporation of ion i together with ion j is given by Eq. (9.15) [53]:

$$d_{ij} = \frac{p_{ij} - e_{ij}}{\sqrt{e_{ij}}} \tag{9.15}$$

where p_{ij} is the number of pairs ij counted in the complete data set and e_{ij} is the number of pairs ij that would be expected in the absence of chemical correlation, given the number of occurrences of each type of ion in all the measured pairs. It should be noted that this value does not give a direct quantification of correlation itself, as it depends not only on the degree of correlation between i and j, but also on the size of the population of i and j. Pairs of ions that are weakly correlated could still give a large value of d_{ij} if the size of the populations of i and j are large, because even a weak correlation could be shown with a large significance. As agreed by the author of the original paper [66], if quantification of the correlation is important, for example, to compare the correlations of two different pairs, or correlations

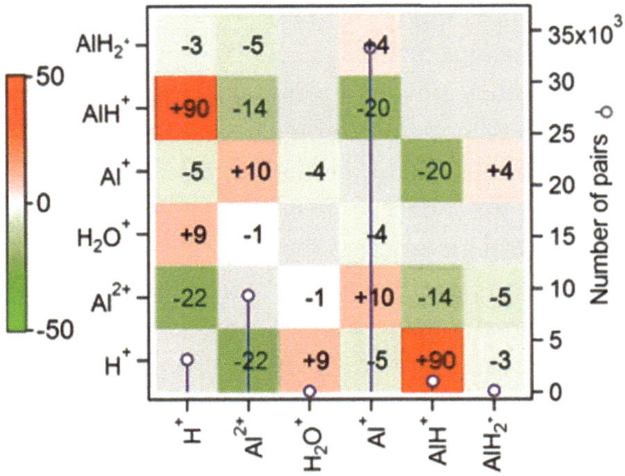

Figure 9.19 Correlation table corresponding to an analysis of pure Al. The values of d are reported for each pair of ion. Positive values indicate correlation [red] and negative values indicate anticorrelation [green]. Strongly significant correlations are indicated by stronger colors and values that are insignificant are not shown. The number of pairs involving each ion species is also shown. The presence of H and H_2O likely corresponds to water adsorbed on the tip surface. *Reprinted from Saxey DW, Ultramicroscopy 2011;111:473–79 with permission of Elsevier.*

from one data set to another, the ratio $(p_{ij} - e_{ij})/e_{ij}$ should rather be used, provided the value of d_{ij} shows that correlation is significant. The different values of d_{ij} are presented in a chart where each row and column corresponds to one ion (Fig. 9.19).

Large positive values indicate a significant correlation, while negative values indicate a significant anticorrelation, which means that two ions are detected together less often than expected from an independent evaporation. Because $d_{ij} = d_{ji}$, the table naturally has a symmetry axis on its diagonal. Ions evaporating in a correlated way often evaporate together in a very small time interval and from the same region on the tip [61]. Hence, their impacts on the detector are very close to each other, and if the ions are identical in nature, they will also reach the detector at the same time, which can make their detection problematic depending on the detectors technology (see Chapter 6: Atom Probe Tomography: Detector Issues and Technology). This can lead to an underestimation of the d_{ij} located on the diagonal of the histogram and conversely an overestimation of the d_{ij} that are off the diagonal. To avoid such bias, the p_{ij} values on the diagonal are set to be equal to e_{ij}. The values of e_{ij} are then recalculated and this is repeated for a few iterations until the values converge. This artificially sets d_{ij} to zero on the diagonal to avoid experimental bias of the rest of the terms. The different correlations that are observed in the histogram could indicate chemical preferences in correlated evaporation or dissociation of molecular ions. The correlation histogram itself does not directly give information about the mechanism that is causing the correlation, but different causes can sometimes be inferred from the rest of the data. In Fig. 9.19, which corresponds to an analysis of Al, the molecular ion AlH_2^{2+} was detected in the mass spectrum, and a positive correlation is observed for AlH^+ and H^+. It can thus be inferred that AlH^+ and H^+ are produced by the dissociation of AlH_2^{2+} [53]. Observation of the behavior of different molecular ions when analysis conditions are changed can also help the interpretation of the observed correlations [63].

Another method that gives information about correlations in multiple events is the correlation histogram or "Saxeygram" [53,64,67,68]. The resulting plot is similar to the correlation table, except in this case the ions are not identified using their range, instead their masses are directly plotted on a 2D histogram. Each count in the correlation histogram corresponds to one pair of ions, with the x and y axes corresponding to the measured mass to charge state ratios of each ion of the pair. In the same way, because a pair AB is equivalent to BA, each pair is taken into account twice in the histogram, which thus is symmetric with respect to its diagonal. Fig. 9.20A is an example of such a correlation histogram obtained from an analysis of pure SiO_2, and Fig. 9.20B is the corresponding mass spectrum including all events. Different peaks can be observed in the correlation histogram, corresponding to different pairs of ions and are themselves connected to tails going in different directions. The largest tails have a positive slope and tend to follow the histogram's symmetry axis. When plotted as a function of corrected time of flight (time of flight corrected for the change of voltage during the analysis and for the differences in flight length) instead of mass to charge state ratio (insert in Fig. 9.20A), these tails all become straight and with a slope equal to unity. This indicates that both ions of the pair are evaporating after the laser pulse, during the cooling of the tip, but with the same delay. This, again, evidences the presence of correlated evaporation, with the evaporation of the first ion during the cooling of the tip triggering the immediate evaporation of the second ion. In a regular mass spectrum that does not contain information about correlation (Fig. 9.20B), these tails are all overlapping and chemical identification of the events present in the tail is not possible. Here, to the contrary, taking into account correlated evaporation adds a second dimension to the mass spectrum, and the different tails

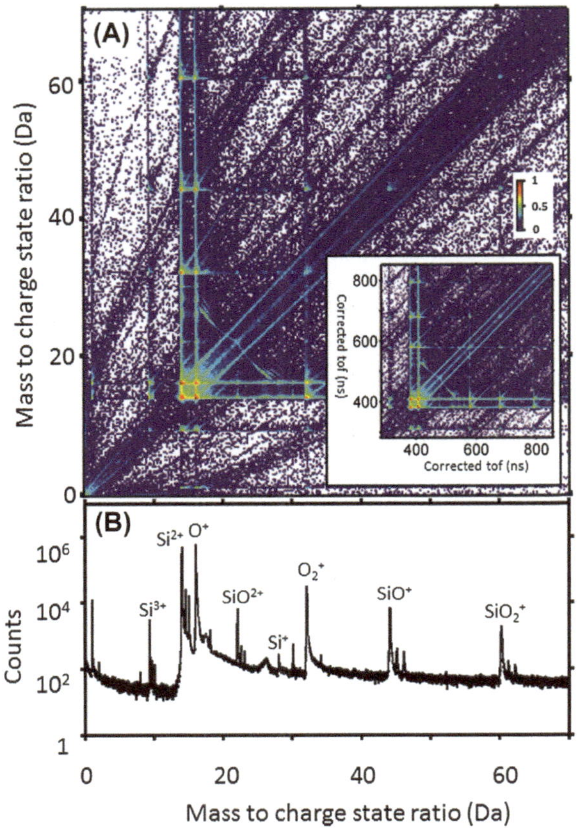

Figure 9.20 Analysis of a sample of SiO_2 using UV laser with (A) the corresponding correlation histogram and (B) the corresponding mass spectrum. The color scale in the correlation histogram represents the logarithm of the number of counts after normalization. *Tof,* time of flight.

can be separated. Thus ions belonging to each tail can be identified provided the tails do not overlap too in the correlation histogram. Interestingly, this is also true for evaporation events that are not correlated to the pulse [53]. In Fig. 9.20A, the different tails can already be seen before the corresponding peaks, evidencing evaporation due to the DC field only. This is possible because, even though these ions evaporate at a random time, they still evaporate together and thus the difference between their times of flight still depends on their chemical natures only.

Other tails are horizontal or vertical, and correspond to the evaporation of one ion at the moment of the pulse and the second ion with a certain delay, with no time correlation with the first ion. Some tracks with a

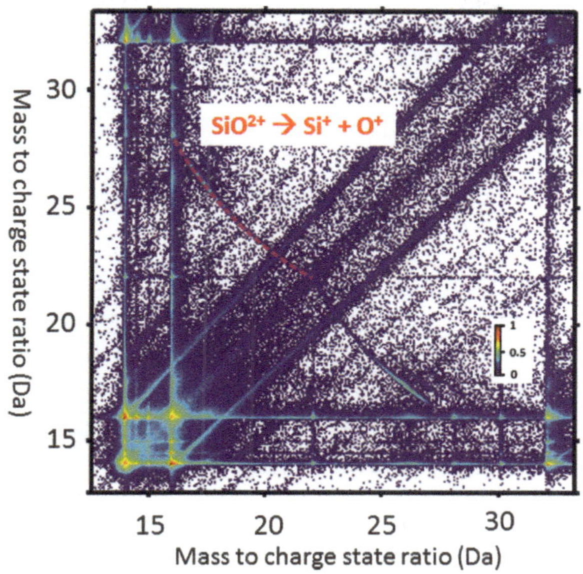

Figure 9.21 Close-up view of the correlation histogram presented in Fig. 9.20. A dissociation track corresponding to the reaction $SiO^{2+} \rightarrow Si^{+} + O^{+}$ is visible. It is fitted in the upper part of the histogram by the purple dotted line using Eq. 9.17. The color scale represents the logarithm of the number of counts after normalization.

negative slope can also be observed, starting at coordinates (16,28) amu and ending on the diagonal of the histogram. A close-up view of this region is given in Fig. 9.21. Such tracks, referred to as dissociation tracks, can be explained by the dissociation of molecular ions [53,64]. Indeed, the dissociation of a molecular ion during its flight in the accelerating field is known to have an effect on the times of flight of the dissociation products [62,69]. This stems from the fact that the total time of flight of an ion depends directly on its mass to charge state ratio during acceleration in the electric field. If dissociation occurs close to the tip surface, the potential drop is traveled entirely by the daughter ions, and the measured mass to charge state ratios will in turn be the ones of each daughter ion. Conversely, if dissociation occurs at the end of the potential drop, far from the tip surface, the measured mass to charge state ratios of both dissociation products will be close to the one of the parent ion. And when a molecular ion dissociates half way during acceleration, the measured mass to charge state ratio will then be an intermediate value between the one of the parent and the one of the daughter ion. They will actually depend on what fractions of the potential drop were traveled by the parent ion and by the daughter ions. Considering conservation of the velocity of the parent ion, the measured mass to charge state ratio of a daughter ion is given by [53]:

$$m_1' = m_1 \left[1 - \frac{V_\mathrm{d}}{V_\mathrm{DC}} \left(1 - \frac{m_1}{m_\mathrm{p}} \right) \right]^{-1} \tag{9.16}$$

where m_1 is the actual mass to charge state ratios of that dissociation product, V_d is the electrostatic potential between the tip and the detector at the location where the dissociation occurred, V_DC is the tip potential, and m_p is the mass to charge state ratio of the parent ion. From this equation, the correlation between the measured mass to charge state ratios of the two dissociation products can be expressed as

$$m_2' = m_2 \left[1 - \frac{m_\mathrm{p} - m_2}{m_\mathrm{p} - m_1} \left(1 - \frac{m_1}{m_1'} \right) \right]^{-1} \tag{9.17}$$

where m_2' and m_2 are the calculated and actual mass to charge state ratios of the second ion. In the correlation histogram, the dissociation track starts at mass to charge state ratios corresponding to Si^+ and O^+. Eq. (9.16) is plotted on the graph for the dissociation path $SiO^{2+} \rightarrow Si^+ + O^+$ and is found to lie on the experimental data, confirming the existence of this dissociation path. In the same way, many dissociation paths can be identified in other materials like GaN [53] or $LiFePO_4$ [64]. Interestingly neutral atoms can also be detected when produced by dissociation of molecular ions [53,64]. Because neutral atoms cannot be accelerated in the electric field near the tip, they are not accelerated toward the detector and do not have enough energy to produce signals on the MCP. However, in the case of a neutral atom formed by the dissociation of a molecular ion, it inherits the velocity of the parent ion and can be detected, and a dissociation track can be observed. For the formation of one neutral atom, Eq. (9.17) becomes simply

$$m_2' = \frac{m_2 m_1'}{m'^1 - mp + m_2} \tag{9.18}$$

with m_1' being the measured mass to charge state ratio of the neutral atom. In general, because dissociation has an important effect on the measured mass to charge state ratios of the dissociation products, the latter often cannot be identified unless dissociation is taken into account [64]. To this day, however, the different analyses of dissociation tracks show that such dissociation events are rare enough to have a negligible effect on composition measurements.

CONCLUSION

In this chapter, various ways to examine APT reconstructions in detail have been presented. The way advanced data mining can be used to track artifacts has also been introduced. Although not explicitly mentioned, the abundance of the methods is a clear indication that APT users are very active for conceiving and testing new algorithmic tools. An inherent drawback coming with this dynamism is that results may differ because of different algorithmic tools. Another limitation may arise from the restriction of the APT user to some methods, which are applied by use and not always because of their relevance to the topic under investigation, whereas more relevant and efficient methodologies have been developed and successfully tested. Insofar as not all the methods presented in this chapter are implemented in commercial softwares, many authors cited here have shared or are ready to share their own published methods.

Things may become more complex when reproducibility suffers from a lack of methodology associated to the choice of parameters to use for data mining. The extreme help provided by the comparison to equivalent randomized data has been highlighted in this chapter. For instance, it is strongly advised to compare the results obtained with a given set of parameters for cluster selection, for instance, with an experimental reconstruction and its randomized equivalent data set. As a conclusion, this chapter and a lot of referenced worked inside provide many keys not to get lost in the obscure world of data mining.

REFERENCES

[1] Stephenson LT, Moody MP, Gault B, Ringer SP. Microsc Res Tech 2011;74:799–803.

[2] Gault B, Moody MP, Cairney JM, Ringer SP. Atom probe microscopy. 2012.

[3] Marquis EA, Hyde JM. Mater Sci Eng R Rep 2010;69:37–62.

[4] Kelly TF, Miller MK, Rajan K, Ringer SP. Microsc Microanal 2013;19:652–64.

[5] Miller MK, Forbes RG. Atom-probe tomography – the local electrode atom probe. 2014.

[6] Larson D, Prosa T, Ulfig R, Geiser B, Kelly T. Local electrode atom probe tomography. 2013.

[7] Moody MP, Stephenson LT, Ceguerra AV, Ringer SP. Microsc Res Tech 2008;71:542–50.

[8] Stephenson LT, Moody MP, Ringer SP. Philos Mag 2011;91:2200–15.

[9] Ceguerra AV, Powles RC, Moody MP, Ringer SP. Phys Rev B – Condens Matter Mater Phys 2010;82:1–4.

[10] Seidman DN, Marquis EA, Dunand DC. Acta Mater 2002;50:4021–35.

[11] Marquis EA. Microsc Microanal 2007;13:484–92.

[12] Stephenson LT, Moody MP, Liddicoat PV, Ringer SP. Microsc Microanal 2007;13: 448–63.
[13] Philippe T, De Geuser F, Duguay S, Lefebvre W, Cojocaru-Mirédin O, Da Costa G, et al. Ultramicroscopy 2009;109:1304–9.
[14] De Geuser F, Lefebvre W, Blavette D. Philos Mag Lett 2006;86:227–34.
[15] Marceau RKW, Stephenson LT, Hutchinson CR, Ringer SP. Ultramicroscopy 2011; 111:738–42.
[16] Christian JW. The theory of transformations in metals and alloys. Elsevier; 2002.
[17] Polmear IJ. Light alloys. Elsevier; 2005.
[18] Guinier A. Nature 1938;142:569–70.
[19] Preston GD. Nature 1938;142:570.
[20] Hellman OC, Du Rivage JB, Seidman DN. Ultramicroscopy 2003;95:199–205.
[21] Hellman OC, Vandenbroucke JA, Rüsing J, Isheim D, Seidman DN. Microsc Microanal 2000;6:437–44.
[22] Stephenson LT, Moody MP, V Liddicoat P, Ringer SP. Microsc Microanal 2007;13: 448–63.
[23] Gault B, Danoix F, Hoummada K, Mangelinck D, Leitner H. Ultramicroscopy 2012; 113:182–91.
[24] Miller MK, Cerezo A, Hetherington MG, Smith GDW. Atom probe field ion microscopy. Oxford University Press; 1996.
[25] Sudbrack CK, Noebe RD, Seidman DN. Phys Rev B 2006;73:212101.
[26] Marquis EA. [Ph.D. thesis]. Northwestern University; 2002.
[27] Haley D, Petersen T, Barton G, Ringer SP. Philos Mag 2009;89:925–43.
[28] Muddle BC, Gable BM, De Geuser F, Gable BM, Muddle BC. 2008.
[29] Hyde JM. [Ph.D. dissertation]. Oxford: University of Oxford; 1993.
[30] Hyde JM, English CA. MRS Proc 2000;650:R6.6.
[31] Kolli RP, Seidman DN. Microsc Microanal 2007;13:272–84.
[32] cerezo LDA. Surf Interf Anal 2007;39:184–8.
[33] Stephenson LT. [Ph.D. thesis]. 2009.
[34] Miller MK, Kenik EA. Microsc Microanal 2004;10:336–41.
[35] Vaumousse D, Cerezo A, Warren PJ. Ultramicroscopy 2003;95:215–21.
[36] Clark PJ, Evans FC. Ecology 1979;60:316–7.
[37] Shariq A, Al-Kassab T, Kirchheim R, Schwarz RB. Ultramicroscopy 2007;107: 773–80.
[38] Xu X, Ester M, Kriegel H, Sander J. Proc 2nd Int Conf Knowl Discov Data Min 1996: 226–31.
[39] Musin OR. Proc Thirteen Annu Symp Comput Geom SCG 2003;97(2):424–6.
[40] Lefebvre W, Philippe T, Vurpillot F. Ultramicroscopy 2011;111:200–6.
[41] Marceau RKW. Design in light alloys by understanding solute clustering processes during the early stages of age hardening in Al-Cu-Mg alloys [Ph.D. thesis]. University of Sydney; 2008.
[42] Lefebvre W, Danoix F, Da Costa G, De Geuser F, Hallem H, Deschamps A, et al. Surf Interface Anal 2007;39:206–12.
[43] Muche L. J Stat Phys 1996;84:147–67.
[44] Deschamps A, Sigli C, Mourey T, de Geuser F, Lefebvre W, Davo B. Acta Mater 2012;60:1917–28.
[45] Karnesky RA, Isheim D, Seidman DN. Appl Phys Lett 2007;91:213–5.
[46] Karnesky RA, Sudbrack CK, Seidman DN. Scr Mater 2007;57:353–6.
[47] Marlaud BBT, Deschamps A, Bley F, Lefebvre W. Acta Mater 2010;58:4814–26.
[48] De Geuser F, Lefebvre W. Microsc Res Tech 2011;74:257–63.
[49] Lefebvre W, Loiseau A, Menand A. Ultramicroscopy 2002;92:77–87.
[50] Liddle JA, Cerezo A, Grovenor CRM. Le J Phys Colloq 1989;50. C8-437–C8-442.

[51] Mazumder B, Vella A, Deconihout B, Al-Kassab T. Ultramicroscopy 2011;111: 571–5.
[52] Tsukada M, Tamura H, McKenna KP, Shluger AL, Chen YM, Ohkubo T, et al. Ultramicroscopy 2011;111:567–70.
[53] Saxey DW. Ultramicroscopy 2011;111:473–9.
[54] Diercks DR, Gorman BP, Kirchhofer R, Sanford N, Bertness K, Brubaker M. J Appl Phys 2013;114:184903.
[55] Mancini L, Amirifar N, Shinde D, Blum I, Gilbert M, Vella A, et al. J Phys Chem C 2014;118:24136–51.
[56] Kingham DR. Surf Sci 1982;116:273–301.
[57] Lefebvre W. [Ph.D. thesis]. University of Rouen; 2001.
[58] de A. Neto AV, de Castilho CMC. J Phys B At Mol Opt Phys 1991;24:2609–15.
[59] Zurlev DN, Forbes RG. J Phys D Appl Phys 2003;36:L74–8.
[60] Riley JR, Detchprohm T, Wetzel C, Lauhon LJ. Appl Phys Lett 2014;104 152102.
[61] De Geuser F, Gault B, Bostel A, Vurpillot F. Surf Sci 2007;601:536–43.
[62] Tsong TT, Liou Y. Phys Rev Lett 1985;55:2180–3.
[63] Müller M, Saxey DW, Smith GDW, Gault B. Ultramicroscopy 2011;111:487–92.
[64] Santhanagopalan D, Schreiber DK, Perea DE, Martens RL, Janssen Y, Khalifah P, et al. Ultramicroscopy 2014;148C:57–66.
[65] Yao L, Gault B, Cairney JM, Ringer SP. Philos Mag Lett 2010;90:121–9.
[66] Saxey DW. Personal communication. 2016.
[67] Xie KY, Breen AJ, Yao L, Moody MP, Gault B, Cairney JM, et al. Ultramicroscopy 2012;112:32–8.
[68] Estivill R, Grenier A, Duguay S, Vurpillot F, Terlier T, Barnes J-P, et al. Ultramicroscopy 2015;150:23–9.
[69] Stepien ZM, Tsong TT. Surf Sci 1998;409:57–68.

CHAPTER TEN

Correlative Microscopy by (Scanning) Transmission Electron Microscopy and Atom Probe Tomography

W. Lefebvre-Ulrikson
Groupe de Physique des Matériaux, University and INSA of Rouen, Normandie University, France

Although a very powerful tool, atom probe tomography (APT) cannot access all categories of information and also suffers from intrinsic limitations. Both drawbacks can be partially overcome by combining APT with another instrument. When this combination is applied to a single specimen, such an approach is commonly referred to as a correlative microscopy approach. Two main goals may be pursued when a correlative approach by APT and (scanning) transmission electron microscopy [(S) TEM] is applied: (1) to improve the APT reconstruction by defining the specimen geometry and/or the distribution and morphology of particular features in the APT specimen prior to analysis; (2) to access additional information in TEM or STEM that are not accessible in APT. However, as far as a correlative approach is foreseen, several questions are raised. First, is the specimen we are looking at strictly the same in both instruments? In other words, how far can we neglect the influence of the specimen exposure to an electron beam? Second, what is the effect of the particular geometry of APT specimen on the data extracted in STEM and TEM? The serial analysis of the same specimen by both techniques has required the development of dedicated methods, which are summarized in this chapter. Examples are provided which illustrate the possible achievements of correlative microscopy by STEM and APT. Then, a last question is asked: what if both instruments would be combined in a same machine?

Atom Probe Tomography
ISBN: 9780128046470
http://dx.doi.org/10.1016/B978-0-12-804647-0.00010-3

MOTIVATIONS FOR A CORRELATIVE APPROACH

Cases exist when input reconstruction parameters are not accessible to the user, for instance, when a new material or a complex heterostructure is analyzed. These parameters may often also evolve as a function of the nature of the feature (eg, layer, particle) emerging at the specimen surface during the sequence of evaporation. It is hence relevant to at least get an idea of how the APT reconstruction should look like. Either prior projections or prior tomography data of the specimen hence become necessary. They can be obtained by TEM, STEM, or scanning electron microscopy (SEM).

When reconstruction parameters are determined trustfully, considering the limited and anisotropic spatial resolution of APT, an important driving force for a correlative approach is to fill the gap of crystallographic information. As an example, dislocations are rarely identified by APT. Segregations of minor elements can hence be strictly related to the presence of a dislocation if this defect has been identified by TEM in a weak beam or bright field mode, for instance [1]. More generally, unambiguous identification and complete characterization of crystallographic defects is hardly accessible in APT, although it might be achieved in some cases [2].

Ways to reveal artifacts in APT have been discussed in previous chapters. Nevertheless, detailed inspection of APT data may sometimes fail to identify artifacts and, a fortiori, to quantify them. Insofar as the specimen surface geometry rules the ion trajectories in APT, stopping an APT acquisition to get a "snapshot" of the specimen surface appears relevant. For instance, it can reveal the existence of a local magnification effect [3] or specimen deformation under the effect of the laser beam used for the field evaporation [4]. Once again, the relevant tool to be combined with APT is an electron microscope.

RELEVANT TECHNIQUES FOR A CORRELATIVE APPROACH

The wavelength of the radiation to be used for imaging must be smaller than dimensions of the feature of interest. As far as the specimen surface or defects need to be observed, the following radiations fulfill the short wavelength requirement: X-rays, and neutron and electron beams. Producing X-rays in a laboratory is an easy task. However, focusing X-rays to enable specimen imaging is not a trivial issue. The use of Fresnel lenses associated to a synchrotron radiation enables performing high-resolution holographic reconstructions with a spatial resolution down to 100 nm [5]. This remains at least two orders of magnitude larger than what the correlative approach under consideration here requires. As for neutron radiation, it is obviously inappropriate for microscopy, insofar as neutron beams cannot be focused and require a source that is complex, extremely expensive, and therefore very rare. Electron beams are hence the more appropriate radiation to be used for a correlation with APT. The relativistic wavelength of electrons is given in the following equation Eq. (10.1):

$$\lambda_r = \frac{\mathrm{h}}{\sqrt{2m_0 eV\left(1 + \frac{eV}{2m_0c^2}\right)}} \tag{10.1}$$

where m_0 is the rest mass of the electron, e is the electron charge, V the acceleration voltage, h the Planck constant, and c the speed of light. The electron wavelength ranges from about 2 to 7 pm for acceleration voltages of 300 or 30 kV, respectively. According to Abbe's limit, the theoretical resolution of electron microscopes is expected to be far below the size of atoms. Experimental limitations linked to aberrations of lenses have for a long time limited resolution in TEM and STEM to about 0.2 nm. The implementation of spherical aberration correctors [6] has decreased spatial resolution down to 0.5 nm in TEM and STEM (the reader can refer to the review by P.W. Hawkes on the topic of aberration correction [7]). Meanwhile, progress in electron optics in SEM has led to improved imaging capabilities. In both techniques, the spatial resolution is now mostly limited by the physics of electron—specimen interactions (cf. next sections).

The various capabilities of APT specimen preparation and characterization in SEM have been discussed in detail in "Chapter 4, Atom Probe Sample Preparation." As for the possibilities of characterization in TEM and STEM, they are also numerous. A schematic illustration of capabilities of STEM and related techniques is given in Fig. 10.1. A long list of acronyms (eg, EELS, ADF, EDS, CDBED, NBD, PED) exist that defines dedicated techniques associated to TEM or STEM. For a deep presentation of these techniques, the reader can look at Refs. [8—10]. The variety of information that can be obtained by analyzing an APT specimen in a (S)TEM is large. It includes: defects structure, elemental mapping (projections), atomic scale imaging, single atom detection (in favorable cases), and oxidation states. For more advanced protocols and techniques, eg, with electron holography, magnetization maps, elastic strains, or electrostatic potentials can be retrieved. Although (S)TEM initially provided two-dimensional data based

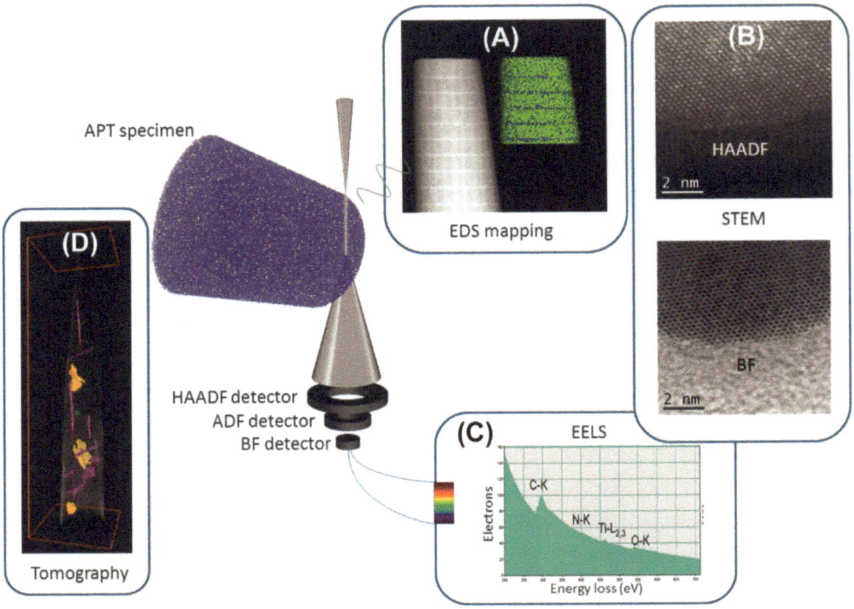

Figure 10.1 Schematic illustration of the multiple modes available in scanning transmission electron microscopy (STEM). An atom probe tip is illuminated by a convergent electron probe below which the transmitted electrons may be collected by several detectors, namely, high-angle annular dark field (HAADF), annular dark field (ADF), and bright field (BF) STEM detectors. (A) X-rays can be collected by an electron dispersive X-ray spectroscopy (EDS) detector to perform elemental mapping. (B) HAADF and BF STEM imaging can be performed simultaneously. (C) Electron energy loss spectroscopy (EELS) spectra and EELS mapping are possible, in the absence of the BF detector. (D) Recording of signals along tilt series may lead to tomographic reconstructions.

on the projections of a three-dimensional (3D) specimen, both (S)TEM and APT now provide data in a 3D space. More interesting for the APT community, electron tomography (ET) can be used to help reconstructing APT. To enable the full characterization of an APT specimen in a (S)TEM, dedicated versatile holders needed to be developed. This is presented in the next section.

EXPERIMENTAL PROTOCOLS

The main specificities of specimen preparation for a correlative approach rely on the following aspects: (1) the analysis of a region of interest; (2) the capability of transferring the specimen from an instrument to the other. Regions of interest are now selected thanks to focused ion beam

(FIB)-SEM. However, the first correlative analyses involving combined use of TEM and APT were reported for the selection of interfaces by electro-polishing [11]. Indeed, the site-specific analysis of grain boundaries in metals and alloys by APT was previously conducted thanks to the prior selection of grain boundaries in TEM (see, for instance, the early work by Krakahauer and Seidman who could investigate grain boundaries by mounting a tip on a double-tilt holder [12]). This was done by gently electropolishing the specimen apex (thanks to very short voltage pulses in an electrolyte) recursively until the interface of interest was close enough from the specimen apex [11]. Additional characterization of the interface could be undertaken thanks to electron diffraction. The site-specific analysis in FIB-SEM enables mounting several tip-shaped specimens on a half-moon grid, as demonstrated by Pfelfer et al. [13] and Gorman et al. [14]. Such a specimen can later be placed on a usual TEM holder. Nevertheless, further analysis of such a specimen by APT requires the use of a local electrode. Without this device, confinement of isopotential electrostatic surfaces close to the tip cannot be obtained and controlled field evaporation becomes impossible. Among the attempts to design reliable holders for correlative analysis in APT and (S)TEM, enabling specimen rotation along two axes, a setup has been developed by Herbig et al. [16], which allows safe manipulation of the specimen from SEM to TEM and APT, among other advantages (Fig. 10.1). The concept of the method proposed by Herbig et al. is to fix FIB made tips of half-moon grid prepared from usual TEM grids. The main force of the method is that the same grid holder can be transferred to the FIB-SEM, the (S)TEM, and the APT, without manipulating the grid, therefore preventing perilous manipulation of the specimen. Another advantage is that the grid can be mounted on a usual TEM double-tilt holder. This enables conventional TEM (CTEM) analysis of dislocations and interfaces [1] in diffraction mode or high-resolution observations.

Insofar as usual TEM holders are designed to support a 3-mm-diameter grid or disc, special holders for tips had to be conceived. Home-made specimen holders were initially made by APT groups [15] but the interest of tip-shaped specimen analysis in ET has motivated the design of commercial holders. The interest here comes from the stability of the projected volumes along tilt series acquisition and from the fact that the tilt axis is nearly identical to the axis of symmetry of the specimen. Thanks to the development of such holders, ET can routinely be applied to tip-shaped specimens prior to their analysis in APT. Such a correlative approach is, however, not free of disadvantages since the specimen is exposed to electron irradiation and to carbon contamination. In addition, using such a holder does not allow specific orientation of the specimen with enough accuracy, since only a single tilt axis is available.

DEGRADATION OF SPECIMEN DUE TO OBSERVATION IN STEM

The main purpose of this section is to seriously consider the following question: is the result of the APT analysis affected by the prior observation of the specimen in a STEM of TEM? To answer this, one has to identify the various potential modifications of the specimen occurring when it is investigated in a (S)TEM. Among the sources of specimen modification, the following have been identified:

1. Specimen contamination under electron beam;
2. Exposure of the specimen to plasma cleaning;
3. Electron beam damage (knock-on displacements);
4. Electron beam heating.

These four points are discussed hereafter.

Specimen Contamination and Plasma Cleaning

Specimen contamination and methods preventing or removing contamination cannot be neglected by who wants to observe an APT specimen prior to an APT analysis. As far as APT specimens are submitted to long and repeated electron beam exposures (eg, for ET acquisitions), high level of contamination by carbon-rich molecules of the specimen apex are currently reached (Fig. 10.2). Such a contamination is detrimental for observations in the (S)TEM and is also unsuitable for further measurements in APT. Preventing specimen contamination may imply the application of plasma cleaning, which may modify the specimen structure in some cases. Moreover, insofar as the lift-out method requires the use of polymerized molecules (see Chapter 4: Atom Probe Sample Preparation), the connection of the APT specimen to its holder may be degraded by the plasma cleaning, which will erode the carbon-rich weld deposited in FIB-SEM. In the following, the issues related to specimen contamination are raised and discussed.

Observation of APT specimens in a (S)TEM often involves surface contamination, whatever the specimen preparation method is. Contamination by carbon-rich polymerized molecules arises by the radiation damage of adsorbed hydrocarbon molecules on the specimen surface [17]. The gaseous hydrocarbons may arise from pump oils, vacuum grease, O-ring seals, or

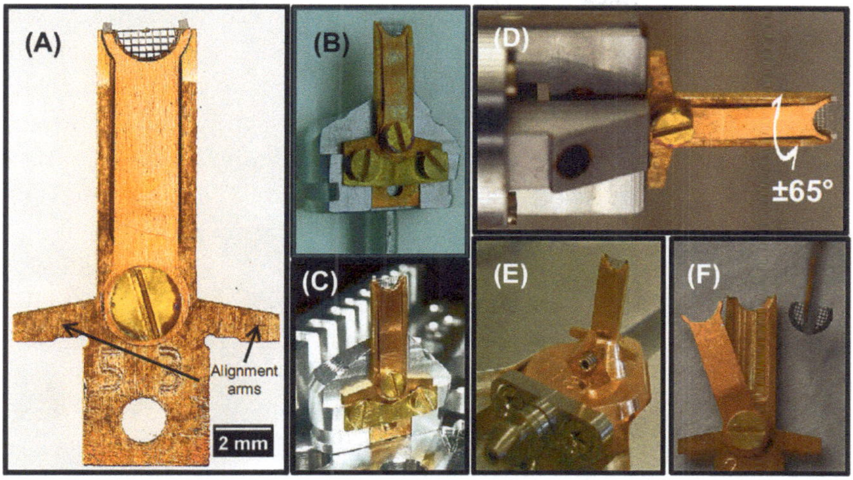

Figure 10.2 Grid holder developed by Herbig et al. [16]. (A) Grid holder. (B,C) Grid holder mounted on a device designed for introduction in the scanning electron microscope. (D) Grid holder mounted on a single-tilt JEOL holder. (E) Holder mounted on a local electrode atom probe (LEAP) instrument. (F) Grid removed from the holder. *Reproduced with permission from Herbig M, Choi P, Raabe D, Ultramicroscopy 2015;153:32–39.*

from the specimen itself [18]. Electropolishing also leads to the deposition of organic molecules. Furthermore, preparation in an SEM may also imply a certain level of carbon deposition. Because of a low vapor pressure and low surface mobility of hydrocarbon molecules [19], the thickness of the contamination layer increases as the specimen is illuminated by the electron beam. This phenomenon may be even more pronounced when high-resolution electron microscopy observations are performed. Indeed, contamination rates depend on the current density [20] and on the temperature of a specimen at the incident electron beam areas [21].

For thin foils, the contaminated layers cover both surfaces of the specimen. Some precautions can be taken to prevent contamination of the specimen. The relative efficiency of methods used to prevent or reduce specimen contamination under electron beam has been investigated in the literature [22,23]. First, one can heat the specimen at a moderate temperature (~100°C or higher temperature if possible) in vacuum (10^{-4} Pa or below) over a night prior to analysis in STEM. This treatment promotes the desorption of hydrocarbon molecules and decreases contamination [24]. However, not all specimens can be submitted to such a treatment, which can lead to evolution of specimen structure, stress relaxation, and modification of the tip shape. Of course, use of a nitrogen cold trap helps improving the vacuum in the vicinity of the specimen. It is, however, not sufficient in all cases, especially when a high electron dose is locally applied. Another way to reduce contamination is to apply a "beam shower." This process consists in bombarding a large area of the specimen by defocusing the illumination and removing the condenser aperture. This way, surface hydrocarbons get fixed on the specimen surface. This later prevents them from diffusing toward the area under investigation. Indeed, hydrocarbons already adsorbed on the specimen surface are otherwise diffusing toward the edge of the irradiated area [25], where they are fixed by polymerization. Cooling the specimen during the observation is an efficient way to reduce the mobility of hydrocarbons [21] and, consequently, the contamination. Let us notice that, instead of growing a layer on the material, the layer could also be eroded under electron beam illumination by an etching of the hydrocarbons. This etching is promoted when water, O_2, or N_2 molecules are adsorbed on the specimen [25]. This last phenomenon may partially explain why contamination levels and rates may strongly vary from one specimen to the other, as a function of the preparation method and of the ability of the material to adsorb molecules promoting the etching of hydrocarbons.

Once the specimen is covered by a layer of carbon-rich molecules, its evaporation in APT might become perilous. Then, two possibilities are offered to the user: either applying ion milling to the contaminated area or submitting the specimen to plasma cleaning. Ion milling in FIB has been widely discussed in "Chapter 4, Atom Probe Sample Preparation". As for the application of plasma cleaning, it is relevant to wonder in which way it might eventually affect the specimen structure, composition, and yield in APT. Plasma cleaning is widely employed in the field of surface processing technologies [26–28]. Its use has been extended to the treatment of (S)TEM specimens and plasma cleaners are currently used as a "black box" tool, for which some parameters can be adjusted (eg, gas species, pressure, specimen exposure time) without deep knowledge of the technique. Before considering the effects of these parameters, a few words about the technique are needed. Generally, artificial plasma is generated by the application of a high-frequency electromagnetic field to a low-pressure gas (in the range of 10^3–10^9 Hz, but plasma cleaners for (S)TEM specimens usually use a radio frequency generator). The electromagnetic field applied is large enough to stress the dielectric gas beyond its dielectric strength, transforming the gas into a conductor, wherein collisions of electrons with atoms generate an increasing number of electrons and ions in a process known as Townsend avalanche. The high-frequency electromagnetic field excites and ionizes the gas atom, which emits typical visible radiations by deexcitation. The color of the plasma is the signature of the species it contains. Ions in the plasma can either be positively or negatively charged.

A plasma hence generates low-energy ions, electrons, and radiations. Plasma cleaning is not expected to heat the specimen. Indeed, the collisions of mobile electrons with much heavier ions let these later at the neutral gas temperature [30]. In commercial systems [29], the energy of ions (acquired thanks to the polarization of the device's sheath) does not exceed 15 eV, which prevents sputtering of the specimen. However, sputtering can be reached if the specimen is polarized above 100 V, which enables accelerating ions of the plasma as demonstrated by Ko et al. [31]. Hence, the specimen is neither heated nor sputtered and its surface modification is controlled by chemical reactions. Short-wave ultraviolet radiations are also generated in the plasma, which are effective in breaking organic molecule bonds (involving a carbon atom with an N, C, or O atom) [32]. This phenomenon facilitates separating complex organic contaminant molecules into smaller pieces. Hence, prior exposure of the specimen to ultraviolet radiation is a possibility to reduce contamination of the specimen under the electron beam.

The nature of the gas used in the plasma cleaner is an important parameter. Use of oxygen is an evidence since disassociated oxygen created by the plasma chemically reacts with the hydrocarbon material present on the specimen surface, converting it into CO, CO_2, and H_2O, which are evacuated by the vacuum system [33]. Use of H_2/O_2 gas is also extremely efficient. If the specimen surface is sensitive to oxidation, nitriding or hydruration, inert gas (eg, He or Ar) shall be used instead of O-, N-, and H-containing gases. In this case, the impact of inert species on the contaminated surface is the process helping the removal of contaminant molecules. The time constant associated to the plasma cleaner efficiency in the reduction of contamination layer has been estimated by Griffiths and Walthers [20]. These authors have shown that the growth of the contamination layer decreases exponentially as a function of plasma cleaning time, with a time constant decay of about 2–4 min. Since contamination arises after a certain time of observation in the TEM, plasma cleaning in a TEM has been foreseen and even built up as reported by Horiuchi et al. [34]. Nevertheless, because of the necessity of introducing gas in the instrument, the instrumental development proposed by Horiuchi et al. has not been spread to commercial instruments. The development of environmental TEMs shall however make evolve this. A plasma cleaning system has also been implemented in APT, in this case in the LEAP system for the cleaning of the local electrode. Fig. 10.2 gives an illustration of the contamination of an APT specimen observed in STEM. The effect of long-time plasma cleaning is also evidenced.

Contamination is not the main limitation of correlative analyses involving STEM and APT. Submitting the specimen to irradiation by a high-energy electron beam implies elastic and inelastic interactions, which will eventually end up in generating significant beam damage. This aspect is the subject of the next section.

Electron Beam Heating

Because of inelastic interactions, a certain quantity of the energy given by incident electrons can be converted into heat. The problem of heat generation due to specimen irradiation by an electron beam has been addressed in detail in Chapters V and XI of Ref. [18] and was also estimated by Egerton et al. [22]. Here, we will consider the transfer of energy per unit length of relativistic electrons to the specimen under the irradiation of a cylinder of radius r_0 and thickness t_h, which is the specimen thickness. The energy transferred dE/dx per electron per unit length can be expressed according to Beth formula:

$$\frac{dE}{dx} = \frac{N_A Z e^4}{4\pi\varepsilon_0^2 A E_0 \beta^2} \frac{\rho}{M_U} \left[\ln\left(\frac{2E_0\beta^2}{J\cdot(1-\beta^2)} \right) - \beta^2 \right] \tag{10.2}$$

where N_A is the Avogadro number; Z is the atomic number of the species in the specimen; E is the charge of the electron; ρ is the density of the specimen (g/cm^3); ε_0 is the permittivity of vacuum; A is the mass number of the species in the specimen; $E = m_0c^2$, with m_0 the rest mass of electrons; $\beta = v/c$, the ratio between the velocity of electrons and the speed of light; $M_U = 1$ g/mol is the molar mass constant; J is the mean ionization energy of species in the specimen, as derived by Bloch: $J = 13.5 \times Z$ in eV.

For a current density j (A/m^2), the heat transferred to the specimen is

$$\pi r_0^2 \frac{j}{e} \frac{dE}{dx} \rho t_h \tag{10.3}$$

In the specimen, a stationary state is reached when the generation of heat is balanced by the radial conduction of heat in the material and the radiation by the top and bottom surfaces. Egerton et al. [19] have established that the radiative term is negligible, even for poorly conductive materials. Consequently, for a material of thermal conductivity λ, a stationary state is reached when the above energy is transferred across cylinders of radius r, with the long axis parallel to the beam:

$$\pi r_0^2 \frac{j}{e} \frac{dE}{dx} \rho t_h = -\lambda \cdot 2\pi r t_h \frac{dT}{dr} \tag{10.4}$$

which simplifies to

$$\frac{dT}{dr} = -\frac{j\rho}{2e\lambda} \frac{dE}{dx} \frac{r_0^2}{r} \tag{10.5}$$

Interestingly, the thickness t_h of the specimen does not influence the temperature rise, which is of importance for APT specimens, which are significantly thicker than standard TEM foils. Integration of the above equation leads to a temperature profile, once boundary conditions have been wisely identified [18]. First, the temperature of the specimen is unaffected and equal to T_0 at a distance R equal to the condenser aperture radius. Second, the temperature rise inside the cylinder of radius r_0 is neglected. The temperature is hence maximal at $r = r_0$. The stationary temperature hence becomes

$$T(r) = T_0 + \frac{j\rho}{2e\lambda} \frac{dE}{dx} r_0^2 \ln \frac{R}{r} \tag{10.6}$$

This enables the estimation of the maximum temperature increase $\Delta T_{max} = T(r_0) - T_0$ as a function parameter related to both the specimen and microscope settings. For an aberration-corrected STEM operating at 200 kV, a probe current ranging from 10^{-11} to 10^{-9} A can be focused on a point with radius of about 0.1 nm. For any kind of metal, such conditions of illumination lead to negligible heating of the specimen insofar as maximum heating does not exceed 10^{-2}K in the illuminated column. Such small values are justified by the high surface over volume ratio of the illuminated area. Even for a carbon thin film, Egerton et al. [22] have established that a 5-nA electron probe focused on a 0.1 nm diameter only leads to a 1.5°C temperature increase. Reimer and Kohl [18] have shown that higher values are obtained in the case of uniform illumination [ie, high-resolution transmission electron microscopy (HRTEM) mode], in particular for poor thermal conductors like SiO_2, for which the temperature increase can reach about 30K. But for metals, uniform illumination does not exceed a few degrees.

Careful analysis of the possible heating of APT specimens has demonstrated that no significant heating occurs either in STEM mode or in TEM mode, except for very particular conditions. Hence, inelastic interactions are not responsible for significant evolution of specimen structure. However, the impact of elastic collisions still needs to be evaluated.

Beam Damage

Any STEM user may have acquired an intuitive knowledge of electron beam damage from his/her own experience and may base this knowledge on the following assertions: (1) a threshold energy exists above which knock-on displacements will occur; (2) the knock-on displacement rate increases with larger magnification. Relevant analyses of APT specimens in STEM often require exposure of the specimen for a long time (eg, for tilt series acquisition) or high magnification observations, which intrinsically increase the possibility of substantial electron beam damage. To ascertain that the structure of APT is not modified by STEM or TEM analysis, it is mandatory to evaluate the parameters controlling the mechanism of knock-on displacement. In the following, the basic concepts related to knock-on displacement are introduced.

In conducting materials (ie, metals and semiconductors), because of the high density of delocalized electrons, knock-on displacement subsequent to elastic scattering is the only damage mechanism [19]. Conversely, radiolysis mechanisms triggered by inelastic scattering of electrons are the dominant beam damage for organic materials and oxides [18,22]. Elastic scattering is the electrostatic deflection of incident electrons by atomic nuclei. Although considered as causing negligible energy transfer, the transfer of energy becomes substantial for large scattering angles θ. For an incident electron of rest mass m_0 and kinetic energy $E_c = e \cdot U$, the maximum transferrable energy T_{max} to an atom of mass M (mass number A) is [35]:

$$T_{max} = 2E_c \frac{E_c + 2m_0c^2}{Mc^2} \tag{10.7}$$

If $T_{max} > T_D$, the threshold displacement energy of the atoms, knock-on displacement may occur. Here, the direction dependence of T_D in crystals will not be considered. The rate at which these displacements will be involved is related to the cross-section of the knock-on displacement.

This quantity is defined in the McKinley—Feshbach approximation by the following equation [35—37]:

$$\sigma_D = \frac{Z^2 4\pi a_0^2 E_R^2}{m_0^2 c^4} \frac{1-\beta^2}{\beta^4} \left\{ \frac{T_{max}}{T_D} + 2\pi\alpha\beta \left(\frac{T_{max}}{T_D} \right)^{1/2} - (\beta^2 + \pi\alpha\beta) \ln\left(\frac{T_{max}}{T_D} \right) - (1 - 2\pi\alpha\beta) \right\} \quad (10.8)$$

where: Z is the atomic number; a_0 is the Bohr radius (53 pm); E_R is the Rhydberg energy (13.6 eV); $\beta = v/c$ is the ratio between the incident electron velocity and the speed of light; T_{max} is the maximum energy transferrable by an incident electron to an atom due to "elastic" scattering; T_D is the threshold displacement energy of the species; $\alpha = Z/137$.

The displacement cross-section, expressed in barn ($1\ b = 10^{-28}\ m^2$) can easily be determined by the two aforementioned equations, if the threshold displacement energy T_D is known for the material under consideration. Then, the concentration of displacements (here, the number of displacements per atom) is obtained via the determination of the electron probe current density j as follows:

$$C_d = \sigma_D \frac{jt}{e} \quad (10.9)$$

where e is the electron charge and t is the time to which an atom is exposed to the electron beam. This last equation exposes clearly how the occurrence of knock-on displacements is linked to the magnification and to the total exposure to the electron beam. In STEM, the pixel size is defined as $P_M = L/(R_{es} \times M_{ag})$ (with L a calibrated image size at a reference magnification, R_{es} the resolution of the image in pixels, and M_{ag} the magnification). Considering the dwell time of the electron probe t_{pix}, for a single scan of an area in STEM, each atom will be exposed to the electron probe for a duration $r \cdot t_{pix}$, with r the ratio between the atom cross-section and the pixel area: $r = \pi r_{at}^2 / P_M^2$. Hence, for a single scan, the concentration of displacements becomes

$$C_D = \sigma_D \frac{j\pi r_{at}^2}{eL^2} (R_{es} M_{ag})^2 t_{pix} \quad (10.10)$$

The dependence of concentration of displacements as a function of the magnification and of the image resolution becomes explicit. Indeed, going

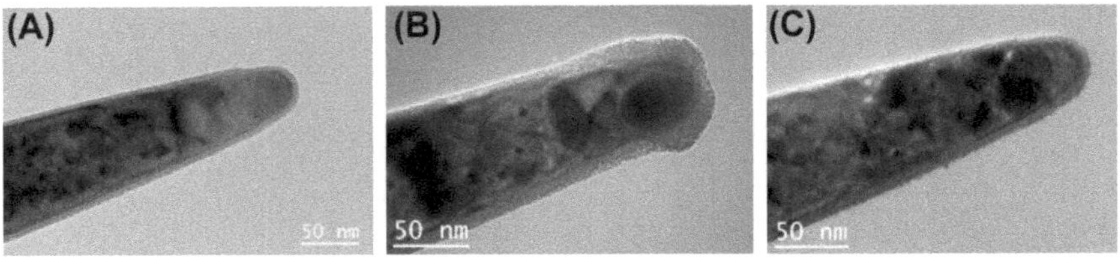

Figure 10.3 Aluminum specimen observed in bright-field transmission electron microscopy. The tip was prepared by standard electropolishing followed by focused ion beam cleaning with 2 keV gallium ions. (A) Specimen before observation in scanning transmission electron microscopy, without prior plasma cleaning. (B) Specimen after high-resolution observations in scanning transmission electron microscopy. A high level of contamination is visible. (C) Specimen after 8′ plasma cleaning under a mixture of O_2 and H_2 gas, which allowed removing completely the contamination layer.

from a magnification of 500k to 10 million increases the concentration of displacements by a factor of 400. This dependence on the magnification is illustrated in Fig. 10.3, which shows a sequence of high-angle annular dark field (HAADF) STEM images at large magnifications.

Materials do not have similar threshold displacement energies T_D, which explains why the concentration of displacements may strongly vary from one specimen to the other. Typically, knock-on damage is considered when $T_{max} > T_D$ or when the knock-on displacement cross-section becomes positive. Eq. (10.10) highlights the critical behavior of the material as a function of the electron beam energy. Identifying $T_D = T_{max}$ indeed reveals a critical value below which no knock-on displacement is expected to occur. T_D is the energy necessary to displace an atom from its normal position through a saddle point to a second stable or metastable position. Displacement energies are related to atomic bonding energies, the amplitude of which varies as a function of the bond character of the solid (Table 10.1).

Table 10.1 Bond Energy per Atom

Bond	Bond Principle	Bond Energy Range (eV)
Van der Waals	Dipole interaction	0.1−1
Metallic	Delocalization of conduction electrons	1−3
Covalent	Partial and directed electron transfer	4−8
Ionic	Electrostatic	8−200

After Hughes AE, Pooley D. Real solids and radiation, Wykeham, London; 1975.

As pointed out by Hobbs [35], the small energies in metallic solids may suggest that atom displacements should be relatively easy to produce. Nevertheless, the high compactness of metals and alloys counterbalance this hypothesis. Indeed, removing an ion core through several saddle point configurations to a stable interstitial position, where there is little room, involves seriously perturbing the bonding of several atoms in the vicinity. Consequently, T_D for metals is more in the range of 10–40 eV, as depicted by the values reported in Table 10.2. For alloys or particular compounds for which data on T_D are unavailable, this quantity has to be calculated either by molecular dynamics or density functional theory calculations [39–42]. Very close values of T_D are reported for Al, Fe, and Cu. Nevertheless, since the maximum energy which an incident electron can transfer to an atom depends on the atom mass (cf Eq.(10.7)), the critical voltage above which knock-on displacement occurs strongly varies from one element to the other. This variation is illustrated by Fig. 10.4. In Table 10.2, the critical acceleration voltage U_C has been calculated for reported values of T_D, using the following equation:

$$U_C = 2m_0c^2\left\{\sqrt{1+\frac{T_D M}{2m_0^2c^2}}-1\right\} \tag{10.11}$$

Finally, the actual rate of displacements will depend on the value of the cross-section, which can be calculated using Eq. (10.8).

Before concluding this section dedicated to beam damage, the particular geometry of APT specimens has to be further considered. Beyond knock-on displacement, displacement of surface atoms can lead to sublimation when they are ejected in vacuum. This process has been presented in detail by Egerton et al. [19]. Insofar as sputtering requires the cut of surface bonds, the authors pointed out the fact that high surface curvatures have a high density of kink and step sites. Such configurations favor sputtering of atoms for incident energies below the threshold for bulk material, as demonstrated for silver nanoparticles [50]. These considerations lead to the conclusion that surface atoms are susceptible to knock-on displacement below the thresholds reported in Table 10.2. Reducing the dose for high-resolution observations of light materials is hence an obvious precaution.

As for the effect of temperature, it is clear that no influence of temperature can be pointed out on the knock-on displacement, which mostly relies on the incident electron energy and the mass of atoms. It is noticeable

Table 10.2 Values for the Threshold Energy for Knock-On Displacement in Pure Materials. Acceleration Voltages in Bold Are Below the Voltages Used Currently in STEM and TEM (ie, 200 of 300 kV)

Element	T_D (eV)	References	Critical Acceleration Voltage U_C (kV)
Al	16–19	[43]	**170–196**
Zn	7.6–11	[44]	**191**
Cd	8.5–11.2	[44]	329
Mg	9.9–13.2	[44]	**100–128**
Ti	23.2	[44]	372
Ni	23–24	[43]	433
Fe	17–24	[43]	329
Cu	19–22	[43]	397
Nb	28	[43]	703
Ta	32	[43]	1210
Pd	34	[43]	883
Mo	34–37	[43]	825
Ag	24–28	[43]	702
Pt	34	[43]	1320
V	25	[43]	414
Zr	25	[45]	640
Au	33–36	[43]	1304
Ga in GaN	73	[46]	1112
N in GaN	32	[43]	**175**
Al in Al_2O_3	24	[43]	**240**
O in Al_2O_3	41	[47]	**242**
Si	12.5	[48]	**140**
C (graphite)	28–31	[35]	**152**
C in SiC	19	[43]	**95**
Si in SiC	42	[49]	389
Ga in GaS	9	[43]	**233**
As in GaAs	9.4	[35]	**257**
Zn in ZnO	57	[35]	903
O in ZnO	57	[43]	328
Mg in MgO	64	[35]	478
O in MgO	60	[43]	331

In case various values of T_D were reported depending on the crystal orientation, only the minimum value is given here. On the right column, we have calculated the minimum value of the acceleration voltage required to initiate knock-on displacements.

that a lower temperature does not reduce the surface sputtering rate of inorganic materials [19]. A lower temperature can, however, reduce the sensitivity of organic specimens to structural damage and mass loss. Indeed, Egerton et al. [51] have demonstrated that cooling of organic specimens

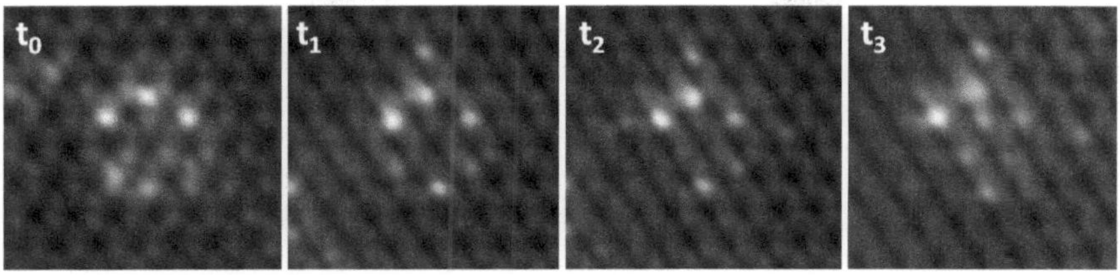

Figure 10.4 Effect of a large electron dose on a Mg-0.51 at.% Nd alloy observed under the following acquisition parameters: semiconvergence angle of 30 mrad, probe current of 54 pA, pixel size of 0.0127 nm, dwell time of 30 ms.

below 100K decreases the rate of mass loss by factors between 3 and 100 and interpreted this phenomenon in terms of lower atomic mobility. Consequently, as far as inorganic APT specimens are investigated, the use of a cooling specimen holder is of no interest for reducing beam damage or beam heating.

CORRELATION OF CONVENTIONAL TEM OR BF-STEM AND APT

In CTEM, the specimen is illuminated by an incident plane wave. CTEM commonly designates imaging modes based on diffraction and diffraction contrasts [ie, diffraction patterns, BF and dark field (DF) images]. High-resolution images are hence not considered, although they are recorded by selecting several waves, including the incident wave by inserting an aperture in the vicinity of the back focal plane of the objective lens, as is made for BF- and DF-TEM. Diffraction contrast helps identifying crystallographic defects such as precipitates, interfaces, and dislocations, most of which cannot be fully described by APT analysis. For instance, dislocations are described by their line and Burgers vectors, which cannot be determined by APT because of the lack of crystallographic information of this technique. Dislocations are visible in BF-TEM thanks to the elastic strain they exert on the lattice. In weak beam mode, Bragg reflections of weak intensity are selected to perform DF-TEM image of dislocations in two wave conditions. The weak beam mode allows identifying their Burgers vectors and the error vector (ie, a reciprocal vector defining the gap to exact Bragg condition) allows identifying their nature. Of course, such a characterization cannot be undertaken using a single-tilt TEM holder. It requires mounting the specimen on a double-tilt or tilt-rotation holder, which can be done thanks to dedicated protocols detailed earlier in this chapter (see section: Experimental Protocols). More information on imaging in weak beam condition can be found elsewhere [10].

The possibility to image dislocation loops in silicon has been evidenced by Larson et al. who could associate a segregation of arsenic to the presence of this defect [52]. Usually, associating unambiguously 3D variations of composition to crystalline defects requires at least the combination of BF-TEM or diffraction and APT analysis as reported in some seminal examples [53–56]. In another study where the authors used BF-STEM, Kuzmina et al. [1] have identified dislocations in an APT tip prior to analysis. BF-STEM provides imaging conditions roughly similar to BF-TEM. This helped them to associate the high levels of Mn concentration revealed by APT to dislocations, which could not be done without a correlative approach. The same group [57] had previously presented an original approach for the characterization of grain boundaries (Fig. 10.5). For this purpose, they have used a nanobeam diffraction in CTEM, which enables identifying with a high degree of precision the orientation of crystals. Doing this, Herbig et al. have been able to correlate the nature of grain boundaries with levels of segregations. This is illustrated in Fig. 10.6.

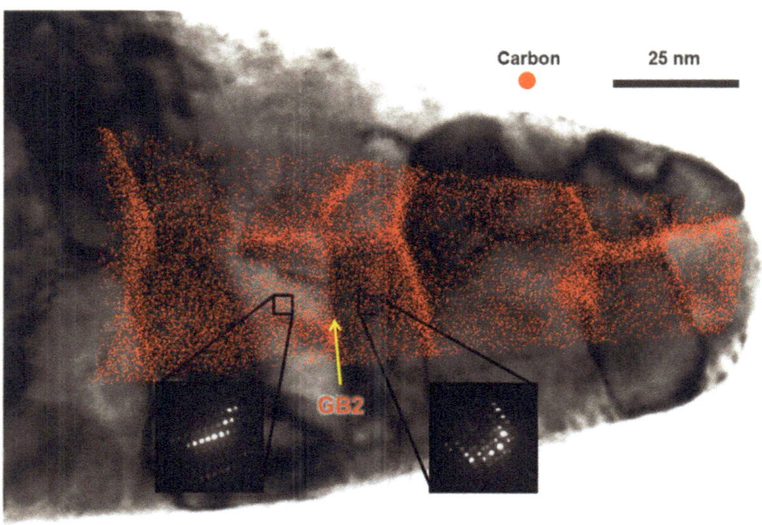

Figure 10.5 Bright field scanning transmission electron microscopy image on which a projection of the carbon map deduced by atom probe tomography is superimposed. The material is pearlitic steel. Grain orientation is determined by nanobeam diffraction. The grain boundary GB2 corresponds to a Σ3 coherent twin, where the level of segregation is lower than the average segregation. *Figure reproduced with permission from Herbig M, Raabe D, Li YJ, Choi P, Zaefferer S, Goto S, Phys Rev Lett 2014;112:126103.*

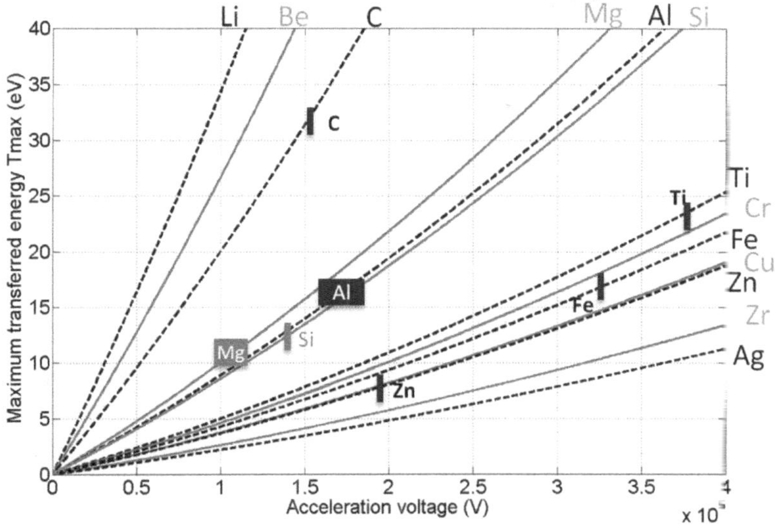

Figure 10.6 Calculation of the maximum transferrable energy of incident electrons for the knock-on displacement. The energy is calculated for several pure elements as a function of the acceleration voltage of the microscope. Threshold energies for the knock-on displacement are superimposed on the curves.

CORRELATION OF STEM IMAGING AND APT: FOCUS ON THE HAADF MODE

The rapid development of spherical aberration correctors (Cs-correctors) and their implementation on commercial (S)TEMs has significantly spread the use of STEM in the early 2000s [7]. The resolution in HAADF STEM commonly reaches 0.1 nm or lower in imaging modes as well as in analytical modes (like EELS and EDS). One of the advantages of STEM is the availability of various signals, which can sometimes be acquired simultaneously, therefore allowing imaging in various modes together with acquisition of analytical signals. Fig. 10.1 provides a schematic illustration of the modes available in STEM. Acquiring analytical signals offers the possibility of identifying specific features in a tip or to compare quantitative data obtained on the same specimen with several techniques including APT [58]. Nevertheless, quantification of concentrations in a tip with EDS or EELS signals is not a trivial issue. Indeed, the thickness has to be determined with accuracy and should be limited to prevent a significant spread of the electron in the specimen, which would enlarge the volume of interaction. It is hence obvious that APT needles do not exhibit a favorable geometry for quantitative composition measurements in STEM.

High-Resolution Imaging in STEM

Influence of Electron Probe Parameters for APT Specimens

HRTEM is not suitable for APT specimens. Indeed, this technique provides a very low depth of focus (below 10 nm), which is far below the size of common APT specimens. Moreover, HRTEM provides complex contrast reversal which may require intensive use of image simulations for the interpretation of images. The task becomes less difficult if the TEM is equipped with a Cs-corrector of the objective lens. Nevertheless, HRTEM remains a relevant technique for thin specimen (thickness in the order of 10–20 nm) and is hence inadequate for APT needles. In opposition to TEM, STEM operates in the convergent illumination mode. Convergence of the electron probe is in the range of 7–40 mrad depending on the size of the condenser aperture. Depending on the microscope voltage, aberration corrections, and other settings, the morphology of the STEM probe can be calculated. Insofar as the probe parameters are of crucial importance for image quantification, they need to be adjusted depending on the type of analysis. Some basics about the electron probe are introduced below.

Without any aberration of the optics of the microscope and assuming that the electron source is a perfect point, the waves arriving on the specimen should be perfectly spherical. Since aberrations exist, the wave slightly deviates from a spherical wave front (Fig. 10.7). The amplitude of the deviation, χ, is a function of two angles: θ, the angular deviation from the optic

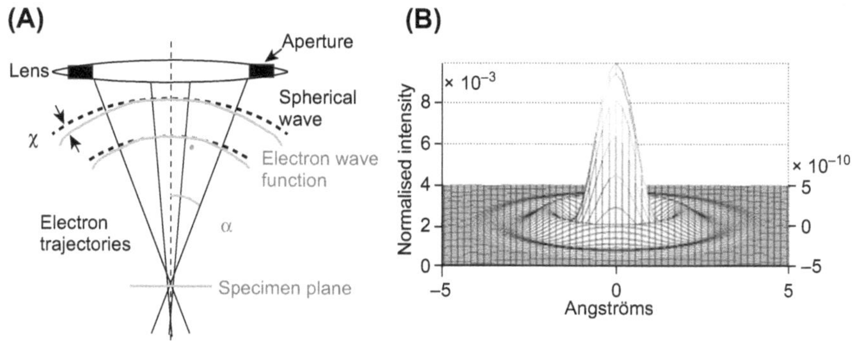

Figure 10.7 (A) Schematic illustration of the deviation of the electron wave front from a spherical wave front. α is the half-convergence angle of the electron probe and defines the opening angle of the condenser (or objective) aperture. Electron trajectories are illustrated in the case of pure spherical aberrations. (B) This later case also corresponds to the calculated electron probe intensity for $C_3 = 0.8$ mm and Scherzer defocus at 200 kV.

Table 10.3 Correspondence Between the Notations of Aberration Coefficients (Nonexhaustive List) Defining the Electron Probe in STEM

Aberration (Symmetry)	Haider's Notation [59]	Krivanek's Notation [60]
Defocus (0)	C_1	$C_{1,0}$ (C_1)
Two-fold astigmatism (2)	A_1	$C_{1,2}$
Axial coma (1)	B_2	$C_{2,1}$
Three-fold astigmatism (3)	A_2	$C_{2,3}$
Spherical aberration (0)	C_3	$C_{3,0}$ (C_3)
Star aberration (2)	S_3	$C_{3,2}$
Four-fold astigmatism (4)	A_3	$C_{3,4}$
Fourth-order axial coma (1)	B_4	$C_{4,1}$
Three-lobe aberration (3)	D_4	$C_{4,3}$
Five-fold astigmatism (5)	A_4	$C_{4,5}$

After Swada H, Sannomiya T, Hosokawa F, Nakamichi T, et al. Ultramicroscopy 2008;108:1467–75.

axis; φ, the azimuthal angle. $\chi(\theta,\varphi)$ can also be expressed as $\chi(\mathbf{k})$, where **k** designates a reciprocal vector in the plane of the condenser aperture. The aberration function is a function of the aberration coefficient listed in Table 10.3 and of the azimuthal angles associated to the directions of the aberrations (if relevant) [9].

$$\chi(\theta,\varphi) = \frac{2\pi}{\lambda}\sum_{mn}\frac{1}{n+1}C_{nm}\theta^{n+1}\cos[m(\varphi-\varphi_{nm})] \tag{10.12}$$

Following Kirkland [9], the point spread function (PSF) or, in other words, the electron probe intensity perpendicular to the optical axis, can be calculated owing to the following equation:

$$h_{\mathrm{ADF}}(r) = A_{\mathrm{p}}\int_0^{\boldsymbol{k}_{\max}} e^{-\frac{2i\pi}{\lambda}\chi(\boldsymbol{k})}J_0(2\pi\boldsymbol{kr})\boldsymbol{k}\cdot d\boldsymbol{k} \tag{10.13}$$

where $k_{\max}$ designates the largest spatial frequency fixed by the condenser aperture ($\alpha = \lambda k$), J_0 is the zeroth-order Bessel function of the first kind and r is the radial coordinate in the plane perpendicular to the optical axis. Simple measurements of the six first aberration coefficients enable the calculation of PSF. An example is provided in Fig. 10.7.

Hexapole correctors allow minimizing/optimizing C_1, A_1, B_2, A_2, C_3, and S_3. It is obvious that decreasing aberration coefficients decreases the probe diameter and improves resolution. Nevertheless, working with the smallest probes may not be the most relevant approach for APT needles. Indeed, minimizing the probe size decreases the depth of focus since the

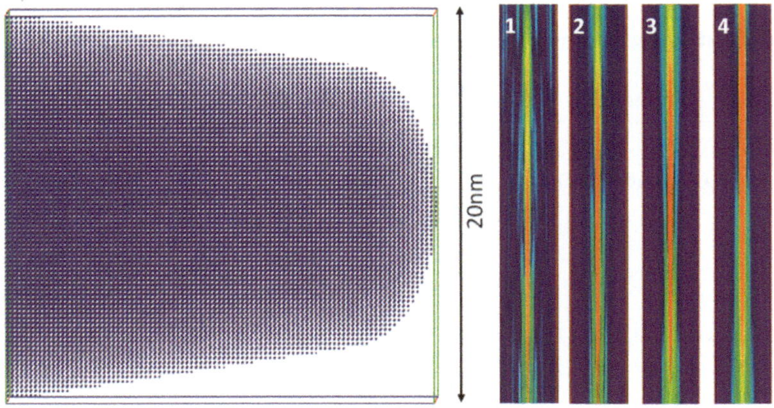

Figure 10.8 (Left) Model atom probe tomography needle. (Right) Effect of the semiconvergence angle α on the electron probe intensity along the direction of propagation. Calculations of electron probe intensity in vacuum were performed for an acceleration voltage of 200 kV, $C_1 = 0$ nm, $C_3 = 0.05$ mm, $C_5 = 20$ mm and other aberration coefficients set to 0. 1: $\alpha = 30$ mrad; 2: $\alpha = 25$ mrad; 3: $\alpha = 20$ mrad; 4: $\alpha = 15$ mrad. Simulations were performed using the methods presented in Ref. [9].

intensity varies more rapidly along the direction of propagation. This phenomenon is illustrated in Fig. 10.8. The depth of focus in aberration-corrected STEM varies as follows [62]:

$$\Delta z = 1.77 \frac{\lambda}{\alpha^2} \tag{10.14}$$

where α is the semiangle of convergence of the electron probe in rads. Here, Δz is defined as the full width at half maximum of the electron probe intensity along the direction of propagation. Values of the depth of focus have been calculated and are reported in Table 10.4.

Table 10.4 shows how the depth of focus decreases drastically as the convergence of the probe decreases. High-resolution observations of thin specimens may be performed with a large convergence (to minimize the probe size), whereas observations of thick specimens will be performed

Table 10.4 Variation of the Depth of Focus as a Function of the Semiconvergence Angle for a 200-keV Electron Beam

Semiconvergence angle α (mrad)	5	10	15	20	25	30	35	40
Depth of focus Δz (nm)	178	44	20	11	7.1	4.9	3.6	2.8

with lower convergence, which implies increasing the probe current by using other settings of condenser lenses. However, this does not mean that high resolution would become impossible. Indeed, only the ultimate resolution of the instrument will not be attainable.

High-Resolution Observation of APT Specimens

Once a specimen has been evaporated in APT using the holder developed by Herbig et al. [16], it becomes possible to perform high-resolution observation by removing the grid and placing the specimen on a TEM double-tilt holder. Such a methodology enabled performing the observations provided in Fig. 10.9. Stopping the APT analysis before tip rupture allows making high-resolution observations of features emerging at the specimen surface at the moment when the acquisition is stopped. By evaluating the morphology of the specimen surface one can estimate the

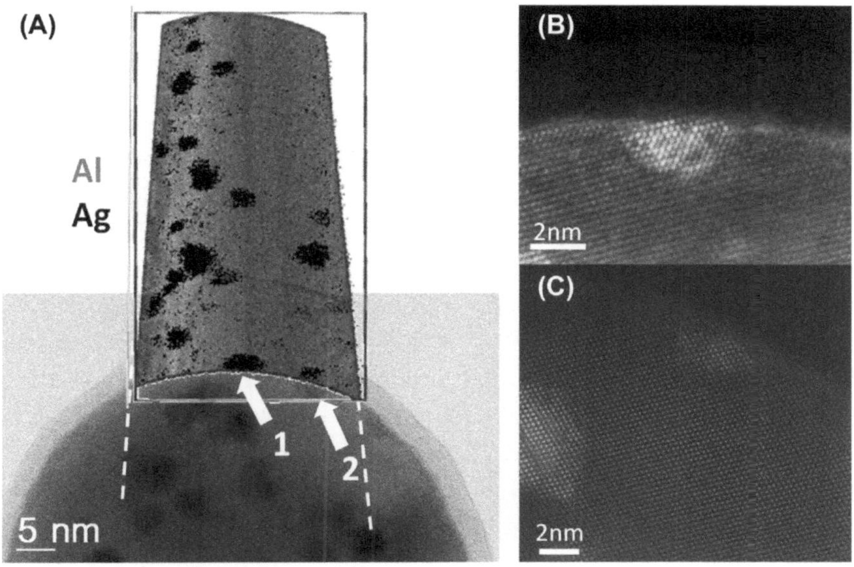

Figure 10.9 (A) Atom probe tomography analysis of an Al-1.7 at.% Ag specimen wherein Ag-rich precipitates are observed. The atom probe tomography reconstruction is superimposed with a bright field scanning transmission electron microscopy image of the specimen observed just after the atom probe tomography analysis has been stopped. The observation in scanning transmission electron microscopy as performed by aligning the specimen along a <110> zone axis of the Al matrix. (B,C) Two precipitates (1 and 2), which were under field evaporation at the end of the analysis, were imaged in high-angle annular dark field scanning transmission electron microscopy at 200 kV with a semiconvergence angle of 15 mrad and a probe size of about 0.1 nm (W. Lefebvre, I. Blum, F. Moyon, F. Vurpillot, unpublished).

respective stabilities of phases under electric field. In Fig. 10.9B, for instance, the Ag-rich particle appears with a slightly protruding position, which is the signature of a larger evaporation field of these particles, where evaporation is known to lead to chromatic aberrations. Such a protocol is expected to help estimating and correcting artifact like local magnification. Moreover, beyond the possibility of performing high-resolution observations, detailed analysis of HAADF STEM image intensities may enable the quantification of atoms, especially for binary systems. This is what is discussed in the next section.

High-resolution images in STEM require orienting the specimen along a zone axis. This involves a well-known phenomenon designated as channeling of electrons. Indeed, in a zone axis orientation, the focused incident beam is "trapped" by atomic potential (mainly tightly bound 1s states of the crystal Bloch waves are excited, leading to high resolution [63]). The channeling phenomenon is illustrated in Fig. 10.10B. A possible limitation in the investigation of APT needles in STEM comes from the fact that, as the specimen thickness becomes large, quantification of image intensity may lead to nontrivial interpretation. Support of image simulations hence becomes compulsory. Fig. 10.10 illustrates how the behavior of the electron probe in APT specimens can be evaluated by means of HAADF STEM image simulations. These simulations were conducted using the multislice algorithm developed by Kirkland [9] and optimized by Robertson [64], including the thermal diffuse scattering intensity in the frozen phonon approximation [65,66]. In zone axis condition and as the specimen gets thicker, it is likely that a significant proportion of the electron probe intensity gets trapped by neighboring atomic columns. This latter phenomenon is known as "cross-talk." It is illustrated in Fig. 10.10C–H. Since APT specimens are significantly thicker than standard STEM foils, this artifact must be considered as soon as quantification of images is pursued [67]. In addition, simulations may help optimizing experimental conditions. Simulations in Fig. 10.10 were performed for two semiangles of convergence. In the case of an Al specimen oriented along [001], it appears that a smaller convergence decreases the proportion of intensity trapped by neighboring columns at a thickness of 20 nm.

Atom Counting in an APT Specimen

The detection efficiency of APT is not 100%. All atoms observed in an (S) TEM are hence not detectable by APT and their quantification might be of

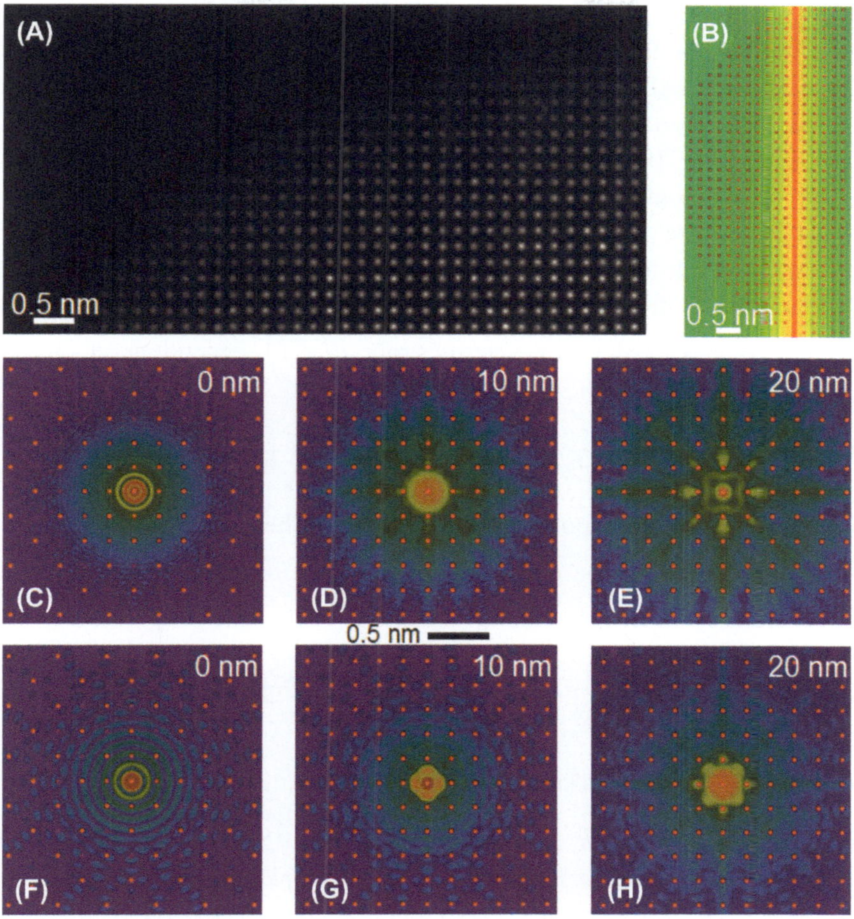

Figure 10.10 (A) High-angle annular dark field scanning transmission electron microscopy image simulation of an Al tip oriented along the [001] zone axis. (B) Side view of an Al tip for a fixed electron probe incident from the top of the image. The channeling effect induces a focusing of the electron probe along the atomic column. (C–H) Simulation of the electron probe intensity map at various thicknesses across an Al specimen oriented along [001]. Probe parameters are set equal except for the semiconvergence angle α which is 30 mrad in (C–E) and 15 mrad in (F–H).

relevance in some cases. Counting atoms in atomic columns by processing the intensity of HAADF STEM images requires a detailed statistical approach. Simply considering that the HAADF intensity grows linearly with a power of Z (the atomic number) or with the number of atoms in the columns is actually a too simplistic approach which holds for very favorable cases. An in-depth investigation of the reliability of HAADF STEM atom counting based on a statistical method has been presented by Van Aert et al. [68]. The first atom counting procedure applied to APT needles analyzed by HAADF STEM has been described in Ref. [67]. This statistical analysis was based on the well-established methodology proposed by Van Aert et al. [68–70], which has led to the first atomic scale reconstructions by discrete tomography in HAADF STEM [71,72]. The main steps of this methodology are recalled here. HAADF STEM images are described based on an incoherent image model and are hence considered as the convolution of an object function $O(\overrightarrow{r_{k,l}};\theta)$, depending on a set of unknown structural parameters, and a PSF of the electron probe $P(\overrightarrow{r_{k,l}})$, depending on a set of probe parameters:

$$f_{k,l}\left(\overrightarrow{r_{k,l}};\theta\right) = O\left(\overrightarrow{r_{k,l}};\theta\right)\cdot P\left(\overrightarrow{r_{k,l}}\right) \tag{10.15}$$

where $r_{k,l}$ is the vector pointing the pixel (k,l). The integrated column intensity should ideally be derived from the function $O(\overrightarrow{r_{k,l}};\theta)$, which is considered as a superimposition of Gaussians.

$$O\left(\overrightarrow{r_{k,l}};\ \theta\right) = \chi + \sum_{m=1}^{M} \alpha_m e^{-\frac{(x-x_m)^2+(y-y_m)^2}{2\rho^2}} \tag{10.16}$$

where χ is a constant background and $\theta = \{x_1, y_1, \ldots, x_m, y_m, \alpha_1, \ldots, \alpha_m, \rho\}$ is the set of unknown structural parameters. To access $O(\overrightarrow{r_{k,l}};\theta)$, a deconvolution would be required but is rarely applied in practice [73]. For each of the M columns ($i = 1,\ldots,M$) identified in the HAADF STEM image, the integrated intensity V_i of a column is obtained by radial integration around the $(x_\mathrm{i},y_\mathrm{i})$ coordinate of its projection. The histogram of column intensities is later fitted with a Gaussian mixture distribution [68–70]:

$$p(V_\mathrm{i}|\Psi_G) = \sum_{g=1}^{G} \pi_g \frac{1}{\sqrt{2\pi\sigma}} e^{-\frac{\left(V_\mathrm{i}-\mu_g\right)^2}{2\rho^2}} \tag{10.17}$$

where $\Psi_G = \{\mu_1,\ldots,\mu_G, \pi_1, \ldots,\pi_G, \sigma, \rho\}$ is a vector containing the values of the parameters defining the distribution. Each value of integrated intensity V_i is hence fitted by a mixture of G Gaussian components, each contributing with given mixing proportion π_g. Insofar as a large amount of various distributions may produce a reasonable fit of the data, based on the literature [74], Van Aert et al. [68] have established that the amount G of

Gaussian components shall be deduced from the drop of the integrated classification likelihood criterion (ICL).

By applying atom counting to simulated images, it was demonstrated that the procedure was reliable only if deconvolution from the PSF was applied in a first step [67]. Fig. 10.11 shows an illustration of atom counting. Applying atom counting to APT needles prior or after their investigation has been demonstrated and opens up many perspectives. However, because of the specific geometry of APT specimens and of the large volumes

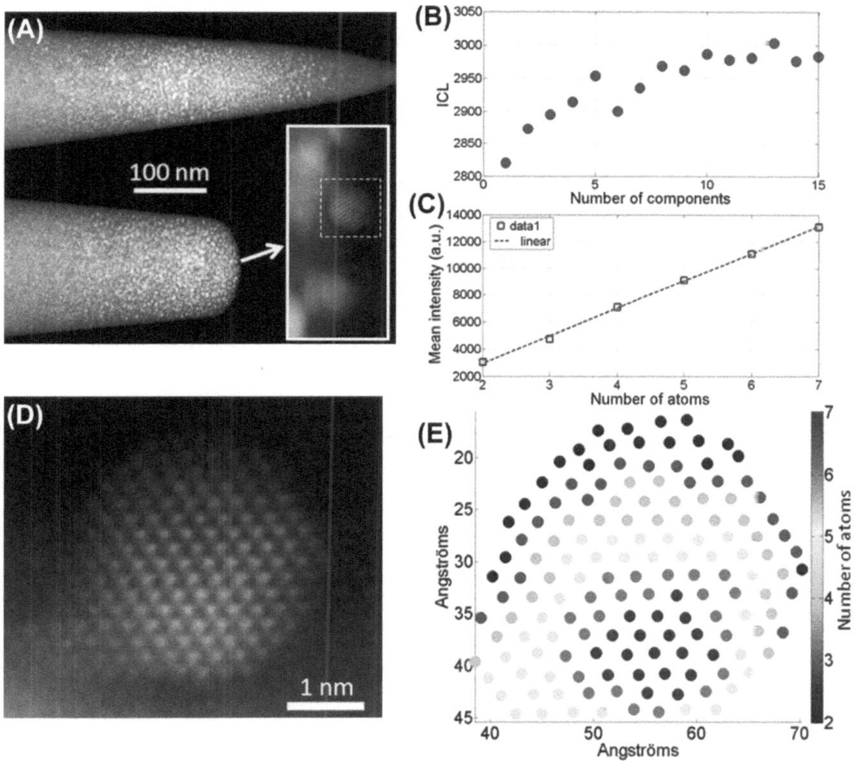

Figure 10.11 (A) High-angle annular dark field scanning transmission electron microscopy images of a CERMET specimen (prepared and analyzed in atom probe tomography by Dr D. Schinde, University of Rouen) containing gold nanoparticles embedded in a MgO matrix. Specimen before (top) and after (bottom) analysis in atom probe tomography. The inset shows the particle selected for atom counting. (B) Variations of the integrated classification likelihood parameter, showing a drop for six Gaussian components. (C) Relationship between column intensity and number of atoms in the nanoparticle. (D) Gold nanoparticle and (E) the result of atom counting procedure applied to this particle. *Results adapted from Lefebvre W, Hernandez-Maldonado D, Moyon F, Cuvilly F, Vaudolon C, Shinde D, et al. Ultramicroscopy 2015;159:4C3–4012.*

investigated, such an approach cannot be applied to all cases and often requires validation by image simulations.

CORRELATION OF ELECTRON TOMOGRAPHY AND APT

The interest of coupling tomographic acquisitions for a same specimen has rapidly been identified by the APT community. One of the main motivation of the approach relies in getting a benchmark for APT reconstructions [75,76]. The first work combining ET and APT was published by Arslan et al. [77]. Performing ET acquisition on APT needles became a routine once dedicated single-tilt holders have been developed. Acquiring tilt series on a needle provides the advantage of larger amplitude in tilt angles, which allows reducing the missing wedge and improving the quality of the reconstruction. Moreover, the particular geometry of APT needles does not lead to variations of thickness along the tilt series. Nevertheless, performing tilt-series acquisitions on an APT needle may be detrimental to the specimen structure. The effect of electron irradiation on the specimen structure has to be considered and acquisition parameters need to be set accordingly (see section: Degradation of Specimen due to Observation in STEM). A compromise must hence be found between the quality of the reconstruction and the preservation of the specimen structure. Improving the signal over noise ratio of images can lead to increase the electron probe intensity, which is detrimental for materials sensitive to knock-on damage. Improving the resolution may also lead to an increase in the number of images by reducing the incremental angle between successive images. However, the total time during which the specimen is irradiated by the electron beam is proportional to the amount of images. Hence, the defect concentration will increase proportionally. Another limitation also comes from specimen contamination (see section: Degradation of Specimen due to Observation in STEM), which deteriorates the quality of images as the contamination layer grows.

No specific acquisition parameters can be provided for the acquisition of APT specimen electron tomograms, since the sensibility to electron beam radiation is strongly dependent on the material (see section: Degradation of Specimen due to Observation in STEM). Only the large specimen thickness is a systematic feature, whatever the APT specimen is. As a consequence, small convergence angles will be used to get better compromise between depth of focus, probe size, and electron probe intensity. Typically, values of α less than 15 mrad are recommended.

Correlative analysis by ET and APT hence offers the possibility to reveal and account for artifacts in APT reconstructions. This is particularly relevant in complex heterostructure materials such as nanoelectronic devices for instance [78]. A seminal example is provided by Fig. 10.12. Here, a super lattice of AlGaN-GaN layers with equal thickness has been investigated in ET after analysis in APT. When fixed compression factors and evaporation fields are set as input parameters for the APT reconstruction (see Chapter 7: 3D Reconstruction in APT: Basics and Advanced Approaches), AlGaN layers are reconstructed with a slightly larger cross-section (perpendicular

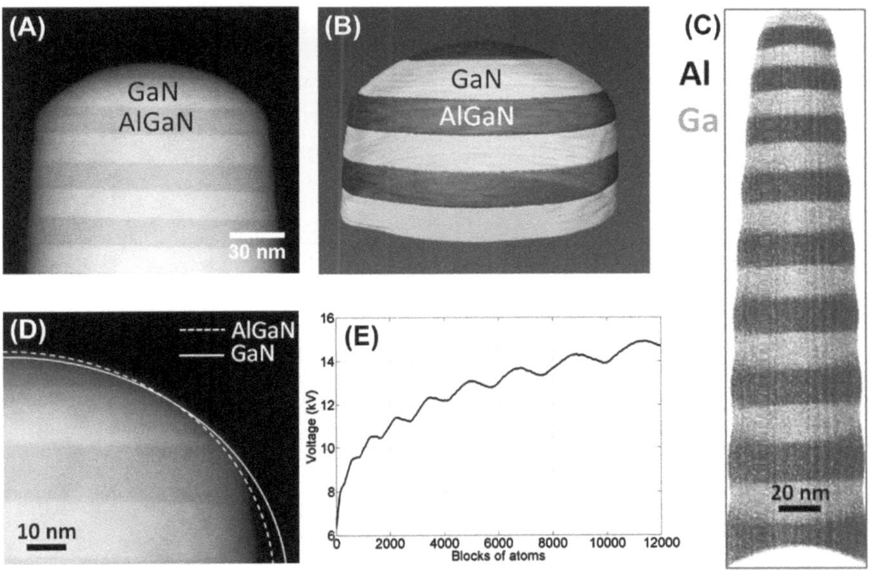

Figure 10.12 Correlative analysis by electron tomography (high-angle annular dark field scanning transmission electron microscopy mode) and atom probe tomography of a supernetwork of AlGaN-GaN. AlGaN layers contained 15 at.% of Al. (A) High-angle annular dark field scanning transmission electron microscopy image of the specimen after analysis in atom probe tomography. (B) Electron tomography reconstruction after atom probe tomography analysis. The acquisition and reconstruction were performed by equally sloped tomography [80] between −79° and +79°. (C) Atom probe tomography reconstruction showing Al and Ga ions. The reconstruction was performed with a single compression factor and a single evaporation field. (D) Schematic illustration of the discontinuity of curvature between emerging AlGaN and GaN layers. (E) Variations of voltage measured during atom probe tomography acquisition for constrained evaporation rate. *Courtesy of Moyon F, Rigutti L, University of Rouen, and Gougneau N, Marcoussis LPN.*

to the depth) than its neighboring GaN layers (Fig. 10.12C). For constrained evaporation rate condition during the APT acquisition, the applied DC voltage is found to increase periodically as an AlGaN layer emerges at the specimen surface (Fig. 10.12E). Such observations are consistent with a higher evaporation field of the AlGaN compound. As shown in Fig. 10.12D, such behavior results in measurable variations of specimen surface curvature in ET. Such measurements are useful for determining input parameters to be used for further simulations of field evaporation. A more straightforward application is the correction of APT reconstructions (not shown here), which can be undertaken using a methodology similar to the one presented in Ref. [79].

Correlating ET with APT on a same specimen may help reaching the best of each technique. Z-contrast, which may allow determining the 3D morphology of nanodevices [78], will fail in measuring 3D levels of segregations of light dopants like boron. This last information is, however, accessible by APT. The correlation of information hence helps going beyond the capabilities of each technique.

CORRELATIVE MICROSCOPY IN A DEDICATED INSTRUMENT

Although the possibility of combining APT and (S)TEM ex situ has been widely demonstrated thanks to dedicated protocols, several considerations justify going beyond this methodology by merging these two techniques in a single instrument. First, sample transfer from one instrument to the other is a limitation. It implies manipulating the specimen, which increases the risk of specimen deterioration. Second, exposure of the material to air after its analysis in (S)TEM might modify the specimen surface by oxidation. Consequently, the structure of surface features observed in STEM might not be correlated with those analyzed in APT. This is particularly true for metals and alloys as evidenced in Fig. 10.9A. Second, although ex situ correlative approaches are sufficient in many cases, (S) TEM observations only provide a snapshot of the specimen surface. If one wishes to understand artifacts of reconstruction in APT by getting geometrical or physical parameters by a correlative investigation, the ex situ methodology remains limited. Indeed, the specimen surface evolves in APT as a function of the field evaporated features. Input parameters for APT reconstructions are hence evolving dynamically with the sequence of field evaporation and it would me more suitable to get a dynamic knowledge of them, which can only be done with the simultaneous imaging by electrons.

If the goal of merging the abilities of electron microscopy with APT is pursued, three possibilities can be foreseen: (1) adapting an electron microscope to an existing atom probe; (2) adapting an atom probe to (S)TEM; (3) thinking a new instrument. The first possibility was proposed by Dr. B. P. Gorman and Dr. D. Diercks at Colorado School of Mines and is presented by Dr. Kirchhofer in her PhD dissertation [81]. The methodology applied here consists in implementing an SEM electron gun on an existing atom probe (LEAP). This enables imaging the specimen in a TEM mode and acquiring selected area electron diffraction patterns to study dynamic phase transformation induced by the use of a laser beam in the atom probe. The second and third methodologies have been inspected in detail in the paper by T.J. Kelly et al. [81]. This article relates the many benefits resulting from merging an STEM and an APT. The authors also introduce the possibility to slightly modify an existing ultra high vacuum cold field emission gun STEM with a modified specimen holder enabling the field evaporation of a He-cooled specimen. This ambitious and interesting perspective might give a clue of what the next generation of instrument will be.

REFERENCES

[1] Kuzmina M, Herbing M, Ponge D, Sandlöbes S, Raabe D. Science 2015 349:1080.
[2] Liddicoat PV, Liao X-Z, Zhao Y, Zhu Y, Murashkin MY, Lavernia EJ et al. Nat Comm 2010;1:63.
[3] Devaraj A, Colby R, Vurpillot F, Thevuthasan S. J Phys Chem Lett 2014;5(8):1361–7.
[4] Sha G, Cerezo A, Smith GDW. Appl Phys Lett 2008;92(4):043503.
[5] Stockmar M, Cloetens P, Zanette I, Enders B, ierolf M, Pfeiffer F, et al. Nat Sci Rep 2013;3:1927.
[6] Zach J, Haider M. Optik 1995;93:112–8.
[7] Hawkes PW. Ultramicroscopy 2015;156:A1–64.
[8] Reimer L, Kohl H. Transmission electron microscopy. Physics of image formation, Springer series in optical sciences. 2008.
[9] Kirkland EJ. Advanced computing in electron microscopy. New York: Springer; 2010.
[10] Williams DB, Carter CB. Transmission electron microscopy, a text book for materials science. Springer; 2009.
[11] Henjered A, Norden H. J Phys E Sci Instrum 1983;16:617–9.
[12] Krakauer BW, Hu JG, Kuo SM, Mallick RL, Seki A, Seidman DN, et al. Rev Sci Instrum 1990;61:3390–8.
[13] Felfer PJ, Alam T, Ringer SP, Cairney JM. Microsc Res Tech 2012;75:484–91.
[14] Gorman BP, Diercks D, Salmon N, Stach E, Amador G, Hartfield C. Microsc Today 2008;16:42–7.
[15] Norden H, Bowkett KM. J Sci Instrum 1967;44.
[16] Herbig M, Choi P, Raabe D. Ultramicroscopy 2015;153:32–9.

[17] Ennos AE. J Appl Phys 1954;5:27.
[18] Reimer L, Kohl H. In: Transmission electron microscopy: physics of image formation, Springer series in optical sciences, Berlin; 2008 [Chapter 11].
[19] Egerton RF, McLeaod R, Wang F, Malax M. Ultramicroscopy 2010;110:991−7.
[20] Griffiths AJV, Walther T. J Phys Conf Ser 2010;241:012017.
[21] Wall JS. Scanning Electron Microsc 1980:99−106.
[22] Egerton RF, Li P, Malac M. Micron 2004;35:399.
[23] Mitchell DRG. Micron 2015;73:36−46.
[24] Ennos AE. J Appl Phys 1954;5:27.
[25] Hren JJ. Introduction to analytical electron microscopy. New York: Plenum Press; 1979. p. 481−505.
[26] Polini W, Sorrentino L. Appl Surf Sci 2003;214:232.
[27] Rider AN, Brack N, Andres S, Pigram PJ. J Adhes Sci Technol 2004;18:1123.
[28] Petersson L, Meier P, Kornmann X, Hillborg H. J Phys D Appl Phys 2011;44:034011.
[29] Isabell TC, Fischione PE, O'Keefe C, Guruz MU, Dravid VP. Microsc Microanal 1999;5:126.
[30] Melliar-Smith CM, Mogab CJ. Plasma-assisted etching techniques for pattern delineation. In: Vossen JL, Kern W, editors. Thin film processes. New York: Academic Press; 1978. p. 497−556.
[31] Ko DS, Park YM, Kim SD, Kim YW. Ultramicroscopy 2007;107:368−73.
[32] Ayako T, Hirose H. Adv Polym Sci 1999;63:441.
[33] Tuszewski M, Scheuer JT, Tobin JA. J Vac Sci Technol 1995;13 A.
[34] Horiuchi S, Hanada T, Ebisawa M, Matsuda Y, Kobayashi M, Takahara A. ACS Nano 2009;3:1297−304.
[35] Hobbs LW. In: Hren JJ, et al., editors. Introduction to analytical electron microscopy. New York: Springer; 1979.
[36] Mott NF, Massey HSW. The theory of atomic collisions. Oxford: Clarendon Press; 1965.
[37] McKinley WA, Feshbach H. Phys Rev 1948;74:1759−63.
[38] Hughes AE, Pooley D. Real solids and radiation. London: Wykeham; 1975.
[39] Gao F, Bacon DJ. Philos Mag A 1993;67:289.
[40] Gao F, Bacon DJ, Newall S. Plios Mag Lett 1998;77:229.
[41] Rossel MD, Erni R, Asta M, Radmilovic V, Dahmen U. Phys Rev B 2009;80:024110.
[42] Werner S, Friis J, Marioara CD, Andersen SJ, Holmestad R. Philos Mag 2015;95: 3524−34.
[43] Jung P. Radiat Eff 1978;35:155.
[44] Whitehead ME, Karim ASA, Loretto MH, Smallman RE. Acta Metall 1978;26: 975−81.
[45] Griffiths M. J Nucl Mater 1989;165:315−7.
[46] Xiao HX, Gao F, Zu XT, Weber WJ. J Appl Phys 2009;105:123527.
[47] Das G. J Mater Sc Lett 1983;2:453−6.
[48] Holmström E, Kuronen A, Nordlund K. Phys Rev B 2008;78:045202.
[49] Lucas G, Pizzagalli L. Nucl Instr Meth Phys Res B 2005;229:359−66.
[50] Braidy N, Jakubek ZJ, Simard B, Botton GA. Microsc Microanal 2008;14:166.
[51] Egerton RF, Crozier PA, Rice P. Ultramicroscopy 1987;23:305−12.
[52] Larson DJ, Thompson K, Alvis RL, Lawrence DF, Ulfig RM, Reinhard DA, et al. Microsc Microanal 2008;14(S2):324−5.
[53] Larson D, Prosa T, Kostrna S, Moe A, Kelly T, Stallybrass C, et al. Microsc Microanal 2006;12(S2):1748−9.
[54] Williams CA, Marquis EA, Cerezo A, Smith GDW. J Nucl Mater 2010;400:37−45.
[55] Baik SI, Yin X, Seidman DN. Scr Mater 2013;68:909−12.

[56] Cojocaru-Miredin O, Schwarz T, Choi PP, Herbig M, Wuerz R, Raabe D. Atom probe tomography studies on the Cu(In,Ga)Se2 grain boundaries. J Vis Exp April 22, 2013;(74).
[57] Herbig M, Raabe D, Li YJ, Choi P, Zaefferer S, Goto S. Phys Rev Lett 2014;112: 126103.
[58] Stender P, Heil T, Kohl H, Schmitz G. Ultramicroscopy 2009;109:612.
[59] Haider M, Uhlemann S, Zach J. Ultramicroscopy 2000;81:163.
[60] Kivanek OL, Dellby N, Lupini AR, 1999;78:1−11.
[61] Swada H, Sannomiya T, Hosokawa F, Nakamichi T, et al. Ultramicroscopy 2008;108: 1467−75.
[62] Nellist PD. The principle of STEM imaging. In: Pennycook SJ, Nellist PD, editors. Scanning electron microscopy. New York: Springer; 2011. p. 91−115.
[63] Nellist PD, Pennycook SJ. Ultramic 1999;78:111−24.
[64] Robertson MD, Raffel K. Microsc Soc Can Bull 2008;36:13−8.
[65] Loane RF, Xu P, Silcox J. Ultramicroscopy 1992;40:121.
[66] Van Dyck D. Ultramicroscopy 2009;109:677.
[67] Lefebvre W, Hernandez-Maldonado D, Moyon F, Cuvilly F, Vaudolon C, Shinde D, et al. Ultramicroscopy 2015;159:403−4012.
[68] Van Aert S, De Backer A, Martinez GT, Goris B, Bals S, Van Tendeloo G, et al. Phys Rev B 2013;87:064107.
[69] De Backer A, Martinez GT, Rosenauer A, Van Aert S. Ultramic 2013;134:23−33.
[70] Martinez GT, Rosenauer A, De Backer A, Verbeeck J, Van Aert S. Ultramic 2014;137: 12−9.
[71] Van Aert S, Batenburg KJ, Rossell MD, Erni R, Van Tendeloo G. Nature 2011;470: 374−7.
[72] Bals S, Casavola M, van Huis MA, Van Aert S, Batenburg KJ, Van Tendeloo G, et al. Nano Lett 2011;11(8):3420−4.
[73] Lefebvre W, Kopp V, Pareige C. Appl Phys Lett 2012;100:141906.
[74] McLachlan G, Peel D. Wiley series in probability and statistics. John Wiley and Sons, Inc; 2000.
[75] Petersen TC, Ringer SP. J App Phys 2009;105:103518.
[76] Petersen TC, Ringer SP. Comput Phys Commun 2010;181:676.
[77] Arslan I, Marquis EA, Homer M, Hekmaty MA, Bartelt NC. Ultramicroscopy 2008; 108:1579−85.
[78] Grenier A, Duguay S, Barnes JP, Serra R, Rolland N, et al. Appl Phys Lett 2015;106: 213102.
[79] De Geuser F, Lefebvre W, Danoix F, Vurpillot F, Forbord B, Blavette D Surf Interf Anal 2007;39:268−327.
[80] Kirchhofer R. Development of a dynamic atom probe [Ph.D. dissertation]. Colorado School of Mines; 2014.
[81] Kelly TF, Miller MK, Rajan K, Ringer SP. Microsc Microanal 2013;19:652.

CHAPTER ELEVEN

Combining Atom Probe Tomography and Optical Spectroscopy

L. Rigutti
Groupe de Physique des Matériaux, University and INSA of Rouen, Normandie University, France

INTRODUCTION

The recent development of laser-assisted atom probe tomography (La-APT) [1] has made it possible to analyze a large variety of nonmetallic nanostructures. Among these structures, there are functional materials of particular interest because of their optical properties: for instance, because they can efficiently emit photons of a given energy and polarization, because they can emit single photons on demand, or because they can selectively absorb a certain interval of wavelengths. In many cases, the optical properties of nanoscale systems result from their structure and composition, which immediately makes it clear why it is interesting to analyze them by APT. The knowledge of the structure and composition of semiconductor quantum wells (QWs), quantum dots (QDs), and nanocrystals or of metal nanoparticles (nps) at the nanoscale and in three dimensions (3D) opens the way to a complete understanding of their optical properties. This consideration is the basis of this chapter, which deals with the problem of connecting optical spectroscopic information with atom probe data at different levels of complexity and accuracy.

If the connection between optical properties and structural APT data is now a reality with growing importance in the domain of nanosciences, there is another perspective interest in connecting optical spectroscopic information and La-APT, still requiring some instrumental development effort. In La-APT, the sample, a sharp tip containing the nanoscale functional structure, is illuminated by a pulsed laser to trigger ion evaporation. It is therefore possible to imagine that the same laser beam could be used to trigger photon emission and field ion evaporation. Similarly, the presence of specifically absorbing structures within the tip could leave a trace in the transmitted

Atom Probe Tomography
ISBN: 9780128046470
http://dx.doi.org/10.1016/B978-0-12-804647-0.00011-5

beam. The realization of an optical bench in situ inside an atom probe would open further possibilities for a more in-depth study not only of the properties of a particular nanoscale structure, but also of the physical processes governing field ion evaporation itself.

RELATIONSHIPS BETWEEN CHEMISTRY, STRUCTURE, AND OPTICS IN NANOSCALE MATERIALS

Under the expression "optical properties" of a given system is included a large amount of information, relative to the interaction of the system with the electromagnetic field of light. This information may concern the fundamental electronic transitions giving rise to photon absorption, emission, or scattering, and also the way the light couples to the system, for instance, through waveguide or antenna effects. This chapter deals with selected optical properties relative to light emission under photon excitation [photoluminescence (PL) or microphotoluminescence (μPL) spectroscopy] of semiconductor heterostructures. The reader should anyway be aware that other systems, such as metal nps, or other optical properties, related to absorption or scattering, could also be addressed in perspective.

By optical emission spectroscopy it is possible to study the following properties of emitted light: (1) energy (wavelength); (2) intensity; (3) polarization. Time-resolved measurements can measure the temporal evolution of such quantities, while parametric measurements can establish how they vary with respect to temperature, applied fields, applied stress, and other quantities that can be varied in a controlled way.

This optical information may be related to the structural properties accessible by APT, and the measurement can be in principle correlated to other complementary microscopy techniques. In the following sections, the way in which structural properties determine optical properties will be made more explicit in the case of semiconductor heterostructures. The information provided here is necessarily succinct, but the provided references guide the reader toward specific and more detailed didactical literature.

Bulk Semiconductors

In bulk semiconductors the fundamental optical transition in the NIR-VIS-UV range is the band-gap transition. Bulk semiconductors strongly absorb photons with energy higher than the band gap, while the absorption of lower-energy photons is weaker. The absorption strength may vary from direct-band-gap semiconductors, in which the photon may be absorbed directly without creating or absorbing lattice vibrations (phonons) and which exhibit a stronger absorption, to indirect semiconductors, in which a phonon must be absorbed or emitted to have the photon absorbed. The dispersion relation (also called E-k diagram) of a bulk direct semiconductor is schematically depicted in Fig. 11.1A. The conduction and valence bands are approximately represented as parabolas, while the distance between the top of the valence band and the bottom of the conduction band is the energy band gap (or shortly, gap) E_g. However, the emitted or absorbed photons may have energies slightly differing from E_g because of many-body effects. Free electron—hole pairs, for instance, may bind together to hydrogen-like states called excitons, and then recombine radiatively producing photons with an energy $h\nu < E_g$ [2].

Polytypes and alloys. The band gap is characteristic of the material, and depends on its chemical composition and on its crystal symmetry. Polytypes with identical compositions but different crystal symmetry will have different band gaps (eg, SiC: 2H, 4H, 3C have $E_g = 3.33$, 3.265, 2.39 eV, respectively [3]). The band gap can be continuously tuned by controlling the chemical composition of binary, ternary, or quaternary alloys (eg, Si_xGe_{1-x}, $In_xGa_{1-x}As$, $In_xGa_{1-x-y}Al_yAs$) during the synthesis of the material. For a large class of alloys, the band-gap dependence on the composition is a simple linear or quadratic function of the alloy fractions [2] but more complex relationships must be considered if the alloys exhibiting spontaneous ordering, such as dilute nitrides (eg, $GaAs_{1-x}N_x$) [4]. The composition of a semiconductor alloy measured by APT is thus a quantity that could be related to its optical emission or absorption properties.

Doping. The insertion of donor or acceptor impurities does not have an important effect on the band gap of a semiconductor, unless the doping density attains particularly high values, giving rise to phenomena such as the Burstein—Moss shift [2], or band-gap renormalization due to the self-energy of free carriers themselves. Such impurity densities depend on the properties of the host material, and can be in principle measured by APT [2].

Optically active deep levels. Further emission mechanisms may arise if specific defects induce the presence of electron states within the gap (deep levels, at energy E_{DL}). Some of these levels may yield radiative recombination. Examples for this behavior are the intrinsic defects giving rise to deep luminescence bands in GaN, or luminescent impurities, such as rare earth ions, which can be inserted into the host matrix during the synthesis [5]. This impurity concentration, if higher than $\sim 10^{18}$ cm^{-3}, can be studied by APT.

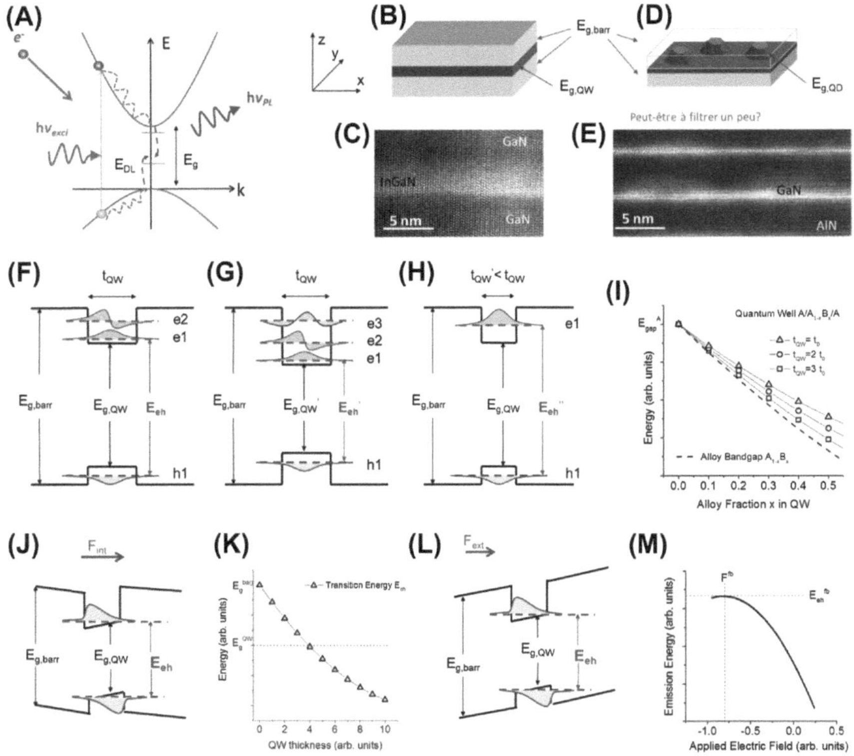

Figure 11.1 (A) Scheme of electronic state excitation by photons or electrons with energy above the band gap of a semiconductor, subsequent relaxation of electron (hole) to the bottom of the conduction (valence) band, and final radiative recombination of carriers yielding the emission of a photon with energy roughly equal to the band gap. (B) Scheme and (C) high-resolution scanning transmission electron microscopy image of a quantum well. (D) Scheme and (E) high-resolution scanning transmission electron microscopy image of Stranski-Krastanov quantum dots. (F) Potential profile and confined electron and hole energies and wavefunctions (subbands) within a rectangular quantum well of a given composition and thickness (expressed as a multiple of an arbitrary t_0). (G) and (H) Show how the energy of the states and of the main radiative recombination evolve with quantum well composition and thickness, respectively. (I) Schematic behavior of the recombination energy within an $A_{1-x}B_x$ quantum well within two barriers A as a function of alloy fraction and quantum well thickness. The alloy band gap is also traced for reference. (J) Potential profile and confined electron and hole energies and wavefunctions in a rectangular quantum well exhibiting the buildup of an internal field. (K) Schematic dependence of the transition energy E_{eh} in a QW in which an internal field is present. (L) Potential profile and confined electron and hole energies and wavefunctions in a rectangular quantum well to which an external field is applied. (M) Schematic dependence of the transition energy E_{eh} on the intensity of the applied field. E_{eh}^{fb} and F^{fb} denote the transition energy at flatband and the field needed to induce a flatband configuration of the quantum well.

Heterostructured Semiconductors and Low-Dimensional Systems

In semiconductor heterostructures, semiconductors with different band gaps are arranged to occupy different spatial regions. The primary goal of this approach is the so-called band engineering, ie, the possibility of designing semiconductor systems with specific electronic properties, which constitute the functional building block of electronic and optoelectronic devices such as high electron mobility transistors, light-emitting diodes, laser diodes, single photon emitters, or photodetectors. In many cases, such specific properties are obtained by confining the electron and/or hole states in extremely narrow planes (several nanometers), cylinders, or islands. This confinement makes wave-like properties of electrons in solids measurable, and is therefore a quantum phenomenon [6]. In the following, we will summarize the properties of selected heterostructure systems and the influence of the structural properties accessible by APT on their optical emission properties.

Quantum Wells

In the simplest picture, such as that represented in Fig. 11.1B, a QW is a thin flat layer (thickness t_{QW}) of semiconductor with a gap $E_{g,QW}$ synthesized between two, generally thicker, layers of semiconductor with band gap $E_{g,barr} > E_{g,QW}$. An example of InGaN/GaN QW observed by high-resolution scanning transmission electron microscopy (HR-STEM) is also reported in Fig. 11.1C. If the bands of the different semiconductor phases align in such a way as that reproduced in the scheme of Fig. 11.1F–H, the band profile across the QW is such that both electrons and holes can be confined within the well. Carriers in a QW are confined only in the growth direction (here denoted as z), but they can freely move along the QW plane (x,y). They thus possess in-plane energy levels which depend on their in-plane wave vector, and are distributed in two-dimensional (2D) bands. The Schrödinger equation of the system may thus be separately solved for the in-plane and the on-axis components. The solution of the Schrödinger equation along the z-direction yields a set of discrete levels, called subbands. For electrons (holes) in a QW, the energy of the ground state subband differs from the minimum of the conduction band (from the maximum of the valence band) of the material constituting the well. This difference is called the confinement energy: it increases with the height of the potential barriers, and decreases for increasing QW thickness [6] as depicted in the schemes of Fig. 11.1F–H. The wavefunctions also become progressively less confined with decreasing barrier gap. Furthermore, a finite number of subbands can be accommodated in a QW with a finite barrier height. The main optical emission taking place within a QW involves the electron and hole ground states, and therefore is affected by the confinement. The transition energy is thus

$$E_{eh} = E_{1e} + E_{1h} + E_{g,QW}$$

The terms appearing in this quantity all depend on the QW composition. The terms $E_{1e,h}$ also depend on the barrier composition and on the QW thickness t_{QW}. The way optical emission energies vary with varying well composition and thickness are schematically depicted in Fig. 11.1G and H for a rectangular QW of composition $A_{1-x}B_x$ within two barriers of composition A.

Quantum Dots

QDs are islands of semiconductor with band gap $E_{g,QD}$ embedded into a semiconducting matrix of larger band gap $E_{g,barr} > E_{g,QD}$. There are different methods for synthesizing such structures, such as Stranski–Krastanov growth, droplet QDs, colloidal QDs, or QDs in nanowires. The scheme reported in Fig. 11.1D and the HR-STEM image of Fig. 11.1E illustrate the configuration of Stranski–Krastanov, whose strain-driven formation occurs on a wetting layer with the same composition of the dot deposited onto a flat substrate. In a QD the carrier wavefunctions are confined in the three dimensions. This no longer gives rise to energy quantization in one direction with the formation of subbands, but to the definition of discrete energy levels, analogously to those of isolated atoms.

In QDs, the electron and hole energies depend on the composition of the QD and of the barrier material and on the QD shape and size, similar to what happens in QWs. This should make clear why it is interesting to study them by APT.

Effect of an Electric Field Across the Heterostructure

In semiconductor quantum confining systems, the PL emission energy as well as the absorption edge may be influenced by the presence of an electric field. This field could be (1) an internal field, such as in polar heterostructures, or (2) an externally applied field. This is known as the quantum-confined Stark effect (QCSE) [7], and it can be exploited in optoelectronic devices such as modulators and tunable absorbers [8].

Case (1) is schematically illustrated in Fig. 11.1J and K, and may occur in semiconductors having crystal directions without inversion symmetry, such as the [0001] in the case of wurtzite crystals. The most studied system exhibiting polar properties leading to QCSE is the III-N system. If the different heterostructure phases have different pyroelectric properties, a static sheet charge builds up at heterointerfaces. If, moreover, a heterostructure phase is coherently strained on another, the strain may set up a piezoelectric polarization. In nitride quantum wells and dots, sheet charges may be of the order of $\sim 10^{-2\circ}\mathrm{C/m^2}$ and internal fields of the order of several V/nm [9]. The effect of the electric field is to reduce the electron and hole energies with respect to the flat-band case, and also to spatially separate the wavefunctions of electron and hole. The radiative recombination energy, intensity, and lifetime are thus affected. The modification to the energy levels also depends on the thickness of the quantum confining structure.

Concerning case (2), illustrated in Fig. 11.1L and M, an electric field can be applied externally on a quantum confining system. Most frequently, this is done by embedding the system in a p-n or Schottky junction, but QCSE induced by contactless techniques has been demonstrated as well [10]. In this way, it is possible to study the parametric dependence of the radiative transition energies and lifetime on the applied field. The maximum transition energy is obtained for the flat-band configuration.

Typically, quantum confining structures exhibit measurable Stark shifts with field values of the order of some fractions of volts per nanometer. The magnitude of such fields should be compared with the surface field applied to an atom probe tip, which is of the order of some tens of volts per nanometer: if only a fraction of this high field would penetrate inside a field emission tip containing an optically active quantum-emitting system, a significant Stark shift would be induced in its transition energies.

EXPERIMENTAL APPROACHES

The connection between optical spectroscopy and atom probe can be established by different means, depending on the available experimental facilities, on the desired degree of accuracy and on the system under study. In this chapter, the experimental approaches have been classified as follows:

- Statistically correlative approaches, in which different (possibly nanoscale) subsystems of a given macroscopic system (a thin film, or a nanowire array, for instance) are studied separately by optical spectroscopy and APT. This approach has become common in semiconductor nanosciences, as it does not present any major blocking point from the experimental point of view.

- Strictly correlative approaches, in which the same nanoscale system is studied before by optical spectroscopy, and after by APT. This approach is challenging, as it is not obvious to prepare an atom probe sample (or a system with size close to an atom probe sample) preserving its optical properties. Also, not any optically interesting nanoscale system is adapted to a correlated study.
- Coupled, in situ approach, in which an optical spectroscopy bench is implemented in an atom probe instrument, allowing for in situ optical study within the atom probe itself. This approach has not been demonstrated yet, but its feasibility is under study.

The main features, advantages, and drawbacks of these three classes of approach are summarized in Table 11.1 and will be discussed in the following sections, reporting and explaining selected examples issued from recent studies.

The Statistically Correlative Approach

By statistically correlative study is meant a set of experiments performed by optical spectroscopy and APT on different parts of the same macroscopic sample. In practice, this could consist, for instance, in studying the macroscopic PL from a large surface of a thin film sample and in the extraction of a field emission tip to study by APT, or, as another example, in the study of a given set of ensemble or individual nanowires by optical spectroscopy, and in parallel by the APT study of another set of nanowires. Thus, the comparative experiment relies on the assumption that the subsystems studied by optical spectroscopy and APT are statistically representative of the overall properties of the macroscopic system studied. This does not mean that it is not possible to draw conclusions on the microscopic causes of the optical properties studied, only that the conclusions will be statistical in nature, and that a more complex approach is needed to really correlate the structural properties of a given microscopic system imaged by APT and its own, individual optical properties. However, the comparative approach has already proven to be very powerful for understanding the influence of structure on the optical properties of a larger and larger class of systems. Up to now, a certain number of statistically correlative studies have been performed on different semiconductor heterostructures [11–15]. One of them will be illustrated in more detail in the following.

Table 11.1 Experimental Approaches for Connecting Optical Properties and Atom Probe Data of Semiconductor Nanostructures

Experimental Approach	Description	Available Information	Limitations
Statistically correlative	Optical spectroscopy and APT performed on different nanoscale parts of the same macroscopic sample	Optical ensemble properties, selected structural properties of nanoscale samples	The connection between optics and structure is statistical
Strictly correlative	Optical spectroscopy and APT performed subsequently on the same nanoscale object	Close correspondence between optical and structural properties In the case of a single emitter, exact correspondence Further correlation with HR-TEM possible	Extremely careful sample preparation needed, high risk of failure due to multiple sample manipulations Possible negative impact of FIB on optical properties Correspondence lost if optical emitter is outside the APT field of view
Coupled[a]	Optical spectroscopy and APT performed on the same nanoscale sample in situ within the atom probe	Compact correlative approach Minimization of manipulation-related specimen fracture risk Possible spectral profiling during tip erosion. Exact correspondence between optical and structural properties even for multiple emitters within a single sample Study of effect of the applied DC electric field on the optical properties	Same limitations as in the correlated approach concerning FIB and field of view Need of matching sample temperature, applied field and incident laser energy to simultaneously perform APT and optical spectroscopy Further correspondence with HR-TEM not possible if optical properties need to be preserved

APT, Atom probe tomography; *FIB*, focused ion beam; *HR-TEM*, high-resolution transmission electron microscopy.
[a]This approach is currently under development.

Si/SiO₂ Nanoparticles

The study presented in this section deals with Si nps embedded in a SiO_2 matrix. The system is schematically depicted in Fig. 11.2A and B. The nanometer-sized particles have a strongly enhanced radiative recombination efficiency with respect to bulk Si: due to quantum confinement, the electron—hole carrier overlap increases and even this indirect band-gap system may yield a significant luminescence signal. Si nps may be assimilated to QDs, but with the important difference concerning the interfaces with the matrix, which in nps are not coherent. In the system under study, nps are obtained upon annealing of Si-rich SiO_x layers, having a thickness t_x, interleaved with stoichiometric SiO_2 layers of thickness t_b. The excess Si tends to segregate at high temperature and to form small clusters. The shape and the size of the nps depend on both layer thicknesses and annealing treatment, and determine the optical properties of the system. Such nps form a crystalline subsystem inside an amorphous matrix, and the crystal orientation is random. Furthermore, the elemental contrast with the matrix is low. These elements make these nps very elusive with respect to transmission electron microscopy (TEM) characterization [16,17].

The PL spectra of selected nanoparticle systems with different layer thicknesses after different annealing treatments are displayed in Fig. 11.2C—E, while the APT reconstruction of a system subvolume of the structures annealed at 1100°C is reported in Fig. 11.2F—H. Fig. 11.2C reports the spectra of a structure having $t_x = 4$ nm and $t_b = 3$ nm. The as-deposited structure produces no luminescence, as expected of uniform SiO_x layers. Annealing 1 h at 1000°C yields a measurable PL signal, which is then greatly enhanced if annealing takes place at 1100°C. The 3D APT reconstruction reported in Fig. 11.2F shows that after annealing at 1100°C the system contains well-shaped and well-defined nps disposed on planes separated by the thickness of the SiO_2 barriers. The structure shown in Fig. 11.2D and G is similar to the previous one, but the barrier thickness is only $t_b = 1.5$ nm. Such thickness is too low to prevent Si interlayer diffusion, and the APT analysis indeed shows that there are still nps, but Si segregation often takes place over different SiO_x layers and inside SiO_2 barriers, so most of the Si clusters are too large to yield a good carrier overlap. Furthermore, nps are larger than in the previous structure. This translates into lower luminescence intensity, and in a slight red-shift of the PL spectra. Finally, a structure with too large SiO_x layers ($t_x = 10$ nm), as shown in Fig. 11.2E and H yields too large nps with no PL signal at all [18].

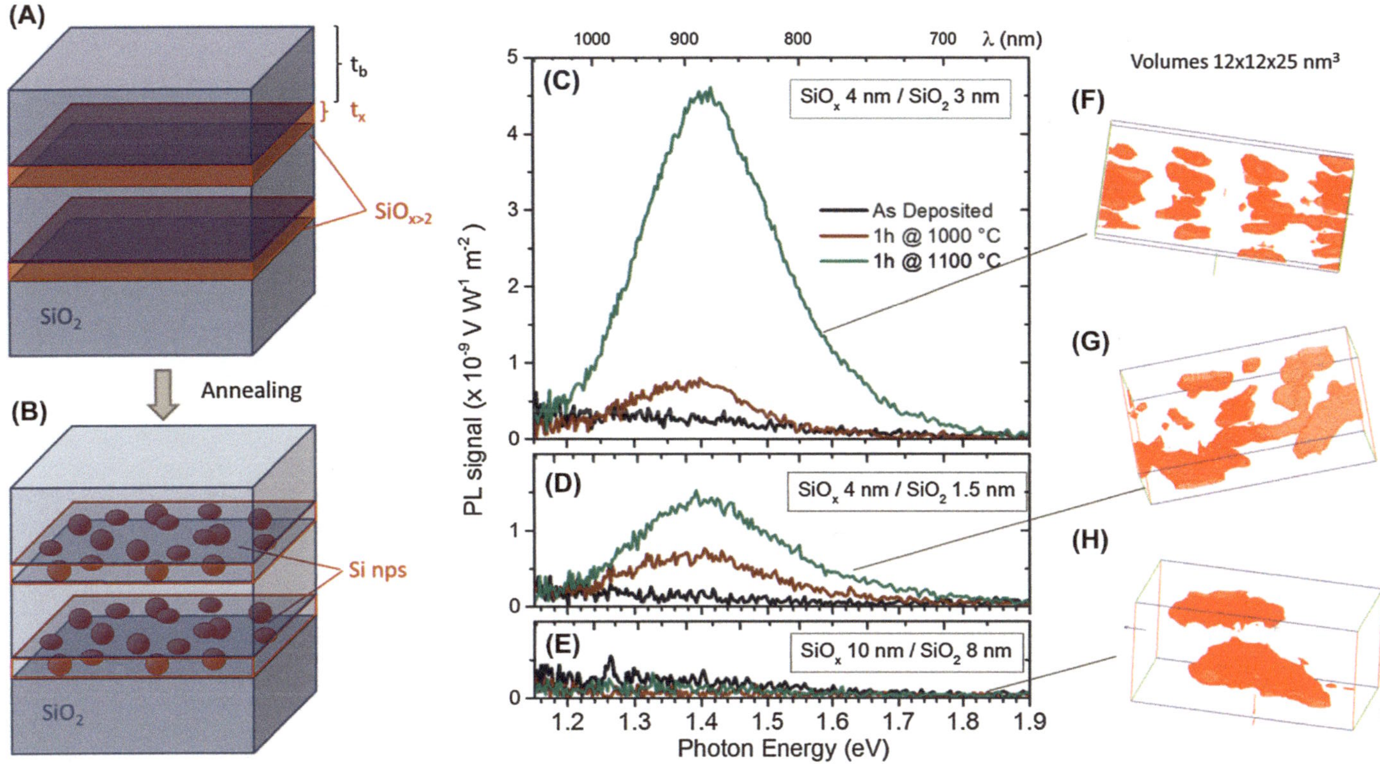

Figure 11.2 (A,B) Scheme of an $SiO_2/SiO_{x>2}$ multilayer before and after annealing. The SiO_x layers (31 at.% oxygen excess with respect to SiO_2) may form SiO_2 layers containing Si clusters or nanoparticles upon thermally induced segregation. Si nanoparticles may yield a photoluminescence signal depending on their density and size. (C–E) Photoluminescence spectra of structures having three different SiO_x and SiO_2 layer thicknesses after 1 h annealing treatments at increasing temperatures and (F–G) the three-dimensional atom probe tomography reconstruction of the Si-rich clusters/nanoparticles samples with three different SiO_x and SiO_2 layer thicknesses after 1 h annealing at 1100°C. *Adapted from Roussel M, Etude et modélisation de la précipitation de nanoparticules de silicium dans des couches de silice nanométriques, [Rouen: Ph.D. thesis]. 2012, Courtesy of Talbot E.*

Beyond the Statistically Correlative Approach

A strictly correlative experiment at the nanoscale consists in the study with different techniques of the same nanoscale object, or of a set of nanoscale objects, each of which is addressed individually by the two techniques. The gain in information that can be achieved by this approach is great: the link between the optical and structural information that can be achieved by comparative approaches is statistical in nature, while a correlative approach establishes individual links at the nano-object level. This is why a lot of different correlative approaches are flourishing in the nanosciences, and this is why there is an increasing number of studies reporting on the correlation between optical emission [19] or optical extinction properties [20] and TEM. However, if one considers correlating an optical study with APT, one becomes rapidly aware that the task is not as straightforward as for a correlation with TEM.

The main reasons for this are the following: on one hand it is difficult to study a given isolated nano-object, subsequently transfer it onto a support tip, and have it analyzed by atom probe; on the other hand it is not obvious to prepare a nanoscale specimen on a support tip that could be studied by optical spectroscopy before and by APT after. The first difficulty is mainly related to nanomanipulation issues, the second one to the use of focused ion beam (FIB) and its heavy impact on the optical properties of nanoscale objects.

The FIB Bottleneck

The main obstacle to the implementation of a correlated approach is related to the use of FIB for the preparation of atom probe semiconducting and dielectric specimens. The preparation techniques based on FIB have been extensively treated in the chapter on FIB and the reader can refer to it for further detail. Fig. 11.3 illustrates how FIB can affect the optical properties of a given sample. The first important consideration is that any exposure of a specimen to FIB, at any typical ion energy, produces some extent of implantation. Fig. 11.3A shows the example of the trajectories of Ga ion of different energy impinging at normal incidence onto a GaN volume, as simulated by the stopping and range of ions in matter (SRIM) [21]. Ions with energy ~2 keV are implanted on a relatively reduced thickness of several nanometers, this thickness significantly increasing to over 10 nm for 30-keV ions, which is the typical energy used for specimen milling operations. It has been clearly shown in

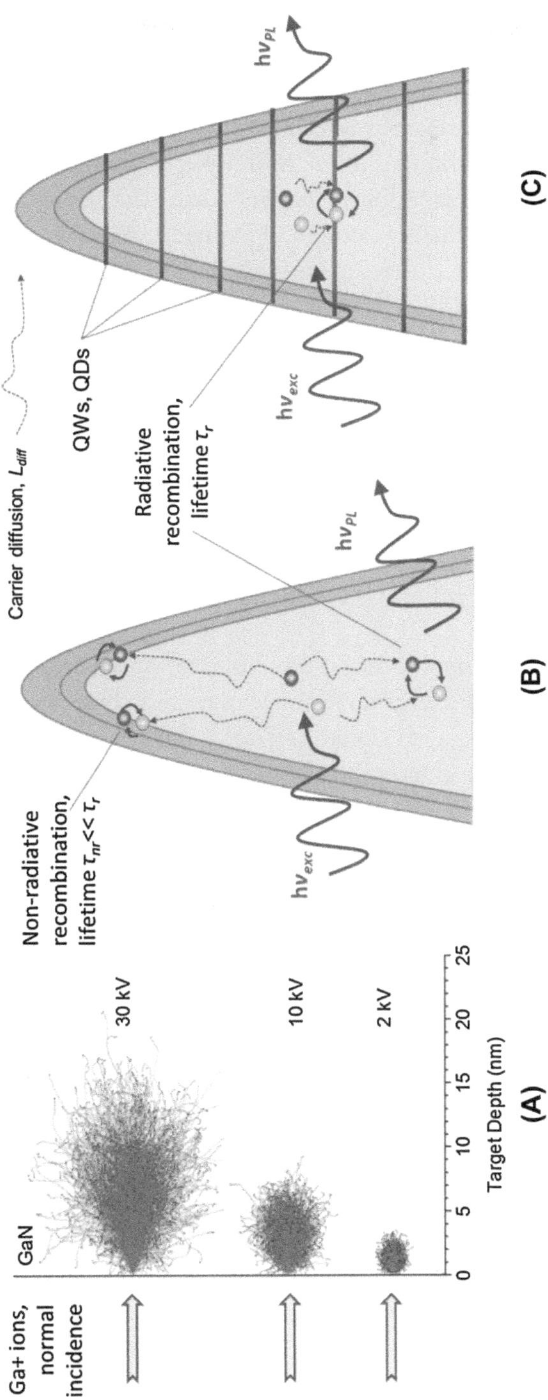

Figure 11.3 (A) Trajectories of Ga ions impinging at normal incidence on GaN and at different energies, as given by SRIM calculation [21]. (B,C) Schematics of carrier generation, radiative and nonradiative recombination processes in a tip sample having a damaged surface: (B) represents a homogeneous tip, (C) a tip including quantum confining structures. *QD*, quantum dot; *QW*, quantum well.

Ref. [22] that the ion doses used for the preparation of Si slices, which do not differ too much from those used in APT specimen preparation, induce the amorphization of an external layer, whose thickness is around 2 nm for 2-keV ions and around 20 nm for 30-keV ions. This latter value is already close to the size of an APT tip specimen. Moreover, ion implantation can take place even further than the thickness of the amorphized region, producing structural damage, extended and point defects. The effect of this on the radiative recombination properties of a sample are illustrated in Fig. 11.3B in the case of a uniformly crystalline sample. As the crystalline properties of the sample are preserved far from the surface, generation of electron–hole pairs is still possible upon absorption of a photon with energy higher than the band gap. However, the created carriers can easily diffuse toward the damaged surfaces, where their nonradiative lifetime τ_{nr} dramatically decreases because of the high density of nonradiative recombination centers. The radiative lifetime τ_r is unaffected with respect to the bulk material, so the relationship $\tau_{nr} \ll \tau_r$ translates into a very efficient decrease of luminescence efficiency.

Two important considerations should guide the operator. The first one, is that independent of the implantation depth—and therefore of the ion energy—the presence of a damaged surface is sufficient to increase the nonradiative recombination rate. Decreasing the ion energy may thus limit the damage, but would never totally resolve the problem of surface degradation of optical properties. The second consideration is that the effect of FIB is strong because APT specimen tips have a large surface-to-volume ratio: the surface is always very close to the region where carriers are generated, and diffusion toward the surfaces is an important process for the enhancement of nonradiative efficiency. So, if a sample contains some structure capable of reducing the diffusion length, this would help prevent the carriers from reaching the surface. Furthermore, if the diffusion-limiting structure also could locally enhance the radiative recombination efficiency, the effect of FIB-induced damage would not have a strong impact. This situation is depicted in the scheme of Fig. 11.3C. In this case, the presence of a multi-QW system is meant to both limit the carrier diffusion [23] and to increase the recombination probability once the carriers are captured by the QWs. Another possibility, not shown here, is the so-called resonant excitation, consisting in creating carriers directly within quantum confining subsystems (quantum wells or dots), from which they could not freely diffuse. In this case, the incident light should have well-determined photon energy, between the e-h1 transition energy and the barrier energy band gap. For the interested reader, a quantitative study of the effect of FIB on the luminescence properties of ZnO microwires is reported in Ref. [24].

What Should Be Kept in Mind if FIB Is Necessary

As a conclusion of this section, some recommendations are given for an implementation of FIB protocols as respectful as possible of the optical properties of the samples under study:

- FIB resistance of the material. Different materials or nanoscale systems are differently resistant with respect to the use of FIB. If this is not known from previous studies, the user should preliminarily test how the FIB affects the optical properties of the system under study. Systems in which the optically active regions are isolated from the surfaces should prove more resistant.
- Ion penetration depth. The operator should keep in mind what is the penetration depth of the FIB in the material under study; such information can accurately be retrieved through SRIM simulation.
- As much protection as possible. The optically active region of the system should be protected as long as possible during sample milling, for instance, through preliminary capping with a passive layer.
- Soft finalization. The final steps of ion milling should be performed with low ion energies (below 10 kV). This will not avoid the presence of a damaged surface layer, but it will limit its impact as much as possible. However, the spatial accuracy of annular milling decreases with decreasing ion energy (the beam becomes less focused and the tip shape develops a larger cone angle) so the operator should find good compromise between sufficient spatial accuracy and reduced ion penetration.

A Correlative Study in InGaN/GaN Nanowire Quantum Wells

This section illustrates an example of correlated optical–structural study of the same nanoscale object by μPL and APT. For sake of completeness, it should be mentioned that the nanoscale object was prepared with a two-step FIB-based protocol: the optical properties of a cylinder of 500 nm diameter have been studied; later on, the cone was sharpened down to a field-emission tip size. Therefore, future studies could still increase the accuracy of the correlation. It also should be mentioned that the presented nanoscale object has undergone a further correlation with HR-STEM, this analysis being performed between μPL spectroscopy and APT. Data relative to HR-STEM and further details can be found in Ref. [25] and in its supplementary information.

Microphotoluminescence

The μPL study of the radial multi-quantum well (MQW) system is described in Fig. 11.4. The microwires studied were grown by metal organic chemical vapor deposition. The growth is performed so that a thinner core is synthesized, then the growth conditions are changed to realize a conformal growth on the upper part of it. During conformal growth, a set of 20 QWs is inserted by modulating the Ga and In precursor fluxes. The external appearance of the wire is displayed in the SEM image of Fig. 11.4A, while Fig. 11.4B reports a scheme illustrating the QW arrangement. Fig. 11.4B

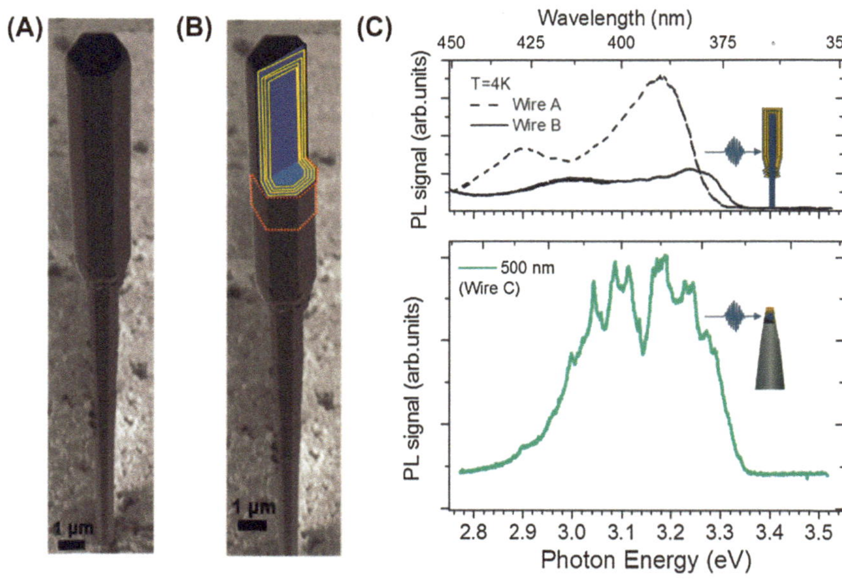

Figure 11.4 (A) Scanning electron microscopy image of one of the analyzed microwires; (B) scheme of the multishell quantum well system, with the *red dashed line* approximately indicating the region from which nanochunk C was extracted; (C) microphotoluminescence spectra of two whole nanowires A and B mounted on a tungsten tip (*black lines*, upper part of the graph) and of nanochunk B, which have been subsequently analyzed by atom probe tomography. *Adapted with permission from Rigutti L, Blum I, Shinde D, Hernàndez-Maldonado D, Lefebvre W, Houard J, et al. Correlation of microphotoluminescence spectroscopy, scanning transmission electron microscopy, and atom probe tomography on a single nano-object containing an InGaN/GaN multi-quantum well system. Nano Lett 2014;14(1):107–14.*

reports the μPL spectra recorded under 244 nm cw laser excitation at 4K. The spectra displayed in the upper part of the graph were collected from whole wires, while those in the lower part were issued by the nanoscale sample, preliminarily milled down by FIB to a cylinder of around 400 nm diameter. The PL spectra of the whole wires consist of a large band, showing two main components but no particular fine structure. The situation is different when analyzing nanochunks. Here narrow lines appear, distributed in the interval 2.9—3.3 eV: this points out that a set of emitters with well-defined energy is contained in the nanochunk. This difference depends on the probed volume: even if tightly focused, the laser illuminates a much larger microwire volume when impinging on whole wires than when it illuminates the nanoscale chunks. The appearance of discrete and narrower peaks points out to the formation of regions with different structural properties, in which the carriers could localize and radiatively recombine. The emission energy would then be primarily determined by the size and composition of the specific emitting region.

Atom Probe Tomography

The MQW system was analyzed by APT in a Lawatap, at temperature $T = 20$K, with an impinging laser energy per unit surface 5—10 μJ/cm^2 and a Ga^{2+}/Ga^{1+} charge state ratio varying from 0.18 to 0.06 throughout the analysis, with a total of 33×10^6 ions detected. Fig. 11.5B displays the position of 10% In atoms in the 3D reconstruction. The leftmost reconstruction image is oriented so as to show the correspondence with the STEM image of part (A). A total of 16 QWs over 20 could be imaged by APT. The STEM image was profitably used to achieve the best possible reconstruction. Remarkably, a good reconstruction could only be obtained by postulating very low detection efficiency, of the order of 0.3. Further details about the III/N compositional measurement and about the efficiency issue in reconstruction can be found in Ref. [26] and in its supplementary information, respectively.

The comparison between STEM and APT shows that the apparently thicker QWs found in the particular zone axis adopted for the STEM appear now as bent QWs in the APT reconstruction, with QWs as thick as the surrounding ones. Fig. 11.5C shows a close-up of a cross-section of QW #4, indicating that the QW interface on the core side is better defined than that on the surface side, a feature probably related to the growth method [27,28]. The QWs are thus asymmetric, and this also has an effect on the electronic band structure.

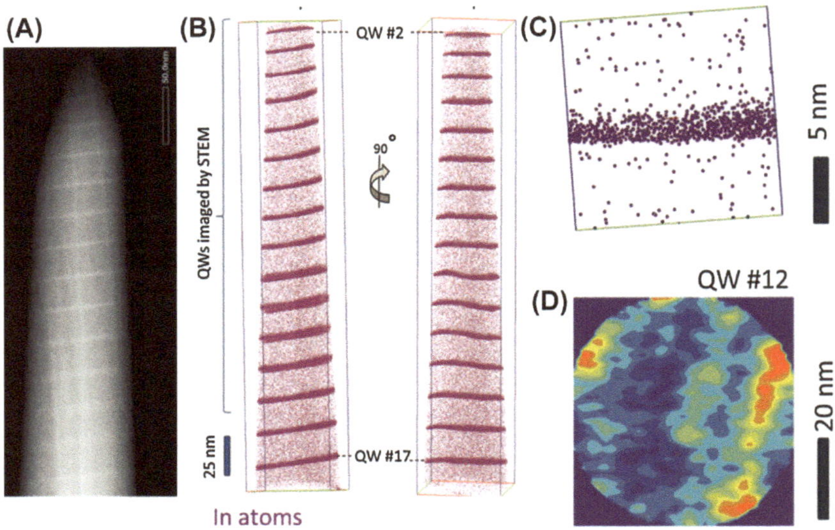

Figure 11.5 (A) High angle annular dark field—scanning transmission electron microscopy (STEM) image of the tip milled out of nanochunk C: the bright contrast nearly *horizontal lines* correspond to InGaN quantum wells (QWs) inside a GaN matrix; (B) atom probe tomography three-dimensional reconstruction of 16 of 20 QWs contained in the tip. Only 10% In atoms are shown for clarity; (C) close-up of a region of QW #4; (D) two-dimensional density plot of the InN fraction calculated in a rectangular box containing QW #12. *Adapted with permission from Rigutti L, Blum I, Shinde D, Hernàndez-Maldonado D, Lefebvre W, Houard J, et al. Correlation of micro-photoluminescence spectroscopy, scanning transmission electron microscopy, and atom probe tomography on a single nano-object containing an InGaN/GaN multi-quantum well system. Nano Lett 2014;14(1):107–14.*

It can be shown that the average In content at the bottom of the QWs is of 8–13%. However, the distribution of In on-plane is not uniform, as visualized in the 2D plot of Fig. 11.5D showing the 2D In/(In + Ga) concentration averaged over the thickness of a rectangular box containing the whole QW. The actual InN fraction can be correctly measured in 3D. The analysis was performed by isolating 1-nm-thick slices within the well, approximately parallel to the m-plane, and calculating the InN fraction for each 2D slice. The measurements indicate that the peak InN fraction is around 20%, which is much larger than what could be obtained by one-dimensional (1D) profiling due to averaging effects between high- and low-In content regions. Furthermore, the distribution of In in the QWs not illuminated by the electron beam in STEM but imaged by APT is similar to the QWs imaged by STEM, indicating that the exposure to the electron beam was sufficiently short not to induce an artificial In rearrangement within the system [29].

Summary: What Can Be Extracted From a Correlated μPL-APT Analysis on the Same Nano-Object

All information described earlier allows thus for a consistent interpretation of the relationship between structural and optical properties of the nanoscale object analyzed. The main conclusions could be drawn by considering the plot in Fig. 11.6, which reports on the ordinate the PL emission energy and on the abscissa the alloy fraction x of InN in $In_xGa_{1-x}N$. The different curves represent the calculation of the PL energy of interband transitions taking place in relaxed m-plane nonpolar rectangular QWs of different thickness as a function of the InN fraction. The

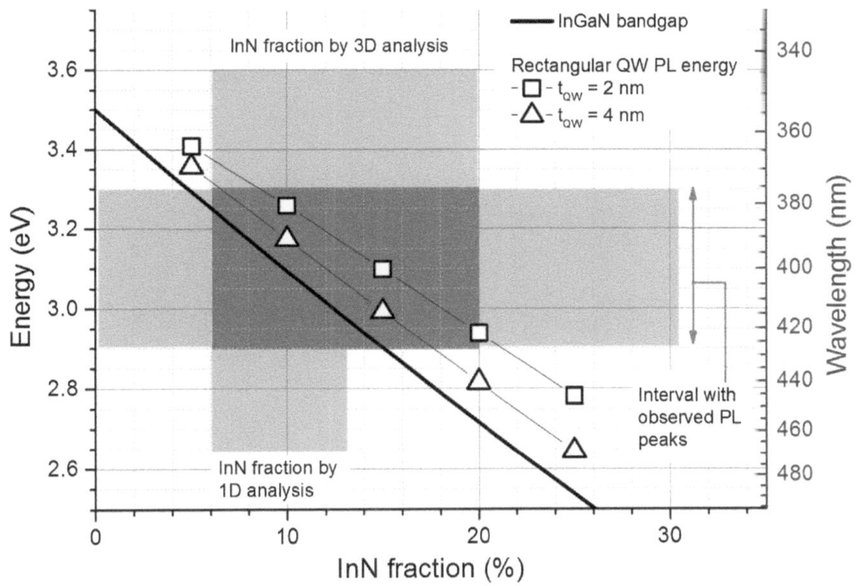

Figure 11.6 Dependence on InN fraction of the InGaN band gap (*black line*) and of the InGaN/GaN photoluminescence (PL) transition energies calculated for rectangular quantum wells of different thickness $t_{QW} = 2$ nm (squares), and $t_{QW} = 4$ nm (triangles). The gray shaded regions correspond to the interval in which the PL narrow lines have been observed in the analyzed nanochunk from wire C, and to the In content found by one- and three-dimensional profiling; the microphotoluminescence spectrum of the nanochunk is also reported on the left-hand side axis (*black line*). *Adapted with permission from Rigutti L, Blum I, Shinde D, Hernàndez-Maldonado D, Lefebvre W, Houard J, et al. Correlation of microphotoluminescence spectroscopy, scanning transmission electron microscopy, and atom probe tomography on a single nano-object containing an InGaN/GaN multi-quantum well system. Nano Lett 2014;14(1):107–14.*

band gap of bulk InGaN is also reported as a reference. In the plot are also highlighted the InN fractions found by 1D profiling (up to 13%) of APT data and by 3D measurement (up to 20%).

The multiple peaks correspond to regions where InN segregation creates local potential wells in which both electrons and holes localize, in a QD-like confining system. The dispersion in the peak energies correspond to the different In content in these regions. The 3D measurement of InN fraction can reasonably account for recorded PL energies, while the lowest energies could not be explained by a simpler 1D profiling. However, it must be underlined that these conclusions neglect the presence of extended defects (stacking faults) which correspond to the nearly vertical bright contrast lines of Fig. 11.5A. These defects are optically active and contribute to a lowering of the energy of e-h transitions of about 0.1 eV [30].

Perspectives: The Coupled, "In Situ" Approach

The scheme of the principle of a coupled μPL and tomographic atom probe instrument is sketched in Fig. 11.7. The basic idea is to exploit the femtosecond laser pulse to trigger the evaporation of single ions from a field emission tip placed in an intense electric field and, at the same time, to generate electron–hole pairs that can recombine radiatively yielding a luminescence

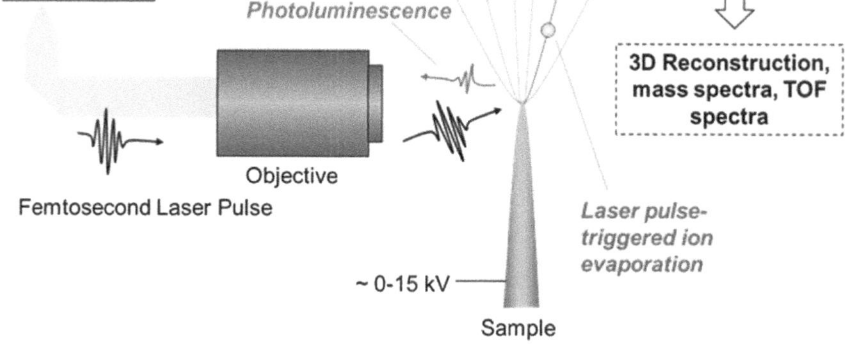

Figure 11.7 Scheme of an in situ microphotoluminescence bench inside a tomographic atom probe.

signal. It must be noticed here that most atom probe experiments are performed keeping the sample base at a temperature varying between 20 and 100K. This is also a favorable interval for the analysis of the PL signal, as radiative recombination is favored at low temperature due to the freeze-out of nonradiative channels. A fraction of the emitted photons, depending on the sample geometry and optoelectronic properties, is collected by the same objective and focused on the entry slit of a spectrometer provided with a spectrally and time-resolved detector, such as a streak camera. It thus becomes possible to analyze the time decay of the PL signal in a selected spectral interval, whose resolution and width may be adjusted through the settings of the spectrometer.

This approach also has the following limitations with respect to a simple correlative approach:

- It is not possible to perform a preliminary study by HR-STEM, as the electron energies used (>100 kV), still preserving the structure of the systems, induce a deterioration of the optical properties, for instance, strongly reducing the radiative recombination efficiency.
- The possibility of performing μPL and APT simultaneously critically depends on the matching of experimental parameters. If it appears convenient to perform both analyses at low temperature and with a laser wavelength exceeding the specimen band gap, further checks must be performed concerning the laser pulse energy and the applied DC bias.

However, the coupled approach may still prove to yield a precise correlation of optics and structure, and a new insight on the physics of field ion evaporation. The main reasons to implement it are the following:

- Measuring μPL in situ in an atom probe allows performing strict correlations between optics and structure in a more compact and straightforward way, and minimizes the manipulation of the specimen between different experiments, decreasing the risk of specimen failure.
- It becomes possible to profile the optical properties of the sample during the evaporation, and thus to attribute spectral features to specific sample regions.
- It would be possible to study the QCSE in an original way, as the field penetrating within the tip is expected to influence the emission properties of localized quantum emitters such as quantum wells and dots. Furthermore, the measurement of the field effectively penetrating within the tip specimen is of particular interest for the understanding of field ion emission physics in dielectrics [31,32].

- It would be possible to verify whether the strain induced by the applied field induces a strong deformation in the quantum emitters close to the surface, which would also have an effect on the optical transition energies.

Summarizing, optical spectroscopy can be carried out in a complementary way to APT in different ways and in different approaches. The correlation between optical and structural information may indeed explain how the optical properties of a given nanoscale optical system depend on the 3D chemical distribution of the different species, an information that only APT is so far capable of producing. On the other hand, future instrumental developments will show also whether optical spectroscopy can be applied to the study of field ion evaporation itself, yielding experimental answers to questions that are still open in the atom probe and field ion emission community.

REFERENCES

[1] Gault B, Vurpillot F, Vella A, Gilbert M, Menand A, Blavette D, et al. Design of a femtosecond laser assisted tomographic atom probe. Rev Sci Inst 2006;77:043705.
[2] Grundmann M. The physics of semiconductors. Berlin: Springer; 2010.
[3] Choyke WJ, Pensl G. Physical properties of SiC. MRS Bull March 1997:25−39.
[4] Mascarenhas A, Francoeur S, Yoon S. Physics of isoelectronic dopants in GaAs. In: Henini M, editor. Dilute nitride semiconductors. Elsevier; 2005. p. 179.
[5] Ziani A, Davesnne C, Labbé C, Cardin J, Marie P, Frilay C, et al. Annealing effects on the photoluminescence of terbium doped zinc oxide films. Thin Solid Films 2014;553: 52−7.
[6] Davies. The physics of low-dimensional semiconductors. Cambridge University Press; 1998.
[7] Miller DAB, Chemla DS, Damen TC, Gossard AC, Wiegmann W, Wood TH, et al. Band-edge electroabsorption in quantum well structures: the quantum-confined Stark effect. Phys Rev Lett 1984;53(22):2173.
[8] Kuo YH, Lee YK, Ge Y, Ren S, Roth JE, Kamins TI, et al. Strong quantum-confined Stark effect in germanium quantum-well structures on silicon. Nature 2005;437(7063): 1334−6.
[9] Bernardini F, Fiorentini V, Vanderbilt D. Spontaneous polarization and piezoelectric constants of III-V nitrides. Phys Rev B 1997;56(16):R10024.
[10] Kudrawiec R. Application of contactless electroreflectance to III-nitrides. Phys Status Solidi (B) 2010;247(7):1616−21.
[11] Benallali H, Cremel T, Hoummada K, Mangelinck D, André R, Tatarenko S, et al. Atomic scale investigations on CdxZn1−xSe quantum dots: correlation between the composition and emission properties. Appl Phys Lett 2014;105(5):053103.
[12] Giddins AD, Keizer JG, Hara M, Hamhuis GJ, Yuasa H, Fukuzawa H, et al. Composition profiling of InAs quantum dots and wetting layers by atom probe tomography and cross sectional scanning tunneling microscopy. Phys Rev B 2011;83:205308.
[13] Watson-Parris D, Godfrey MJ, Dawson P, Oliver RA, Galtrey MJ, Kappers MJ, et al. Carrier localization mechanisms in InGaN/GaN quantum wells. Phys Rev B 2011;83: 115321.
[14] Mancini LEA. Three-dimensional nanoscale study of Al segregation and quantum dot formation in GaAs/AlGaAs core-shell nanowires. Appl Phys Lett 2014;105(24):243106.

[15] Jeon N, Loitsch B, Morkoetter S, Abstreiter G, Finley J, Krenner HJ, et al. Alloy fluctuations Act as quantum dot-like emitters in GaAs-AlGaAs core—shell nanowire. ACS Nano 2015;9(8):8335.

[16] Roussel M, Talbot E, Gorbilleau F, Pareige P. Atomic characterization of Si nanoclusters embedded in SiO_2 by atom probe tomography. Nanoscale Res Lett 2011;6:164.

[17] Roussel M, Talbot E, Nalini RP, Gourbilleau F, Pareige P. Phase transformation in $SiOx/SiO_2$ multilayers for optoelectronics and microelectronics applications. Ultramicroscopy 2013;132:290—4.

[18] Roussel M. Etude et modélisation de la précipitation de nanoparticules de silicium dans des couches de silice nanométriques [Rouen: Ph.D. thesis]. 2012.

[19] Rigutti L, Jacopin G, Largeau L, Galopin E, De Luna Bugallo A, Julien FH, et al. Correlation of optical and structural properties of GaN/AlN core-shell nanowires. Phys Rev B 2011;83:155320.

[20] Billaud P, Marhaba S, Cottancin E, Arnaud L, Bachelier G, Bonnet C, et al. Correlation between the extinction spectrum of a single metal nanoparticle and its electron microscopy image. J Phys Chemistry C 2008;112:978.

[21] Ziegler JF, The stopping and range of ions in matter. [Online] Available: http://www.srim.org/; [accessed 12.01.15.].

[22] Mayer J, Giannuzzi LA, Kamino T, Michael J. TEM sample preparation and FIB-induced damage. MRS Bull 2007;32(05):400—7.

[23] Zagonel LF, Rigutti L, Tchernycheva M, Jacopin G, Songmuang R, Kociak M. Visualizing highly localized luminescence in GaN/AlN heterostructures in nanowires. Nanotechnology 2012;23(45):455205.

[24] Rigutti L, Vella A, Vurpillot F, Gaillard A, Sevelin-Radiguet N, Houard J, et al. Coupling atom probe tomography and micro-photoluminescence spectroscopy: exploratory results and perspectives. Ultramicroscopy 2013;132:75—80.

[25] Rigutti L, Blum I, Shinde D, Hernàndez-Maldonado D, Lefebvre W, Houard J, et al. Correlation of microphotoluminescence spectroscopy, scanning transmission electron microscopy, and atom probe tomography on a single nano-object containing an InGaN/GaN multi-quantum well system. Nano Lett 2014;14(1):107—14.

[26] Mancini L, Amirifar N, Shinde D, Blum I, Gilbert M, Vella A, et al. Composition of wide bandgap semiconductor materials and nanostructures measured by atom probe tomography and its dependence on the surface electric field. J Phys Chem C 2014; 118(41):24136—51.

[27] Köster R, Hwang JS, Durand C, Le Si Dang D, Eymery J. Self-assembled growth of catalyst-free GaN wires by metal—organic vapour phase epitaxy. Nanotechnology 2010;21:015602.

[28] Köster R, Hwang JS, Salomon D, Chen X, Bougerol C, Barnes JP, et al. Nano Lett 2011;11:4839.

[29] Bennett SE, Saxey DW, Kappers MJ, Barnard JS, Humphreys CJ. Atom probe tomography assessment of the impact of electron beam exposure on InGaN/GaN quantum wells. Appl Phys Lett 2011;99:021906.

[30] Mancini L, Hernández-Maldonado D, Lefebvre W, Houard J, Blum I, Vurpillot F, et al. Multi-microscopy study of the influence of stacking faults and three-dimensional in distribution on the optical properties of m-plane InGaN quantum wells grown on microwire sidewalls. Appl Phys Lett 2016;108(4):042102.

[31] Silaeva EP, Karahka M, Kreuzer HJ. Atom probe tomography and field evaporation of insulators and semiconductors: theoretical issues. Curr Opin Solid State Mater Sci 2013;17(5):211—6.

[32] Silaeva EP, Arnoldi L, Karahka ML, Deconihout B, Menand A, Kreuzer HJ, et al. Do dielectric nanostructures turn metallic in high-electric DC fields? Nano Lett 2014; 14(11):6066—72.

APPENDIX A

EXERCISES

1. Calculate the magnitude of the electric field of a 50-nm radius tip, subjected to an electrical potential of 10 kV. Field factor: β or $k_F = 5$.
2. Give the classical expression of the magnification of the APT and calculate its value for an ICF $= m + 1 = 1.6$ and a tip to detector distance (L) of 10 cm, tip radius being $R = 50$ nm.
3. We consider a Ni sample (FCC, $a = 0.35$ nm) in FIM. Calculate the physical radius r_1 of the first 001 terrace of Ni (pole radius) in FIM for a tip radius $R = 50$ nm, assuming a single atom remaining at the center of the terrace. We give $r_n \sim (2\,R \cdot d_{hkl} \cdot n)^{1/2}$, with n the index of the nth terrace. Demonstrate this equation. What is the radius of the second and third terrace?
4. Calculate the time of flight for an Al^{++} ion ($m = 27$ amu), in a standard AP, with $V = 5$ kV, $L = 10$ cm. In these conditions give an order of magnitude of the maximal pulse repetition rate to ensure a correct detection of this element.
5. What are the parameters restricting the detection rate (in atoms per second)?
6. Give the classical expression linking the evaporation rate K of a single atom at the sample surface of a field emitter, maintained at a temperature T, situated in a potential well $Q(F)$ with F the applied electric field (assuming the frequency of vibration of the atom at the sample surface).

   If Q is linearly dependent on the surface field F, and F_{EV} the critical field of evaporation, give an expression of the evaporation rate. Give a demonstration of this expression using the Müller Schottky model of field evaporation.
7. Give a definition of the evaporation flux at the tip surface, and of the detection rate. What are the differences between these expressions?
8. Why do we choose a detection rate <0.1 atom/pulse in the general case?
9. Explain why we need pulses to field evaporated a sample. What are the two kinds of evaporation modes? Explain this need using a graph giving the threshold field of evaporation as a function of the base temperature for a given detection rate.

10. Explain what are preferential evaporation and evaporation retention? Why these phenomena may bias the measurement of composition? In voltage mode, what are the conditions that can prevent these biases?
11. We investigate steel (BCC ferrite -$a = 0.29$ nm) containing 0.5 at.% carbon (X) and we want to study the carbon concentration fluctuations at the nanometer scale. Calculate the number of ions detected in an elementary volume sampling of 8 nm^3 (consider a detection efficiency $Q = 0.5$). How many carbon ions are detected in this volume (in average)? Give the probability to detect n carbon atoms in this volume. Using the average value, what is the uncertainty due to statistical sampling fluctuations? Repeat the calculation for iron containing 20% Cr (stainless steel). What are the conclusions?
12. An APT volume obtained from the analysis of an alloy composed of two species A and B presents some atomic planes (FCC, $a = 0.5$ nm). Using the desorption map, two poles are clearly seen and indexed ((002) and (111)). What is the angle between normal directions to the two poles? Lateral and depth precisions are $\Delta z = 50$ pm, and $\Delta L = 400$ pm. Is it possible to observe atomic planes from both sets of atomic planes (justify)? It is not possible to observe both sets of planes at the same location in the volume; can you explain why? What should be the required lateral resolution to image both sets at the same location?

PROBLEMS

Müller Model in *W*

We consider the Müller model (image force model) for the evaporation of a tungsten tip in atom probe tomography. Physical parameters related to *W* are provided below:

Element	Λ (eV) Sublimation Energy	I_1 (eV) Ionization Energy (1 Electron)	I_2 (eV) Ionization Energy (2 Electrons)	I_3 (eV) Ionization Energy (3 Electrons)	ϕ (eV) Work Function
W	8.66	7.98	17.64	24	4.54

1. Calculate tungsten evaporation field for each charge state following the Müller model. According to the model under which majority charge status should tungsten ions be observed? Experimentally, the evaporation field is estimated at 57 V/nm (80K for 0.1 atomic layer/s) and the charge is 3+. Is the model consistent with the experience?

   Measurements carried out by Kellogg give the variation of Q_n, the activation energy, as a function of the applied field F; data are given in the following graph:

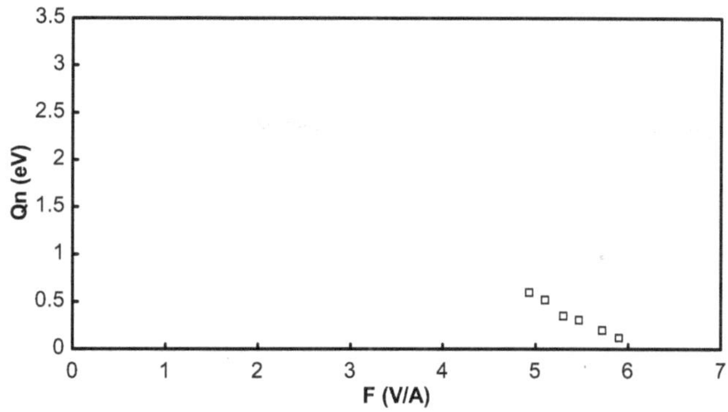

F(V/A)	Q_n(eV)
4.93	0.6
5.1	0.52
5.3	0.35
5.47	0.31
5.72	0.2
5.9	0.12

2. Demonstrate that if we neglect polarization and field penetration terms, Q_n can be written according to the Müller model in the form:

$$Q_n = Q_0\left[1 - \left(\frac{F}{F_e}\right)^{1/2}\right],$$

3. If F_e is the evaporation field at 0K, demonstrate that this equation can be simplified to

$$Q_n \approx Q'_0\left[1 - \left(\frac{F}{F_e}\right)\right] \quad \text{with} \quad Q'_0(F) \sim Q_0/2$$

4. Using the graph, what is the value of Q'_0? What is the conclusion?

Temperature Under Laser Illumination

A tungsten sample prepared in the form of a tip of 50 nm radius of curvature is field evaporated by applying an electrical potential V_0 and laser pulses focused on the end of the sample. The tip temperature is 20K; the detector 8 cm in diameter is placed 10 cm from the sample. Q_D, its detection efficiency, is 50%. Laser energy is here equivalent to 25% of the DC potential required for evaporation without laser.

1. Assume that ICF $= m + 1 = 1.6$, estimate the analyzed diameter and surface on the tip.
2. It is assumed that the fraction of atoms with a significant risk of evaporation represents a thickness of 0.05 nm on the tip surface. The average atomic volume of an atom is 1.6×10^{-29} m^3. What is the number of atoms N in a position to be field evaporated on the surface?

   We measure the evaporation flux for two different base temperatures depending on the laser energy E_p (nJ/pulse) (Figure below).

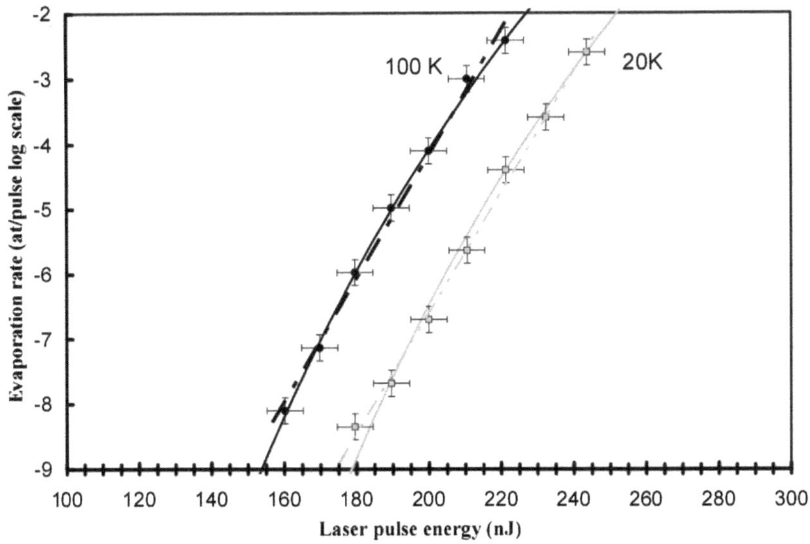

3. It is assumed that the laser increases the sample temperature during a duration τ, with a temperature rise linearly depending on E_p, ie, $T = T_0 + \alpha E_p$, such that the atoms are evaporated during the duration of this temperature pulse. Demonstrate using the evaporation law $K = \upsilon \cdot e^{-\frac{Q_n}{kT}}$ that the evaporation flux Φ (atom/pulse) can be written as

$$\ln(\Phi) = A - \frac{Q_n}{k_B(T_0 + \alpha E_p)}$$

where $A = \ln(\upsilon N Q_D \tau)$

4. Demonstrate that, for the same evaporation flux Φ, the value of α may be deduced from the gap between the two curves, such as $\alpha = \frac{T_1 - T_0}{E_0 - E_1}$ with E_0 the laser energy required to evaporate the sample at T_0 and E_1 the laser energy required to evaporate the sample at T_1. Deduce the temperature-induced rise for an energy of 220 nJ/pulse (value for 25% pulse fraction, 0.01 atom/pulse at 20K).
5. The value of $A = 13.8 \pm 1$ is derived from the adjustment of the two curves. ν is assumed between 10^{12} and 10^{13} s^{-1}. What is the estimated duration τ of field evaporation?

FIM

Consider the following FIM micrograph obtained on tungsten, for a potential 8 kV, the screen 8 cm in diameter is located 7 cm from the tip. Tungsten is a BCC metal of lattice parameter of 0.316 nm. Stereographic projection is given hereafter. This projection points to some centers of low indices <*hkl*> directions.

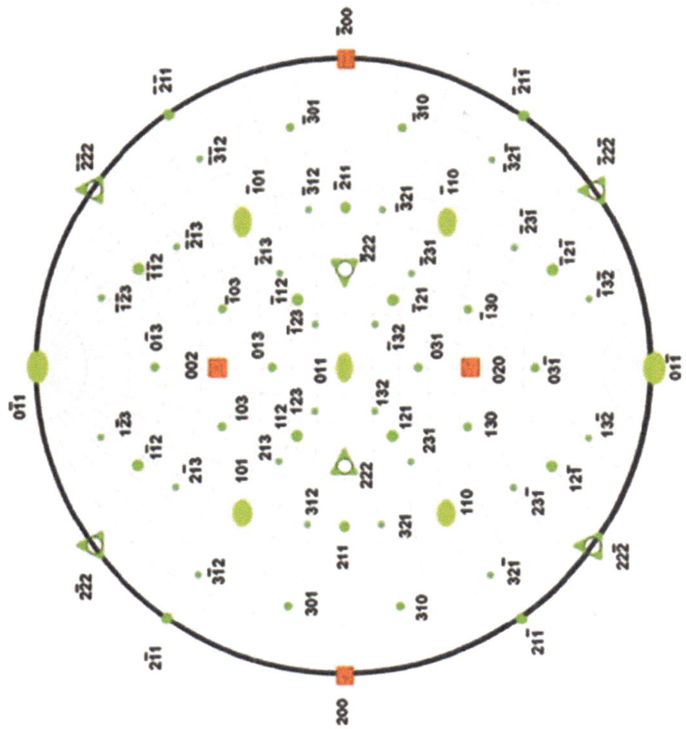

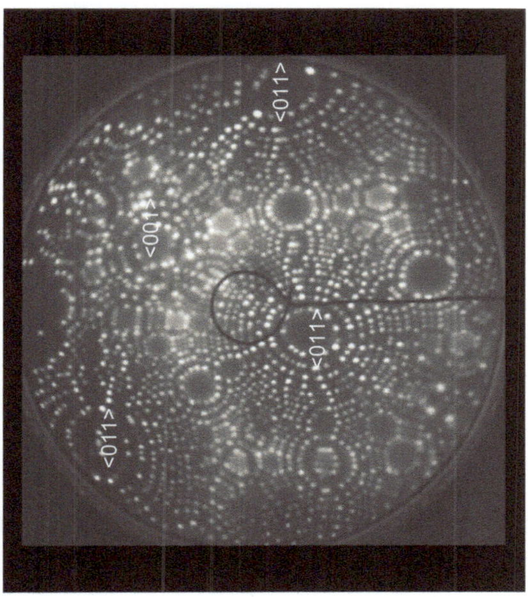
<011>
<001>
<011>
<011>

1. We recall the formula to calculate the angle between two crystallographic directions index <h,k,l> and <$h'k'l'$>:

$$\cos(\theta) = \frac{hh' + kk' + ll'}{\sqrt{h^2 + k^2 + l^2} \times \sqrt{h'^2 + k'^2 + l'^2}}.$$

The distance between the center pole of the (011) (near the central area analysis) and the center pole of (001) is 2.9 cm. Using simple geometric considerations, estimate the angular compression setting of the projection, as $\theta \sim (m + 1) \cdot \theta'$, with θ the crystallographic angles and θ' the projection angles on the screen.

The pole of the <011> direction is elliptical. We measure the maximum and minimum radii (in centimeters) of successive terraces around the pole crystallographic <011>, these measures are given in the following table:

N	R_{max} (cm)	Angle on Tip	cos (θ)	R_{min} (cm)	Angle on Tip	cos (θ)
1	0.43867244			0.36363636		
2	0.68686869			0.54256854		
3	0.86002886			0.72727273		
4	0.99278499			0.86580087		
5	1.13708514			1.02164502		

2. Find the angles of the terrace radii (complete or reproduce the provided table).
3. Using geometric considerations, demonstrate that the radius of a terrace of index n follows $\cos(\theta_n) = A - n \cdot \frac{d_{hkl}}{R}$, where R is the radius of curvature of the tip, d_{hkl} the distance between two successive crystallographic planes, and A a constant.
4. On a chart draw the corresponding laws and estimate the slope of the lines obtained. Deduce the tip radius in both cases. Deduce the approximate surface of analysis.
5. k_F is estimated at ~7. The potential is 8 kV. What is the estimated electric field above the surface?
6. What is the imaging gas used (justify)?
7. Assume the tungsten evaporation field at 0K of $F_e \sim 60$ V/nm, and write the variation of the energy barrier Q_n evaporation as a linear law of the field with $Q_0 \sim 2$ eV $\left(Q_n \approx Q_0'\left[1 - \left(\frac{F}{F_e}\right)\right]\right)$. What is the estimated value of the energy barrier in the experimental cases?

8. The vibration frequency is about 10^{12} Hz; the number of surface atoms is about 5000. At 80K, determine the evaporation rate. Is the image of the tip stable?
9. What is the magnitude of field required to evaporate 1 atom per pulse on a 0.5-ns-wide pulse? What is pulse fraction and field reduction (or pulse contribution) if the DC field is about 45 V/nm?
10. What should be the temperature of a 0.5-ns temperature pulse to achieve the same evaporation rate?
11. The electric field induces a surface electrostatic constraint equal to $F_S = \varepsilon_0 F^2/2$ (N/m^2) ($\varepsilon_0 = 8.84 \times 10^{-12}$ F/m). The $R_{0.2}$ elastic limit is about 1000 MPa. Calculate the stress for the case of continuous imaging (45 V/nm) and in the case of continuous evaporation (0.01 terraces per second), and where electric pulses. What about the risk of rupture in each case?

Artifacts

G. Sha et al. (Ultramicroscopy 102 (2005) 151) analyzed by APT and FIM an aluminum-based alloy (7050 alloy) in which exist the small quasi-spherical particles of composition Al_3Zr ($L1_2$ structure) of a few nanometers in diameter. The goal is to determine the composition of the phases present and in particular the location of the minority solutes (Zn, Cu, Mg). It is stated in the publication that the evaporation field of the Al matrix is estimated at 19 V/nm.

Consider the following ion micrographs obtained on Al 7050, for a potential of 3.23 kV (A) and 4.5 kV (B) under an argon atmosphere. Argon has a BIF (best image field) of about 22 V/nm.

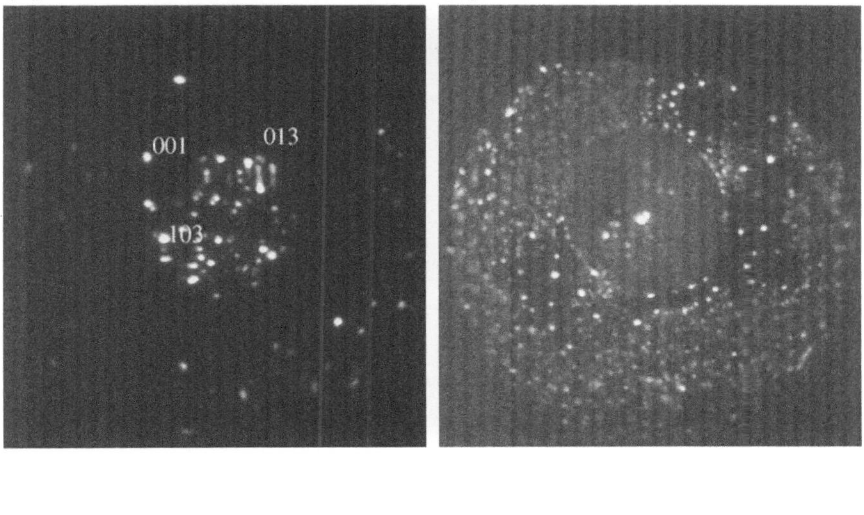

1. Explain why the author used argon as an imaging gas (instead of neon or helium).
2. Explain the differences between the two pictures, knowing that the particle Al_3Zr is visible in the center. In particular, what do you think of the field evaporation of the Al_3Zr phase (relative to the aluminum)?
3. Draw a schematic diagram of the shape of the tip (in transverse section).
4. The following figure shows another particle for which it is possible to observe the surface crystallography, including the terraces of the poles of the <001> and <102> directions of the FCC structure $L1_2$. The author indicates that he observed 13 terraces (001) between the two separate directions of 26.6°. The structure has a lattice parameter $a = 0.408$ nm. What is the distance between atomic planes along the <001> direction?
5. Give the formula used to calculate the angle between the directions?
6. G. Sha estimates the curvature radius at the particle surface to be 23 nm. With a diagram, explain the method. Note that the estimated tip radius is about 68 nm.

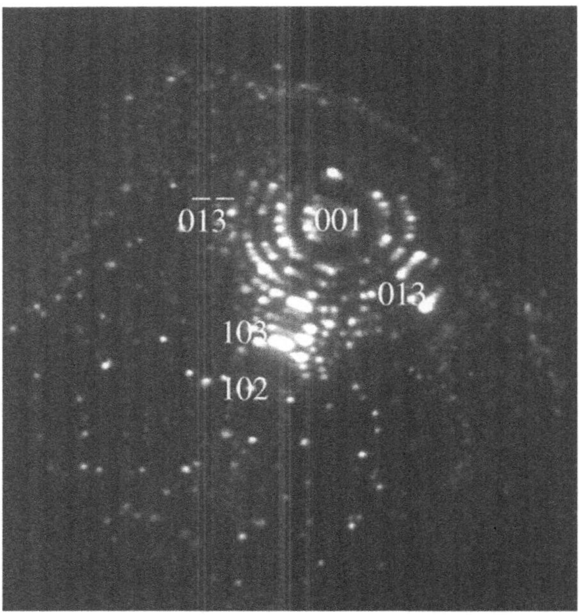

7. The mass spectra obtained by atom probe on the tip surface are presented above. It gives in (A) the matrix of the spectrum and (B) the spectrum of

the particle. What is the physical phenomenon to observe multicharged species (Al^{++}, Zn^{++})?

8. Explain the difference between the figures.

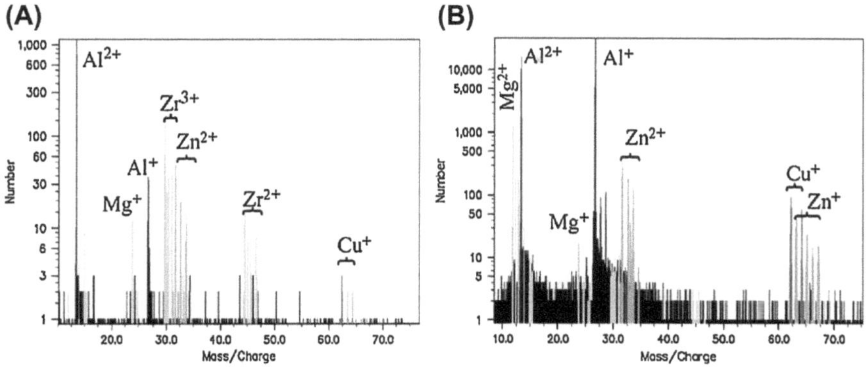

9. The following image shows the impacts on the detector for a small number of atoms collected during the analysis. Al is represented by black dots, the other species by small gray spheres. Give an interpretation of the spatial distribution of impacts (name of the observed artifacts).

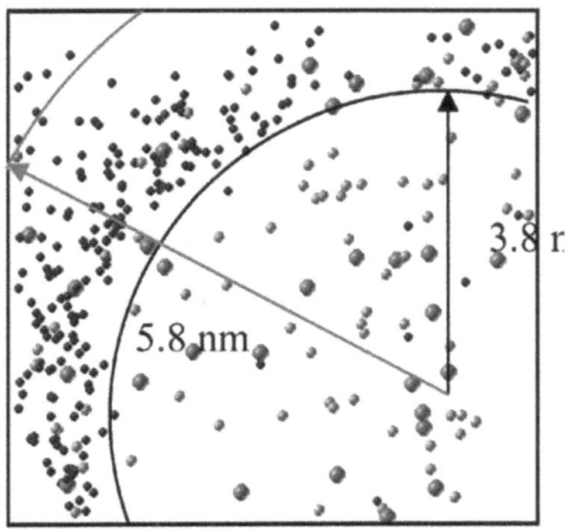

APPENDIX B

FIRST STEPS IN DATA PROCESSING

INTRODUCTION

The aim of this section is to put into practice some of the basic concepts introduced in the previous chapters. These "exercises" are some practice proposed to beginners who are in the training period and starting their first APT analyses and corresponding data processing. APT is a technique that could be used to address a large number of issues in various research fields on a large variety of materials. There are also various kinds of atom probes, with different tunable parameters and different data processing software. Since it would have been impossible to design specific exercises to fit any specific situation or problem, only general ideas and problems are reviewed in the following exercises, keeping the approach as general as possible. All aspects are addressed: atom probe and field ion microscopy, mass spectrometry, three-dimensional reconstructions, composition measurements, and statistics and data filtering. Most of the work proposed here relies on data processing, but some useful ideas are also proposed to check the influence of analysis or imaging parameters. In principle, any proper data set could be used to make these exercises, it is simply advised as a start, if possible, to avoid a complex situation with a material containing too many different elements. And, of course, a data set containing nanoscale features (eg, clusters, segregations, different phases) might help to explore deeper the proposed options.

MASS SPECTROMETRY

Definitions of Intervals

Any local composition measurement done with the APT relies on mass spectrometry, the calibration of the mass spectrometer, the optimization of the mass resolution, and the definition of mass intervals. This last step is usually fully controlled by the user who has to select the proper range for each peak. To understand how this step might affect significantly your measurements, the following are proposed:

1. Define mass intervals significantly wider than each peak—compute the composition.
2. Define mass intervals significantly thinner than each peak—compute the composition and compare with (1).
3. Define mass intervals that fit the best the peak width in your opinion—compute the composition.
4. Change the resolution of the mass spectrum (do not forget that a mass spectrum is a histogram; changing the resolution could give another design of the peaks); define mass intervals that fit the best the peak width in your opinion; compute the composition and compare with (3).

Influence of the Background Noise

The background noise will mainly affect minor elements. It is proposed here to check its influence. If there is no minor element in your material (impurity or low-concentration element), take one isotope with a low abundance.

1. Define all mass intervals; compute the composition.
2. If your data processing software has an option to automatically remove the background noise, then apply it. Compare the composition with (1). Note that the lower the concentration the larger the deviation. If your data processing software does not have an option to remove the background noise, export the mass spectrum, remove the contribution of the noise in the mass interval of interest and compare.

Checking Natural Abundances of Isotopes

The natural abundance of elements is often used to identify without ambiguity the ions on mass spectra. Besides, it could be useful to check the natural abundance at least on one majority element, to check if multihit events on the detector did not affect significantly the measurements.

Influence of Analysis Conditions

If possible, carry out two analyses at very different evaporation rates.

1. Check for the lowest evaporation rate how the background noise affects the measured concentration on minor elements.
2. Check for the highest evaporation rate if the ratio between isotopes is still near the natural abundance (if not, it is because of a large fraction of multihits on the detector).

If possible, carry out two analyses at different temperatures and/or laser energy.

3. Check a possible change in charge states of ions and note that it could be favorable to avoid peak overlaps.
4. Check if there is a change in the proportion of multihits on the detector and a change in the proportion of isotopes.
5. Check for any change in the overall composition (preferential evaporation may occur only in some specific conditions).

THREE-DIMENSIONAL RECONSTRUCTIONS

There are several procedures that can be used to compute three-dimensional reconstructions from APT raw data. They all involve some input parameters and in any case it is worth trying to check the influence of a little change of these parameters on the final result.

1. If available on your software, modify the detection efficiency and observe the changes in the dimension of the reconstructed volume.
2. If available on your software, try the standard reconstruction using the compression factor and evaporation field. Modify these values by about 10% and observe the changes in the dimension of the reconstructed volumes.
3. If possible, identify some crystallographic poles in the cross-section of your analyzed volume and try to image atomic planes in the vicinity. Measure the evolution of the interatomic distance as a function of reconstruction parameters.
4. Perform analyses at various temperatures and check the influence on the spatial resolution (atomic planes should appear a more distinct way at lower temperatures).

COMPOSITION PROFILES AND STATISTICS

The easiest way to explore local composition gradients in reconstructed volumes is first to compute some composition profiles. It is basically done with a sampling box moved across the volume and for each position the concentration is computed. Due to statistical fluctuations, the number of atoms involved (ie, the size of the box) matters.

1. Compute a composition profile across the volume using different sampling volume sizes and compare the fluctuations on the profiles.
2. Find an object of interest in the volume (particle, segregation for example) and compute a profile across. Check the influence of the sampling box size

on the gradient and on the local concentration. Check also the influence of the box orientation relative to the interface or the gradient direction.

3. If available, check the evolution of the number of atoms in the sampling volume. A change of the local composition often leads to a change in the local evaporation field and thus to the local magnification (local magnification effects). Evaluate an estimate of the local change of the evaporation field.

FILTERING AND STATISTICS

There are a huge number of different advanced processing routines to filter data and sort out some statistical information. If they rely on local composition measurements, then the statistics matter. Limitations could be checked on random distributions, such as the distribution of a specific isotope in a homogeneous region (for example ^{54}Fe).

1. Select one minor isotope and apply some filtering or cluster identification with a composition criterion. As soon as you see clusters, it indicates that your criterion is not optimized since they do not exist. The cluster search only exhibits the statistical fluctuations.
2. If possible, plot the first nearest neighbor distance distribution on one minor isotope. If you are using the maximum separation distance for cluster search, interpret the results of question (1) based on the histogram.

FIELD ION MICROSCOPY

Field ion microscopy provides atomic scale images of the needle-shaped samples used for APT. In case of crystalline materials, ion trajectories (compression factor) and the local geometry of the tip can be computed directly from images.

1. From an FIM image where a good contrast of crystallographic poles appears, compute the local radius on various poles and conclude eventually about faceting. By measuring the evaporation voltage under ultrahigh vacuum, estimate the evaporation field (note that under the gas pressure used for FIM the evaporation field is significantly reduced).
2. Carry out imaging at different temperatures to check the influence on spatial resolution. Estimate the evaporation field at different temperatures.

INDEX

'*Note*: Page numbers followed by "f" indicate figures and "t" indicate tables.'

D

E

F

FSC
www.fsc.org
MIX
Papier | Fördert gute Waldnutzung
FSC® C083411

Kolibri 360 GmbH
Ferdinand-Jühlke-Straße 7
99095 Erfurt, Deutschland
produktsicherheit@kolibri360.de